AuthorHouse™ UK Ltd.
500 Avebury Boulevard
Central Milton Keynes, MK9 2BE
www.authorhouse.co.uk
Phone: 08001974150

First published by AuthorHouse   1/5/2011.

ISBN: 978-1-4520-8594-4 (sc)

NONCOMMUTATIVITY AND ORIGINS OF STRING THEORY

E.B. TORBRAND DHRIF

ABSTRACT. In this book we derive (parts of) string theory from quantum field theory in half the dimension and geometry with clifford and group valued coordinates interpreted in terms of half-densities(Spinors). This gives a glimpse of Grand Unified Theory as the author understands it. We work in a general mathematical framework but the bulk of this thesis is in theoretical physics.

The front cover: The sun, a raging inferno of surface temperatures 6000 K, the source of admiration, contemplation and fear since the beginning of mankind, as illustrated by heliofigurative deities such as Amon-Re (the , according to myth, supposed to be first king of Egypt), Helios (Greek sungod) and Huitzilopochtli (Aztec god of war and sun). Despite it's considerable temperatures, it is but a part in the shivering carcass of the early universe. A large number of it's features are amenable to modern theoretical physics, inclusive some of it's chaotics. In the string field theoretic perspective of this thesis, it consists of $D3$-branes, dual to particles in space-time, creating and annihilating worlds as particles disintegrate and annihilate in the raging plasma, living as subspaces in $D = 10$ noncommutative stringy space-time. This has parallels in ancient mythology, e.g in Shiva, who dances worlds to their destruction and likewise creates them, or Quetzalcoatl, the great feathered serpent of the Aztecs, who organized the original cosmos and participated in the creation and destruction of various world periods. Quetzalcoatl ruled the fifth world cycle and created the humans of that cycle. The story goes that he descended to Mictlan, the underworld, and gathered the bones of the human beings of the previous epochs. Upon his return, he sprinkled his own blood upon these bones and fashioned thus the humans of the new era. One wonders, somehow, if the hypothesis in this thesis is more accurate than the thoughts of these ancient peoples.

The back cover; An eclipse of the sun, already ancient people knew how to predict these. Today Quantum Field Theory and String Theory can be used to generate predictions in cosmology and astronomy, such as e.g. Hawkings area theorem.

1. A Philosophical Preface

As I began studying quantum theory in early 96 I became more and more fascinated with the subject. I remember specifically walking home one day with a CD disc when the sun shined abundantly past the roof of a tall building and was reflected back from the disc in a straight ray to the shaded side of the building, leaving a wiggling spot of light as I continued to walk. What hit my mind then was what a brilliant piece of physics that was— Fermat's principle just emanating from photons statistically occurring more often in the line of constructive interferences in the quantum sum over histories. This principle of extremal langrangians is one of those that have dominated our view of physics, indeed many of our old friends like Newton's first law or the Hamilton-Jacobi, Schrodinger, Klein-Gordon, Dirac and Yang-Mills equations are simple consequences of this search for extremals, may it be at different level and quantization, and it is a typical example of one of the reasons for why this thesis was written the way it was; Unity in diversity.

Quite opposite to this search for this unity in diversity are the opinions of those who relish on the extremes, may they be practical, theoretical or simply pertaining to one very fixed idea or the other— exemplified by those who believe in the wide spread opinion that quantum gravity is irrelevant to experimental science and hence to physics. The latter seems to be an example of the very high price we have paid as a consequence of 'just see to that the experimental predictions are correct'- point of view prevailing in those parts of physics— this from the perspective of this *assumed* continuing and engrossing unity. Physics is a science of concepts, perhaps like mathematics at it's best, and we- but only probably- need to discern a set of important 'concepts' so that we can achieve this supposed unity. Linked to achieving this supposed 'unity' is the 'why?' question— we believe it to be important in physics because we think it may be possible to answer; the structures of reality interact with each other in a, to physical science, highly interesting manner, probably giving rise to each other simultaneously and often in a, to the human mind, exhausting complexity—to achieve consistency between them is often initially a difficult matter. So the 'why?' question can be asked and there are some means to make a (first) theoretical check of the answer. This is important to whatever efforts we can do in the strive for any unity we might achieve. In the opposite vein, misuse of the Occam razor, in the *don't* ask 'why?' way, in part as a consequence of the 'Let's just see to that the experimental predictions are correct-philosophy' in some parts of physics— mostly the practical ones and the theoretical subsciences pertaining to them—

has led to much loss of global comprehension in science. Again negating but in another circumstance, we are predisposed to think that the opposite is imperative in the most theoretical part of physics, i.e the strive for physical interpretation and prediction in string theory etc, for it deblindes us at our frontier, forcing us to find the errors we undoubtedly make, and gives us precious intuition on how to proceed from one point to another. The practical man that does not ask why is cut away from theory and hence from deductive thought itself in physics, and the theorist that does not ask how much is cut away from reality and thus the very essence of physics.

Perhaps a framework that would incorporate the better of what we have already done, mixing some (theoretical) extremes, would be too much to hope for, but if there was such a thing we would at least guess it to be a very good and natural start. That is the credo that motivates the developments below. A modest credo, or hope, in not only one hypothesis from one scientific culture or scientist being right, but many, and that the main task of present physics may be to interrelate these is thus what drives us. Of course the *true* hope is then that when we see the interrelations we also see the answers to the questions "Why?" and "How much?".

This thesis aims at giving such a framework, an incomplete blueprint of physics on foundations of noncommutative geometry (both in the sense of noncommutative coordinates in particular and in the more general sense of Connes), twistor geometry and string theory. We hope that it achieves this goal in some sense, leaving it to the reader to make his own opinion after having read the text. And it does attempt answers to the questions "Why?" and "How much?". Actually it does at least seven or eight experimental predictions out of string theory.

How does it do this? To put it briefly, it puts together string theory with noncommutative geometry and twistor, or rather spinor, geometry and it derives string theory out of field theory. D-branes and AdS-CFT correspondences, in particular the proof of Maldacenas theorem, are an integral part of this unification, as well as the below explained notions of N-admissibility and generalized supersymmetry. The string theory that comes out of field theory in this sewing of physics has a noncommutative space-time, with e.g. $U(N)$ gauge degrees of freedom and Chan-Paton factors, thus permitting us to directly identify it with M-theory strings in $D = 10, \cdots$, and we end up in a scenario where there are no extra dimensions in $D = 10$ string theory, which is the critical dimension for the original type II theory.

This critical dimension turns out to interrelate directly to Weyl invariance of Yang-Mills and Hilbert-Einstein lagrangeans in $D = 4$, and is what makes possible two other quantizations of gravity other than the usual string theoretic in $D = 10$. The four dimensional Euclidean space-times come out hyperkahler as suitable backgrounds in most cases due to physicality conditions, and this induces related structures on subspaces in $D = 10$. As some mere simple examples the Minkowski property in $D = 4$ falls as a matter of stability whilst the spatial sets in $D3$-branes evolve temporally on the branes in the growth direction of a harmonic dilaton field which defines the time after Wick rotation to the Euclidean region. We can mention, as a simple and illustrative example of a relation to noncommutative geometry, algebraic geometry, and Morse theory, that this evolution can be obtained via a homomorphism from a C^*-algebra acting on the left of initial conditions on a space-like slices on the $D3$-branes, which can also be described as smooth varieties with the time being smooth deformations of these varieties and topological fluctuations of the space-like slices to zeros of the gradient of the dilaton field. It ends with a map of physics and comments illustrating how the parts of physics are supposed to interrelate to each other, as well as enumerations of the conjectures necessary.

Some things can be improved in this thesis, including both the conceptual point of view and mathematical style, although the author has worked on this project from some time now it has really seemed never-ending, and his originally high ambitions have not been met in satisfactory way. He has tried to set up some basic guidelines, mostly rooted in his own philosophical beliefs, and to follow these as much as possible.

We have tried to achieve highest possible generality in all mathematics done and when doing a mathematical argument to be as simple and abstract as possible. On the other hand when doing physical arguments we attempt to concrethisize and give examples, spelling things out rather then letting them being taken for granted. That has contributed to the volume of this thesis. The reader is presupposed to be at least acquainted with most of modern everyday mathematics of physics like (non-real)algebraic, differential and general topology, complex and real differential geometry through the theory of fibre bundles, in particular spin geometry, complex and real analysis in one and several variables, linear and nonlinear functional analysis, group theory, advanced PDE theory, algebra and so on. Even though quite a lot of mathematics background has been included (for a master thesis), both for sheer beauty and to help the reader get comfortable with the authors conventions, it cannot be expected to account for the enormous literature in the branches of mathematics relevant to this thesis. We also presuppose knowledge of most of the basics of modern theoretical physics, at least up to and inclusive quantum field theory and non-abelian gauge fields but hopefully a little bit more. Some knowledge in string theory is helpful but not necessary. The first book of J.Polchinski and the lecture notes of E.D'Hoker from the 1996-1997 year at the Institute for Advanced Study should suffice more than abundantly.

As we touch the work of many others when we finally reach our objective of putting together physics, we have chosen to , out of respect of e.g. the many string theorists that have been dealing with string theory without necessarily realizing that it is the inevitable consequence of field theory, we have decided to deal with this by simply referring to such sources when they were known to us, to maintain and emphasize other authors work, as well as saving us from increased work labor. It is not without courage that a generation of string theorists developed string theory, and the author, having seen how his unification was greeted in Stockholm, feels only stronger and more compassionate in the cause of protecting the interests of those few individuals who dared propose string theory and have maintained struggle since then.

I apologize in advance for my shortcomings. This is a work written by a young theorist for mathematicans and physicists alike, and as such it suffers from some inevitable compromises. My age, knowledge, and time simply did not suffice for more. However, the perfect obviously exists but as an illusion, so perhaps this sketch of physics will, within obvious limitations, do.

But firstly; Our gratitude to our teacher and supervisor J. Mickelsson, who has been with us from the very beginning of our undergraduate time— and a little bit before— at the Royal Institute of Technology, Stockholm, Sweden, and B.S. DeWitt, whom we asked to comment this galactic size and non-trivial material on gravity, this in view of his unique experience in gravity, which is probably unsurpassed in generality. His support at a critical moment of this project enabled it's completion.

CONTENTS

7

11

"Omnibus ex nihil duscendis sufficit unum."
G.W. VON LEIBNIZ, 1646-1716.

"All beginnings are obscure. Inasmuch as the mathematician operates
with his conceptions along strict and formal lines, he, above all, must
be reminded from time to time that the origins of things lie in greater
depths than those to which his methods enable him to descend.
Beyond the knowledge gained from the indivdual sciences, there
remains the task of *comprehending*. In spite of the fact that the views
of philosophy sway from one system to another, we cannot dispense
with it unless we are to convert knowledge into a meaningless chaos."
H. Weyl, Space, time, Matter, 1918.

"... there are "mystics", among whom I count myself, who hope for a
synthesis which embraces aspects of all the rival theories. As a
mathematician I would find it a pity if God had not found some use
for all the beautiful ideas that have been put forward.
Clues, indicating that such a synthesis is not totally hopeless, include
the key role of integrable systems, solitons, duality, holomorphic
geometry and supersymmetry."
Sir Michael Atiyah, 1998.

1.1. **Organization.** This theoretical(mathematical) physics thesis is subdivided into 4 parts. Firstly comes one mathematical part on noncommutative mathematics in finite and infinite dimension. Then comes Part II, which concerns itself with the origins and foundations of gravity. In Part II we totally let go on any requirements of rigor or suchlike, as we undertake one of the most difficult tasks that can be undertaken—a bare handed quantization of gravity out of no foundations at all—except for elementary quantum field theory/mechanics and general relativity. Throughout Part II we strive consciously towards string theory, noncommutative geometry, and twistor geometry, and the aim is to derive enough intuition to derive these concepts out of field theory. In Part III we use some rigor again, and this time we try to make checks, describe phenomenology and if possibly make clear statements of exactly which our hypothesis are after that the dust of Part II has settled to some extent. Finally in Part IV we mention some more or less interrelating ideas that did not fit into the main body and finish off with mind maps and diagrams of physics and the various conjectures.

E. B. Torbrand

ABSTRACT. In this paper we firstly extensively survey some hyper-complex methods that find use in gravity. After this we briefly apply this to computations and conjectures in gravity and also make a generalized Hypercomplex Maldacena Conjecture. The mass spectrum of states is computed for the simplest case for D3-branes.

1. INTRODUCTION TO HYPERMATHEMATICS

Hyperanalysis is the science that deals with non-real analysis from an intrinsic point of view, and it' historical foundations rest on complex analysis in one and several variables, Lie analysis, Clifford analysis and hypercomplex dittos. The work of Riemann on complex function theory and the discoveries of Cayley, Hamilton, Clifford and Lie set up, provided one is willing to take a point of view that is general enough, a commond ground for examining the analysis of spaces of objects with some specified structures. As much as geometry and topology involve global questions hyperanalysis also does, and following the usual convention of mathematics, we shall commence with local studies and then extend to global cases. Thus, in a sense, we shall first be examining the *local* analysis of objects in some certain structural categories and then move on to the whole objects. The final step of this globalization shall be to regard an entire structural category, and this might be called "quantum" or "statistical" hypergeometry. It is "quantum" hypergeometry— In which we shall later naturally associate the objects to D-branes —that is the driving reason for mathematical escapades in this thesis, and in order to understand it we must first have a local understanding of structural categories and manifolds.

It is common in hyperanalysis that some things are valid for large spaces of structural categories, but also that the individual structural categories and structural objects have specific phenomenology. We shall be concentrating a little bit on the generalities, and try to spot them in specific examples.

Hyperanalysis in one variable has been to some part previously investigated in science and of these developments not all have been accessible, rigorous and intrinsic, however there are some books that touch these topics. The well known parts of hyperanalysis are complex analysis and in some extent several complex variables. In one complex variable the books by L.Ahlfors, H.Cartan and R.Remmert can be recommended. In several complex variables the books by Hormander, Krantz and Gunning can also be recommended. As for noncommutative examples there does not seem much mentionable in book format, however Manin's book on complex geometry/gauge field theory contains some such material and I believe that Riesz has written a book in Clifford analysis-a book that I unfortunately have not been able to get hold of. As for the rest the reader is referred to Fueter's work from the 30's and a review article by Dunford from the 70's. Hyperanalysis is thus in some sense a dormant field; indeed it might be too large a field for us ever to cope with.

The simplest example of hyperanalysis in several variables is several complex variables-pluricomplex analysis- and this simplest commutative example is known to be quite non-trivial at times. Perhaps it is wise to understand that a tree can have many leaves yet each leaf be very individual. Thus in respecting the individuality allowed by generality we commence our exposition.

2. Pluricomplex Analysis

2.1. Introduction to Several Variables.

The general theory of several complex variables was formulated considerably later than the one dimensional counterpart. Some of the fundamental facts were elaborated by K.Weierstass in the late nineteenth century but otherwise the field laid mostly dormant until the shift of the century when some problems, for example the problem of extension of holomorphisms from the boundary of the unit ball throughout the ball was solved in the affirmative by Hartogs and Poincare in 1903-1906 and an investigation of the problems now known as the th 1:st and 2:d Cousin problems was undertaken by Cousin. By then some of the differences between several variables and the one variable case became apparent, such as the nonexistence of biholomorphisms from the unit ball to the polydisc in $\mathbb{C}^n$, $n > 2$, contrary to the one variable case which essentially reduces the question of biholomorphic equivalence to topological equivalence by the Riemann mapping theorem. Work by Bergman and others, such as the famous Bergman kernel giving an integral representation on the unit ball and the Martinelli-Bochner formula, giving an integral representation on arbitrary domains, gradually evolved into the more function theory like of one variable aspects of several variables[1]. But it is in no way to be believed that those were the only developments, indeed the work which has set its most indistinguishable mark is the non function theoretic. This tradition in several variables got its seeds from the work of K.Oka who systematically solved problems with some brilliant new ideas based on work by the genius H.Cartan, who realized the scope of Oka's work and laid the foundations of concepts like sheaf theory together with it's father Leray. Today work by Grauert, Remmert, Leray, Dolbeault and many others has continued to enrich this active field of mathematical research.

The aim of the present section is to introduce the reader to the basics of several complex variables. Prerequisites are almost none, and indeed arguments are kept extremely simple, withstanding the surge for use of advanced analytical, geometric or topological methods, with some exceptions like an explicit use of the calculus of forms and elementary homological calculus. The presentation is made as close to the one

---

[1]The latter formula has a non-holomorphic kernel and this can be annoying. The author is presently considering a new integral representation which is valid without modifications on arbitrary domains in $\mathbb{C}^n$ and does not suffer from this drawback. His integral representation also renders the result by Narasimhan concerning continuation of meromorphisms throughout the ball trivial. A version is reviewed later in this thesis.

variable case as possible in order not to scare away the student in (probably) his/hers first confrontation with this beautiful theory.

So why study several complex variables? Mathematically the answer is selfevident-simply for fun. However there are more reasons- it turns out that a lot of questions are linked to several variables theory and can be successfully resolved within the framework provided, the vast usefulness charecterizing the one variable theory can often be percepted with equal or greater force in this larger framework. It is not probable that anyone with an applied interest will ever read this essay, however the use of several variables in practical applications can be made immense. Some examples - Classical dynamical systems can be treated as hamiltonian manifolds which can, modulo integrability, be attacked as complex manifolds (actually Kahler!)and the state of a C.D system can be obtained by exponentiation on a holomorphic vector bundle(See Bishop and Goldberg or Novikov et al [7]) - Penrose twistor theory with applications to P.D.E's and gravity - P-branes & string theory (TeichmÃ$\frac{1}{4}$ller spaces,moduli spaces ,Hodge theory and what not)-quantum field theory(Euclidinization(i.e in practice often a Wick rotation) and the then needed analytic continuations, evaluation of integrals,etc) - Reformulating differential equations as Cauchy Riemann equations - Projective embeddings of all kinds of stuff that people want to embedd, like varieties(One of the basic tools here being the Kodaira embedding theorem) and in string compactification purposes.

2.2. **The Basic Facts.** Let D, Ω be compact subsets of $\mathbb{C}^n$ if not otherwise stated and reserve the letter $\alpha \in \mathbb{Z}^n$ to denote a tuple in multi-index notation, with$|\alpha| = |\alpha_1| + |\alpha_2| + \cdots$ where we also reserve n, pronounced n, to mean the tuple $(1, 1, 1, ...)$. Call a function f analytic on Ω if it is described locally on a neighbourhood of Ω by the power series expansion $f(\zeta) = \sum_{|\alpha|=0}^{\infty} a_{(\alpha)}(\zeta - z)^{\alpha}$, where we use the standard metric topology on $\mathbb{C}^n$ given by the euclidean distance function $d(\zeta, z) = ||\zeta - z|| = \sqrt{\sum_{i=1}^{n} |\zeta_i - z_i|^2}$. We denote the set of analytic functions on Ω by $O(\Omega)$ and for starters we will differ between holomorphisms on Ω, which we define to be the set of smooth functions satisfying $\bar{\partial} f = 0$ on Ω ,and analytic functions on Ω, $\bar{\partial}$ being the Dolbeault operator defined pointwise on $C^{\infty}(\Omega)$ by $\bar{\partial} f = \frac{\partial f}{\partial \bar{\zeta}_{\sigma}} d\bar{\zeta}_{\sigma}$. The topology of $O(\Omega)$ is defined to be the one generated by the supremum norm on compact subsets of Ω. Now since the set of holomorphic and antiholomorphic coordinates in $\mathbb{C}^n$ are orthogonal and describe $\mathbb{C}^n$ just as well as any other coordinate choice we note the split $d = \partial + \bar{\partial}$, where $\partial f = \frac{\partial f}{\partial \zeta_{\sigma}} d\zeta_{\sigma}$ again acting pointwise on $\mathbb{F}(\Omega) \equiv C^{\infty}(\Omega)$, d being just

17

the ordinary exterior derivative,(if you do not know what an exterior derivative is, just think of it as the differential operator). We denote by $\Lambda(\mathbb{C}^n) \equiv \Gamma(\mathbb{C}^n, \Lambda T^*\mathbb{C}^n)$ the space of smooth functions(sections) $\omega : \mathbb{C}^n \to \Lambda T^*\mathbb{C}^n$, where $\Lambda T^*\mathbb{C}^n$ denotes the space of totally antisymmetric covariant tensors in $\mathbb{C}^n$, which are naturally genertd by $d\zeta_i$ and $d\bar{\zeta}_i$. Such a function(also named section) is called a form. We use a special symbol, called a wedge,to denote the totally antisymmetric product between two tensors in $\Lambda T^*\mathbb{C}^n$. To clarify things we take some examples;

Example 2.1. *Fix $n \geq 4$ in this example. $d\zeta_1 \wedge d\zeta_2 \wedge d\bar{\zeta}_3 \wedge d\zeta_3 \in \Lambda T^*\mathbb{C}^n$ and if we expand we have $d\zeta_1 \wedge d\zeta_2 \wedge d\bar{\zeta}_3 \wedge d\zeta_3 = d\zeta_1 \otimes d\zeta_2 \otimes d\bar{\zeta}_3 \otimes d\zeta_3 - d\zeta_2 \otimes d\zeta_1 \otimes d\bar{\zeta}_3 \otimes d\zeta_3 + \forall$ other antisymmetric permutations, similarly $d\zeta_1 \wedge d\zeta_2 = d\zeta_1 \otimes d\zeta_2 - d\zeta_2 \otimes d\zeta_1$. $(\cosh(\zeta_1\zeta_2 + 2i) + \zeta_4)d\zeta_1 \wedge d\zeta_2$ is a function taking values in $\Lambda T^*\mathbb{C}^n$.*

We remember from elementary multivariable calculus Gauss and Stokes theorem. Their natural generalization is contained in the above machinery, for let us define $\int_M \omega = \int_M f(x)dx_1 dx_2 \cdots dx_n$ where $\omega = f(x)dx_1 \wedge dx_2 \wedge \cdots dx_n$, $\omega \in \Lambda(M)$, M being some smooth manifold. If you do not know what a manifold is, just think of it as a compact subset of $\mathbb{C}^n$ which is locally homeomorphic to some lower dimensional space $\mathbb{R}^j$, $0 \leq j \leq 2n-1$, this will be correct in some sense as long as you are dealing with compact manifolds (Whitney's theorem). Stokes theorem states then $\int_{\partial M} \omega = \int_M d\omega$, ∂ being the boundary operator. Elementary analysis, for example the Riesz representation theorem or the Radon-Nikodym theorem, states a pattern of peculiar relationship of duality between sets and functions via integrals under some circumstances. Let us in the spirit of those lines of reasoning define a 'product' $\langle \, , \, \rangle : C_n(\mathbb{C}^n, \mathbb{C}) \times C^n(\mathbb{C}^n, \mathbb{C}) \to \mathbb{C}$ where we set $C^n(\mathbb{C}^n, \mathbb{C}) \equiv \Lambda^n(\mathbb{C}^n) \equiv \Lambda(\mathbb{C}^n)|_{forms\ of\ order\ n}$, and call the elements in that space *cochains*, the corresponding sets in $C_n(\mathbb{C}^n, \mathbb{C})$ we integrate over are then called chains.[2]

Excercise 2.1. *Familiarize yourself with this 'product'.*

Note that in the above 'product' form we have have $d^\dagger = \partial$, $\dagger$ denoting adjoint and that the boundary of a boundary is nill, hence $d^2 = 0$ thus the Dolbeault split of the exterior derivative gives $\partial^2 = \bar{\partial}^2 =$

---

[2] The reader should not believe that duality between these two spaces is always the case, nor take the proof of such a matter lightly. However, such an assertion can be proved under some circumstances, most notably compactness of the manifold considered. In positive cases this implies deRham's theorem $\frac{ker(d|_{C^n(M,\mathbb{C})})}{Ran(d|_{C^{n-1}(M,\mathbb{C})})} \simeq \frac{ker(\partial|_{C_n(M,\mathbb{C})})}{Ran(\partial|_{C_{n+1}(M,\mathbb{C})})}$.

$\partial\bar{\partial} + \bar{\partial}\partial = 0$ when identifying terms in different

$$\Lambda^{(p,q)} \equiv \Lambda|_{p \text{ holomorphic and } q \text{ antiholomorphic indices}}.$$

Graded Leibnitz rules hold for both the exterior derivative and the boundary operator, namley $d(\omega \wedge \eta) = d\omega \wedge \eta + (-1)^p \omega \wedge d\eta$, $\omega \in C^p$, $\eta \in C^q$ and $\partial(\omega \wedge \eta) = \partial\omega \wedge \eta + (-1)^p \omega \wedge \partial\eta$ $\omega \in C_p$, $\eta \in C_q$ and we also have $\omega \wedge \eta - (-1)^{pq}\eta \wedge \omega = \omega \wedge \omega = 0$. We define $Z^p(X, \mathbb{R}) = Z^p(X) = ker(d|_{\Lambda^p})$, $Z_{\bar{\partial}}^{q,p}(X, \mathbb{C}) = Z_{\bar{\partial}}^{q,p}(X) = ker(\bar{\partial}|_{\Lambda^{q,p}})$, $B^p(X, \mathbb{R}) = B^p(X) = Ran(d|_{\Lambda^{p-1}})$, $B_{\bar{\partial}}^{q,p}(X, \mathbb{C}) = B_{\bar{\partial}}^{q,p}(X) = Ran(\bar{\partial}|_{\Lambda^{q,p-1}})$ [3] and call these sets (Dolbeault)cocycle group and (Dolbeault)coboundary group. We are now finally in a postion to start our field trip.

Lemma 2.1 (Abel's lemma). *Assume $a_{(\alpha)} \in \mathbb{C}^n$ is given for $\alpha \in \mathbb{N}^n$, and for some $\zeta \in \mathbb{C}^n$ we have $\sup |a_\alpha \zeta^\alpha| = M < \infty$; Then falls*

$$\sum_{|\alpha|=0}^{\infty} a_{(\alpha)} \zeta^\alpha < \infty$$

in the polydisc defined by $P(0, r) = B(0, |\zeta_1|) \times B(0, |\zeta_2|) \cdots \times B(0, |\zeta_n|) = B(0, |\zeta|)^n$, convergence being normal.

Proof. Pick $\lambda \in (0, 1)$ and $K \subset B(0, \lambda|\zeta|)^n$. Note $|a_\alpha z^\alpha| = |a_\alpha \zeta^\alpha| \lambda^{|\alpha|}$, $z = \lambda\zeta$. Thus

$$|a_\alpha z^\alpha| \leq |a_\alpha \zeta^\alpha| \lambda^{|\alpha|} \leq M\lambda^{|\alpha|}, \quad \forall \alpha \in \mathbb{N}^n, \ z \in K$$

Since $\sum_{\alpha \in \mathbb{N}^n} \lambda^{|\alpha|} = (\sum_{i=o}^{\infty} \lambda^i)^n < \infty$, the result follows. $\square$

[4]

Corollary 2.1.

$$O(\Omega) \subset H(\Omega)$$

Proof.

$$\bar{\partial}f = \bar{\partial} \sum_{\alpha \in \mathbb{N}^n} a_\alpha (\zeta - z)^\alpha = \sum_{\alpha \in \mathbb{N}^n} \bar{\partial}a_\alpha (\zeta - z)^\alpha = \sum 0 = 0, \zeta \in B(z, r)^n$$

---

[3]When a field is an argument in any of these symbols it simply denotes what field the (vector)space in question is defined over.

[4] Remember that in a footnote on previous pages we spoke of a group named the deRham cohomology, if that group is trivial, you can always inverte d modulo the numerator of the quotient. One defines $Z^n \equiv ker(d|_{C^n(M,\mathbb{C})})$, calling the elements in that set cocycles, similarly $B^n \equiv Ran(d|_{C^{n-1}(M,\mathbb{C})})$ are called coboundaries. Thus if one is lucky one may get a cocycle to integrate and hence $\partial^2 = 0$ implies under trivial cohomology that one may look at the integration domain modulo cycles. This is actually precisley what happens when we speak of conservative fields in physics. There is another kind of Cohomology, called Dolbeault cohomology which is basically the same construction except the conjugate Dolbeault operator replaces the exterior derivative.

[5] follows using the previous corollary and noting that uniform convergence of the differentiated partial sum most definitely is at hand. $\square$

We sometimes set a little symbol as an index to the space we are considering to fix the symbols for the coordinates used, for example $\mathbb{C}_\zeta^n$ denoting $\mathbb{C}^n$ with $\{\zeta\}$ as coordinates.

Theorem 2.1 (General Cauchy integral formula on Riemann surfaces).

$$f(z) = \frac{1}{2\pi i}\{\int_{\partial\Omega} \frac{\omega}{\zeta - z} - \int_\Omega \frac{\bar\partial\omega}{\zeta - z}\} \quad z \in \Omega, \ \omega = f(\zeta,\bar\zeta)d\zeta \in \Lambda^{(1,0)}(\mathbb{C}_\zeta)$$

Proof. By compactness Ω has a finite cover. For simplicity let us consider the case when Ω is simply connected and connected. Then, letting $D_{z,R}$ denote a disc around z of radius R, we have

$$\langle \partial\Omega, \tfrac{\omega}{\zeta-z}\rangle = \langle \partial\Omega \sim D_{z,R}, \tfrac{\omega}{\zeta-z}\rangle + \langle \partial D_{z,R}, \tfrac{\omega}{\zeta-z}\rangle$$
$$= \langle \Omega \sim D_{z,R}, d\tfrac{\omega}{\zeta-z}\rangle + \langle \partial D_{z,R}, \tfrac{\omega}{\zeta-z}\rangle$$
$$= \lim_{R\to 0}\langle \Omega \sim D_{z,R}, \tfrac{\bar\partial\omega}{\zeta-z}\rangle + \lim_{R\to 0}\langle \partial D_{z,R}, \tfrac{\omega}{\zeta-z}\rangle$$
$$= \int_\omega \tfrac{\bar\partial\omega}{\zeta-z} + \lim_{R\to 0}\int_0^{2\pi} \tfrac{f(\zeta,\bar\zeta)}{Re^{i\theta}} dRe^{i\theta}$$
$$= \int_\omega \tfrac{\bar\partial\omega}{\zeta-z} + 2\pi i f(z)$$

and thus by linearity of $\langle\,,\,\rangle$ in the first argument under disjoint unions the assertion falls on arbitrary 2-chains $\Omega \in \mathbb{C}$. $\square$

Corollary 2.2 (Cauchy's Main Theorem).

$$f(z) = \frac{1}{2\pi i}\int_{\partial\Omega} \frac{\omega}{\zeta - z}, \quad z \in \Omega, \omega \in Z_{\bar\partial}^{(1,0)}(\mathbb{C}, \mathbb{C})$$

Proof.

$$\omega \in ker(\bar\partial)$$

$\square$

Corollary 2.3 (The Polydisc Cauchy Main Theorem). *Assume* $\omega \in \Lambda^{(n,0)}(\Omega_\zeta^n)$
$\Omega^n = \Omega_1 \times \Omega_2 \times \cdots \Omega_n, \ \omega = fd\zeta^n \equiv f\bigwedge_{i=1}^n d\zeta, \ f \in H(\Omega^n)$. *Then*

$$f(z) = \frac{1}{(2\pi i)^n}\int_{\partial\Omega^n} \frac{\omega}{(\zeta - z)^n}, \quad z \in \Omega^n$$

---

[5] Since the previous corollary states quite nice properties in the local representations we are studying we will use = rather than $\sim$, and will have to live with the small inadequacy. Later we will prove uniqueness of this expansion thus fully justifying the use of an equivalence in the sense of a transitive, reflexive and symmetric relation.

Proof. By $\bar{\partial} f = 0$ it falls that f satisfies a Cauchy-Riemann condition in each variable separatly. Thus by

$$\int_{\partial\Omega^n} \frac{\omega}{(\zeta - z)^n} = \int_{\partial\Omega_1} \frac{d\zeta_1}{(\zeta_1 - z_1)} \int_{\partial\Omega_2} \frac{d\zeta_2}{(\zeta_2 - z_2)} \cdots \int_{\partial\Omega_n} \frac{d\zeta_n}{(\zeta_n - z_n)} f(\zeta)$$

it suffices by the previous corollary to show that iterated integration holds. By compactness of Ω^n and continuity on the same the Lebeque criterion for Riemann integrability holds, hence Lebesgue integrability. The Lebesgue measure being complete the Fubini theorem now holds(See Royden[5]), and so we assert the corollary. $\qquad\square$

Excercise 2.2. *Differentiate under the integral sign and formulate a residue theorem on $\Omega^n \subset \mathbb{C}^n$. What form must the quotient analytic set $g = 0$, where we are considering $\frac{f}{g}$, $g, f \in H(\Omega^n)$ as an integrand have for you to be able to use this theorem? Examine $g = \zeta_1^2 + \zeta_2^2 + \zeta_3^2$, $1 + \zeta_1 + \zeta_2 + \zeta_3 + \zeta_1\zeta_2 + \zeta_1\zeta_3 + \zeta_2\zeta_3 + \zeta_1\zeta_2\zeta_3$ and $\cos(\zeta_1 + 3i)$- try to think about how the applicability dependens of dimensionality and numerator. The residue you have obtained is called the Cauchy residue, and is a special case of a more general residue called the Grothendieck residue(See Dolbeault[6]).*

Excercise 2.3. *Evaluate*

$$\int_{\mathbb{R}^3} \frac{\cos(\sqrt{x+y}+z)dxdydz}{(x^2+1)(y^2+1)(z^2+1)}$$

6

We define the locus z of a singularity to be the point specified by the denominator factorization $(\zeta - z)^\alpha$, and α to be the order of the locus. If this $\alpha = \infty$ we say that we have an essential singularity at z.[7]

Theorem 2.2 (Polydisc Laurent Theorem). *Assume f to be holomorphic on the polyannulus defined by*
$(B(z, R) \sim B(z, r))^n, r < R$. *Then*

$$f(\zeta) = \sum_{\alpha=-\infty}^{\infty} a_{(\alpha)}(\zeta - z)^\alpha \quad \zeta \in (B(z, R) \sim B(z, r))^n, r < R.$$

convergence being normal and unique.

---

6 Answer:$\frac{\pi^3 c\acute{o}s(1)}{e^2}$

[7]This is in no way sufficient to describe singularities in several variables. Rather the situation is highly interesting, leading to many twists and turns. See Chirka[6]. The reason for this definition is that we are always able to use the W. preparation theorem (See section 4) locally when fixing all coordinates but one.

Proof. Thus assume for simplicity $z = 0$, and set Ω_i to be a ϵ-parted annulus, letting all contours be positive, and defining the chains $C_\alpha = \chi_{\mathbb{N}^n}(\alpha)C_2^\epsilon - \chi_{\mathbb{Z}_-^n}(\alpha)C_1^\epsilon$,

$$f(z) = \frac{1}{(2\pi i)^n} \int_{\partial\Omega^n} \frac{\omega}{(\zeta-z)^n}, \; z \in \Omega^n$$
$$= \frac{1}{(2\pi i)^n} \left(\int_{C_1^\epsilon} + \int_{C_2^\epsilon n} \right) \frac{\omega}{(\zeta-z)^n} + O(\epsilon)$$
$$= \epsilon \to 0 = \frac{1}{(2\pi i)^n} \int_{C_1^0 n} \frac{\omega(-1)^n}{(z(1-\zeta/z))^n} + \frac{1}{(2\pi i)^n} \int_{C_2^0 n} \frac{\omega}{(\zeta(1-z/\zeta))^n}$$
$$= \frac{1}{(2\pi i)^n} \int_{C_1^0 n} \frac{(\sum_{\alpha\in\mathbb{N}^n}(\frac{\zeta}{z})^\alpha)\omega(-1)^n}{z^n} + \frac{1}{(2\pi i)^n} \int_{C_2^0 n} \frac{(\sum_{\alpha\in\mathbb{N}^n}(\frac{z}{\zeta})^\alpha)\omega}{\zeta^n}$$
$$= \sum_{\alpha\in\mathbb{N}^n} \left(\frac{1}{(2\pi i)^n} \int_{C_1^0 n} \frac{((\frac{\zeta}{z})^\alpha)\omega(-1)^n}{z^n} + \frac{1}{(2\pi i)^n} \int_{C_2^0 n} \frac{((\frac{z}{\zeta})^\alpha)\omega}{\zeta^n} \right)$$
$$= \sum_{\alpha\in-\mathbb{N}^n} \left(\frac{1}{(2\pi i)^n} \int_{(-C_1^0)^n} \frac{\omega}{\zeta^\alpha} \right) z^{\alpha-n} + \sum_{\alpha\in\mathbb{N}^n} \left(\frac{1}{(2\pi i)^n} \int_{C_2^0 n} \frac{\omega}{\zeta^{\alpha+n}} \right) z^\alpha$$
$$= \left(\sum_{\alpha\in\mathbb{Z}_-^n} + \sum_{\alpha\in\mathbb{N}^n} \right) \left(\frac{1}{(2\pi i)^n} \int_{C_\alpha n} \frac{\omega}{\zeta^{\alpha+n}} \right) z^\alpha$$
$$= \sum_{\alpha\in\mathbb{Z}^n} \underbrace{\left(\frac{1}{(2\pi i)^n} \int_{C_\alpha n} \frac{\omega}{\zeta^{\alpha+n}} \right)}_{a_{(\alpha)}} z^\alpha$$
$$= \sum_{\alpha\in\mathbb{Z}^n} a_{(\alpha)} z^\alpha$$

where the exchange of summation and integration (a sum is an integral under the counting measure) holds from the Fubini theorem (See Royden[5]) by integrability(Summability) of the function considered and completeness of both the Lebesgue and counting measure. Uniqueness is obvious, and applying the previous theorem on geometrical series normality follows. $\qquad\square$

Corollary 2.4 (Polydisc Taylor theorem). *Assume f to be holomorphic on $N(z) \subset \mathbb{C}^n$. Then on the largest polydisc $B(z,r)^n \subset N(z)$*

$$f(\zeta) = \sum_{|\alpha|=0}^\infty a_{(\alpha)}(\zeta - z)^\alpha, \; \zeta \in B(z,r))^n$$

convergence being normal and unique. [8]

Proof. Since f is a holomorphism in each variable all $a_{(\alpha)}; \alpha_\sigma < 0$ vanish, thus the corollary falls in view of the above statement by noting the formula given for the Laurent series coefficients. $\qquad\square$

Theorem 2.3 (Generalized Vitali Theorem). *The natural restriction homomorphism $r : O(B(0,1)^n) \to O(\Omega^n)$ has compact action for $\Omega^n \subset\subset B(0,1) \subset \mathbb{C}_\zeta^n$.*

Proof. Note that since we are dealing with a holomorphism on the ball we have a taylor series expansion available throughout it, thus

---

[8]i.e in some sense $O(\Omega) = H(\Omega)$.

$$
\begin{aligned}
|f(z) - f(z')| &= \tfrac{1}{(2\pi)^n} |\int_{\partial \Omega^n} \tfrac{f(\zeta)}{(\zeta-z)^n} - \tfrac{f(\zeta)}{(\zeta-z')^n}| \\
&= \tfrac{1}{(2\pi)^n} |\int_{\partial \Omega^n} \tfrac{f(\zeta)}{(\zeta-z)^n} - \tfrac{f(\zeta)}{(\zeta-z')^n}| \\
&\leq \tfrac{1}{(2\pi)^n} \mu(\partial \Omega^n) \sup_{\zeta \in \partial \Omega^n} |\tfrac{f(\zeta)}{(\zeta-z)^n} - \tfrac{f(\zeta)}{(\zeta-z')^n}| \\
&\leq \tfrac{1}{(2\pi)^n} \mu(\partial \Omega^n) \sup |\sum_{\alpha \in \mathbb{N}^n} \tfrac{f(\zeta)(z^\alpha - z'^\alpha)}{\zeta^{\alpha+1}}| \\
&\leq \tfrac{1}{(2\pi)^n} \mu(\partial \Omega^n) \sup |\tfrac{f(\zeta)}{\zeta^{\alpha+1}}| |\sum_{\alpha \in \mathbb{N}^n} (z^\alpha - z'^\alpha)| \\
&\leq M \|z - z'\|
\end{aligned}
$$

for some $M \in \mathbb{R}_+$ for close enough z, z' [9]. Thus noting to each $\delta > 0$ there is a ϵ defined by $\delta = \frac{\epsilon}{M}$ satisfying

$$
|f(z) - f(z')| < \epsilon \; \forall z, z'; \; \|z_i - z_i'\| < \delta
$$

we have uniformly continuous f. Now Ω_i is compact, thus Ω^n is compact. Let $\mathfrak{F} \subset O(B(0,1)^n)$ be bounded, then the M in the above inequalities can be chosen to be valid over the entire family since it only depends on $diam(\mathfrak{F})$ for a fixed $\Omega^n \subset\subset B(0,1)$, but then we have a bounded equicontinuous family of functions $\mathfrak{F}$, the Ascoli-Arzela theorem (See Royden[5]) now asserts that the range of restriction is a normal family of functions. $\qquad \square$

Corollary 2.5. *The natural restriction $r : O(\mathcal{O}) \to O(K)$ is compact for compact $K \subset\subset \mathcal{O}$, $\mathcal{O}$ open in $\mathbb{C}^n$.*

Proof. Since the topology we are considering is equally well generated by balls or polydiscs and we are dealing with a topological matter the assertion has to fall. $\qquad \square$

Theorem 2.4. *There are no biholomorphisms from the polydisc to the unit ball in $\mathbb{C}^n$, $n \geq 2$*

Proof. For simplicity we will assume that that this assumed biholomorphism f extends holomorphically to the boundary. On the boundary it would be necessary that $\|f(\zeta)\|^2 = f(\zeta) \dagger f(\zeta) = 1$. Let q_n be fixed and situated on the boundary of $\partial \Omega_n$ for $q = (q_1, q_2, \cdots, q_{n-1}, q_n)$. It is seen that $(N(q) \sim q) \cap (\mathbb{C}^n \sim \partial \Omega^n) \neq \emptyset \; \forall N(q)$, Ω_σ open, hence q is in the topological boundary of this polydisc. A biholomorphism is among other things a homeomorphism mapping boundaries to boundaries, but then individual coordinate $\zeta_\rho, \rho \neq n$ might be used on a subset of the boundary for $n > 1$. Hence follows $0 = \sum_\sigma \partial_\rho \bar{\partial}_\rho f_\sigma \bar{f}_\sigma = \sum_\sigma |(\partial_\rho f_\sigma)|^2 \Rightarrow 0 = |\partial_\rho f_\sigma|^2$. Thus $\partial_\rho f_\sigma = 0$ and since the derivative of a holomorphism is again a holomorphism the maximum modulus principle yields that

---

[9]We are assuming implicitly that the origin is not on the boundary of the polydisc considered.

this derivative is a map attaining the value 0 throughout the polydisc when considering components of it. Reordering components this will hold in every component ρ, thus f is singular, contradicting the bijective property of a biholomorphism. $\square$

To get a little bit of fresh air we introduce the Bergman formula

$$f(z) = \frac{(n-1)!}{(2\pi i)^n} \int_{\partial B(0,1)_{\mathbb{C}^n} = S_{\mathbb{C}}^n} \frac{f(\zeta)\omega'(\bar{\zeta})\wedge\omega(\zeta)}{(1-\zeta^\dagger z)^n},$$
$$\omega(\eta) = d\eta n$$
$$\omega'(\eta) = \sum_{k=1}^n (-1)^{k-1}\eta_k d\eta_1 \wedge \cdots d\hat{\eta}_k \cdots \wedge d\eta_n$$

and say goodbye to n.[10]

Theorem 2.5 (Hartogs 1906). *Assume $f \in O(B(0,1+\epsilon)_{\mathbb{C}^n} \sim B(0,1-\epsilon)_{\mathbb{C}^n})$, $1 > \epsilon > 0$. Then f has a unique holomorphic continuation throughout the ball.*

Proof. Thus use the Bergman formula and obtain

$$f(z) = \frac{(n-1)!}{(2\pi i)^n} \int_{\partial B(0,1)_{\mathbb{C}^n} = S_{\mathbb{C}}^n} \frac{f(\zeta)\omega'(\bar{\zeta}) \wedge \omega(\zeta)}{(1-\zeta^\dagger z)^n}$$

clearly being unique(which is seen by for example subtracting another different function g from f which is obtained from the bove integral), defined throughout the ball and(Differentiation under the integral sign being permitted by the standard theorems of calculus.) by

$$\bar{\partial}|_{\mathbb{C}_z^n} f(z) = \frac{\bar{\partial}|_{\mathbb{C}_z^n} (n-1)!}{(2\pi i)^n} \int_{\partial B(0,1)_{\mathbb{C}^n}=S_{\mathbb{C}}^n} \frac{f(\zeta)\omega'(\bar{\zeta})\wedge\omega(\zeta)}{(1-\zeta^\dagger z)^n}$$
$$= \frac{(n-1)!}{(2\pi i)^n} \int_{\partial B(0,1)_{\mathbb{C}^n}=S_{\mathbb{C}}^n} \bar{\partial}|_{\mathbb{C}_z^n} \frac{f(\zeta)\omega'(\bar{\zeta})\wedge\omega(\zeta)}{(1-\zeta^\dagger z)^n} = 0$$

a holomorphism. $\square$

We denote by $H(X,Y) = O(X,Y)$ the set of holomorphisms from the complex manifold X to the complex manifold Y. A holomorphism from one complex manifold to another complex manifold is a smooth map for which the local coordinate components f_σ satisfy Cauchy-Riemann conditions $\bar{\partial} f_\sigma = 0$, $1 \leq \sigma \leq n, n = dim_{\mathbb{C}}(X)$ on each chart. The following theorems will treat some topological properties of holomorphisms.[11]

Theorem 2.6 (Generalized Liouville Theorem I). *Assume*

$$f \in H(X,Y)$$

---

[10]Puih!-I use a much better symbol in private, but hey- that's what Latex had to offer.

[11]Please note that a manifold cannot have a boundary, this is important when considering the following theorems.

for X a compact manifold. Then f is either surjective or singular.

Proof. Since X is compact $f(X)$ is compact hence closed. A nonsingular holomorphism is an open map hence $f(X)$ is open since X is in the topology of X. But then $f(X)$ is open,closed and non-empty, the only set in Y with such a property is Y itself. $\square$

Corollary 2.6 (Generalized Liouville Theorem II). *Assume $f \in H(X,Y)$, X compact, Y non-compact. Then f is a constant.*

Proof. Since $f(X)$ is compact and Y is not f cannot be surjective, hence the corollary falls in view of the above statement. $\square$

Theorem 2.7. *Assume that $f \in H(X,Y)$, X a compact manifold. Then f is a biholomorphism IFF f is injective.*

Proof. Since f is injective f is non-singular hence surjective. f is obviously continuous, hence it now suffices to note that every closed subset K in the compact X is again compact and so $f(K)$ is compact hence closed and so f is closed as map. These properites combined yield that f is a biholomorphism.

$\square$

Example 2.2. *Here comes an application of several complex variables.*

Theorem 2.8 (On Solutions to the Laplace Equation.). [12] *Let $M \subset X$, X a complex manifold of dimension n and M a manifold with boundary. Assume further M to be biholomorphically equivalent to a set with boundary given by*

$$D = \{\zeta \in \mathbb{C}^n_\zeta \cong \mathbb{R}^{2n} | Im(\zeta_n) = 0\}.$$

Then $\square\phi = 0$ with $\phi|_{\partial M} \equiv B_1$ and $d_c\phi|_{\partial M} \equiv B_2$ for ϕ real given, d_c denoting codimensional exterior derivative, has a unique solution given by

$$\phi = \frac{\Phi + \bar{\Phi}}{2} + \frac{\int_{\bar{\zeta}}^{\zeta} d_c\Phi}{2i}$$

where $\square = \sum_{1 \le \sigma \le 2n} \partial^2_\sigma$ and Φ is the codimensional continuation of ϕ from D to $\mathbb{C}^n_\zeta$, where $\phi = Re(\Phi)$. [13]

---

[12]Note that the powers of algebra and analytic continuation permit us to reduce the massless Dirac and Klein-Gordon(i.e d'Alemberts equation in the K.-G. case) to the treated case. Please also note that the inhomogeneous equations are essentially only a question of shifting boundary values.

[13]Note the elementary relation $v = \int *du$ for $f = u+iv, f \in O(\mathbb{C})$,* being Hodge star.

Proof. Note that $ker(\partial\bar{\partial})$ is invariant under biholomorphisms. Also $\frac{\Phi+\bar{\Phi}}{2} \in (O^+ + O^-)(\mathbb{C}^n) = ker(\partial\bar{\partial})$. We note $\frac{\int_{\bar{\zeta}}^{\zeta} d_c\Phi}{2i}|_{\partial M} = Im(\Phi|_{\partial M}) = 0$ since ϕ was real and so B_1 holds.

$$d_c\frac{\Phi+\bar{\Phi}}{2} + \frac{d_c\int_{\bar{\zeta}}^{\zeta} d_c\Phi}{2i} = o + Im(id_c\phi) = d_c\phi$$

follows from noting that $\frac{d\phi(x+it)}{dt}|_{t=0} = \phi'i$, since then a infinetisimal change of value of ϕ in codimensional direction relative to D has to be imaginary implying $d_c\frac{\Phi+\bar{\Phi}}{2}|_{\partial M} = Re(d_c\phi)|_{\partial M} = 0$ and

$$\frac{d_c\int_{\bar{\zeta}}^{\zeta} d_c\Phi}{2i}|_{\partial M} = \frac{d_c\int_0^{\zeta} d_c\Phi + d_c\int_{\bar{\zeta}}^0 d_c\Phi}{2i}|_{\partial M} = \frac{d_c\phi(x+it)i + d_c\phi(x-it)i}{2i}|_{t=0} = d_c\phi$$

thus also B_2 is true. Uniqueness now follows from the standard theorems of PDE's. $\qquad\square$

Excercise 2.4. *Find the Taylor series expansions of $g_1 = \cosh(\zeta_1\zeta_2 + 3i)$, $g_2 = \frac{1}{\zeta_1\zeta_2 - 2i\zeta_2 - 3\zeta_1 + 6i}$ around $(1+i,1)$ and $(1,0)$ in $\mathbb{C}^2$. Is it possible to find a Taylor series expansion for g_2 at $(2i,3)$? Can you find a point $z \in \mathbb{C}^2$ such that there is no expansion for g_2 in any neighboorhood of that point? What if you consider the one point compactification $\mathbb{C}^2 \cup \infty = \widehat{\mathbb{C}^2}$, how many points are omitted by g_1 ?*

When doing these problems it might be worth remembering that $\frac{1}{(\zeta-z)^n} = \sum_{|\sigma|=0}^{\sigma=\infty} \frac{(z-w)^\sigma}{(\zeta-w)^{\sigma+n}}$, however you will have to find the convergence polydisc on your own!

14

2.3. Power Series of Several Complex Variables.

In this short section we shall look into the properties of power series - mostly where they converge and a special kind of set called Reinhardt domain. Define a Reinhardt domain to be an open set Ω such that $\zeta \in \Omega$ implies $(e^{i\theta_1}\zeta_1, e^{i\theta_2}\zeta_2, \cdots, e^{i\theta_n}\zeta_n) \in \Omega$, $\theta_i \in [0, 2\pi)$. One says that a set $D \in \mathbb{R}^{2n} \cong \mathbb{C}^n$ is logarithmically convex if $\log\tau(D) \equiv \{\xi \in \mathbb{R}^n : \xi = (\log|z_1|, \log|z_2|, \cdots, \log|z_n|), z \in D\}$ is convex. We have

Lemma 2.2. *The convergence domain of power series $\sum_{\alpha\in\mathbb{N}^n} a_\alpha z^\alpha$ is logarithmically convex.*

Proof. Assume ξ and η are points in $log\tau(D)$ then we are to show $t\xi + (1-t)\eta \in log\tau(D)$. Choose $p, q \in D$, $\lambda \in (0,1)$ for $t \in (0,1)$. By convergence at λp and λq we have $\exists M$;

$$|a_\nu|\lambda^{|\nu|}|p^\nu| \leq M, \quad |a_\nu|\lambda^{|\nu|}|q^\nu| \leq M, \quad M < \infty$$

---

[14]By now the reader has seen more than one example of a singularity, and probably correctly realized that they are extended objects in $\mathbb{C}^n$, $n \geq 2$.

$\forall \nu \in \mathbb{N}^n$. It follows

$$|a_\nu||\lambda^{|\nu|}||p^\nu|^t|q^\nu|^{1-t} \leq M$$

and thus by Abel's lemma we have convergence in a neighbourhood of $a_t = (|p_1|^t|q_1|^{1-t}, |p_2|^t|q_2|^{1-t}, \cdots, |p_n|^t|q_n|^{1-t}) \in D$, i.e $t\xi + (1-t)\eta \in log\tau(D)$. $\square$

Lemma 2.3. *Let D be Reinhardt. If D is logarithmically convex, then D is convex with respect to monomials in the coordinates z_i.*

Proof. We must prove that the hull [15] of every compact set K contained in D is strictly contained in D. By compactness of $\hat{K}$ it suffices to prove $\hat{K} \cap \partial M = \emptyset$. Since K has a finite cover Q of polydiscs Q_i it suffices to prove that $\hat{Q} \cap \partial M = \emptyset$. Let $p \in \partial D$ with all components nonzero. Then $p^* \equiv log|p| \in \partial log\tau(D)$ and thus by convexity $\exists$ a linear $L(\xi) = \sum_{j=1}^n \mu_i \xi_i, \mu_i \in \mathbb{R}_+$ satisfying $L(\xi) < L(p^*), \xi \in log\tau(D)$. Let Q^* be the finite set of points in $log\tau(D)$ which corresponds to the polycentres in $Q = \{Q_i\}$. One can find rational $\alpha > \mu > 0$ sufficently close to μ_i such that for $\tilde{L}(\xi) = \sum_{j=1}^n \alpha_i \xi_i$ one has $\tilde{L}(\xi) < \tilde{L}(p^*), \xi \in Q^*$. The above equation has to be true after multiplication with the positve common denominator of $\alpha_1, \cdots, \alpha_n$, so we may assume $\alpha_i \in \mathbb{Z}_+^n$. Then the monomial $m_\alpha(z) = z^\alpha$ satisfies $|m_\alpha||_Q < |m_\alpha(p)|$ i.e $p \notin \hat{Q}$. For any remaining points $p \in \partial D$ we can always reorder the coordinates so that for some $1 \leq l < n$ one has $p_1 p_2 \cdots p_l \neq 0$ while $p_{l+1} = \cdots = p_n = 0$. If $\pi_l : \mathbb{C}^n \to \mathbb{C}^l$ is the projection then $log\tau(\pi_l(D)) \subset \mathbb{R}^l$ satisfies the convexity requirement and so $\pi_l(D)$ is convex. The preceding argument applied to $\pi_l(p)$ now gives a monomial m in the coordinates satisfying $|m||_Q < |m(p)|$, so $p \notin \hat{Q}$ even in this case.

$\square$

Theorem 2.9. *Let D be Reinhardt centered at the origin. Then the following are equivalent 1) D is the convergence region of a power series. 2) D is logarithmically convex. 3) D is monomially convex. 4) D is holomorphically convex. 5) D is a domain of holomorphy.*

Proof. 1) $\Rightarrow$ 2) $\Rightarrow$ 3) follows in view of previous lemmas and 3) $\Rightarrow$ 4) is trivial. 4) $\Rightarrow$ 5) is discussed in the next chapter and 5) $\Rightarrow$ 1) is likewise trivial by Taylors theorem. $\square$

Example 2.3.

$$\sum_{\alpha \in \mathbb{N}^n} \frac{|\alpha|^{|\alpha|} z^{2\alpha}}{\alpha^\alpha}$$

converges precisely in $B(0,1) \subset \mathbb{C}^n$.

---

[15]See the Oka-Cartan section, i.e. section 5, for the notion of hulls.

2.4. The Weierstrass preparation theorem. Let us define a function holomorphic at $a = (a_1, a_2, \cdots, a_n) = (a', a_n)$ with $f(a) = 0$ to be $z_n - regular$ of order $k \in \mathbb{N}$ at a , if $g(z_n) = f(a', z_n)$ has a zero of order k at $z_n = a_n$, i.e if

$$g(a_n) = g'(a_n) = g''(a_n) = \cdots g^{(k-1)}(a_n) = 0$$

Furthermore we say that a function has total order k at some fixed point if the last nonvanishing homogeneous polynomial is of order k in the homogeneous expansion of the function around the point.

Lemma 2.4. *If f is a holomorphic function of total order $k < \infty$ at a point w , then after a suitable nonsingular linear change of coordinates in $\mathbb{C}^n$ the function will be regular of order k in z_n at the point w .*

Gunning and Rossi. Set w to be the origin, the function has then the homogeneous expansion around it. Select any point $a = (a_1, a_2, a_3, \cdots) \neq 0$ such that $P_k \neq 0$, P_k denoting the homogeneous polynomial of order k, since a is not nill there are constants such that the linear change of coordinates

$$z_i = a_i \zeta_n + \sum_{j=1}^{n-1} b_{ij} \zeta_j$$

is nonsingular. In these new coordinates our function $g(\zeta) = f(z(\zeta))$ still has total order k, and moreover $g_k(0, \cdots, 0, 1) = f_k(a_1, \cdots, a_n) \neq 0$; but then g is regular of order k in ζ_n at the origin, as desired. $\square$

Lemma 2.5. *Suppose f is holomorphic at the origin and $f(0) = 0$ and furthermore is $z_n - regular$ of non-vanishing order k. Then for sufficiently small $\delta_n > 0$ there is $\delta' > 0$, such that for each fixed $z' \in B(0, \delta')^n$ the equation f(z',z)=0 has precisely k solutions (counted with multiplicities) in the disc $|z_n| \leq \delta_n$.*

Gunning and Rossi. By hypothesis, for each sufficiently small $\delta_n > 0$, $g(z_n) = f(0', z_n)$ is holomorphic on $|z_n| \leq \delta_n$, g has a zero of order k at 0, and $g(z_n) \neq 0$ for $0 < |z_n| \leq \delta_n$. By continuity of f and Rouche's theorem, there is $\delta' > 0$ such that the conclusion of the lemma holds for all $z' \in B(0, \delta')^n$. $\square$

When dealing with analytic sets of the type $f = 0$ one might want to have a factorization of the function in question in the variable z_n and an algebraic unit u, i.e in this context a function $u \neq 0$ in some neighbourhood in consideration. To be more precise, given a function f of nonzero total order at a point z_0 one would like to write $f = u(z_n - \phi)^{\alpha}, u \neq 0 \; \forall z \in N(z_0), \; \phi = \phi(z_1, \cdots, z_{n-1})$. Can this be done and are these zeros considered in the variable z_n holomorphic functions? Unfortunately the answer is in general negative if we require the zeros to be holomorphic. However it turns out that when looking at the maximally symmetric functions of the zeros ϕ_i that are generated when expanding the above expression we discover that they are holomorphic. These expanded polynomials have a fancy name - Distinguished Weierstrass pseudopolynomials.

Theorem 2.10 (Weierstrass preparation theorem). *Assume that f is holomorphic at 0, $f(0) = 0$, and suppose f is $z_n - regular$ of order $k \geq 1$. Then there is a unique factorization*

$$f = \omega u$$

on some polydisc $B(0, r)^n$, where $\omega \in O(B(0_n, r_n))$ is a distinguished pseudo-polynomial of degree k at the origin and u is a unit in the polydisc considered.

Gunning and Rossi. Let f be a function holomorphic in a neighboorhood of the polydisc $B(0, r)^n$ and regular of order k in z_n at the origin. By a previous lemma, there is a polydisc $B(0, \delta)^n \subset B(0, r)^n$ such that for every point $(z_1, \cdots, z_{n-1}) \in B(0, \delta)^n$ f has k zeros in $|z_n| < \delta_n$; These zeros will be denoted by $\phi_1, \cdots,$ and for these $\phi_i(0, 0, \cdots) = 0$, $|\phi_i(z_1, z_2, \cdots)| < \delta_n$ will hold. Set

$$h = z_n^k + a_1 z_n^{k-1} \cdots + a_k$$

where the a_j' are the elementary symmetric functions, which actually are holomorphic in the values ϕ. It is easy to see that the a_j's are holomorphic in $B(0, \delta)^n$, holding $(z_1, z_2, \cdots)$ fixed(See Ahlfors[4], page 154)

$$\sum_i \phi_i{}^r = \frac{1}{2\pi i} \int_{|\zeta|=\delta} \frac{\partial f}{\partial \zeta} \frac{\zeta^r}{f}$$

The function $f(\cdots, z_{n-1}, \zeta)$ is nonzero on $|z_1| < 1, |z_2| < 1, \cdots |z_n| = 1$, hence the power sum above is holomorphic on $B(0, \delta)^n$. Since the elementary symmetric functions are polynomials in the power sum, they are also holomorphic in the same polydisc. Moreover, $a_j(0, 0 \cdots, 0) = 0$ since $\phi_i(0, \cdots, 0) = 0$. Consequently, the function represents a Weierstass polynomial. The polynomial h is clearly the unique Weierstass polynomial having the same zeros as the function f in $B(0, \delta)^n$. To complete the proof of the theorem, we need to show that the quotient $u = \frac{f}{h}$ is holomorphic and non-vanishing in $B(0, \delta)^n$. This quotient is by construction holomorphic and non-vanishing in $|z_n| < \delta_n$. Setting M to be a least upper bound of $|f|$ in $\bar{B}(0, \delta)^n$ and setting $m > 0$ to be the greatest lower bound of $|h|$ on $|z_1| \leq \delta_1, |z_2| \leq \delta_2, \cdots, |z_n| = \delta_n$ it follows from the maximum principle in one variable z_n that $|u| \leq \frac{M}{m}$ $\forall \, z \in \bar{B}(0, \delta)^n$. Then by the Riemann extension theorem [16] u is indeed holomorphic and thus the non-vanishing property already being noted the theorem is concluded. $\qquad\square$

Corollary 2.7 (Implicit Function Theorem). *Assume $f \in O(B(0, r)^{(n)})$ and regular of order 1 in z_n at z_0; Then $\exists \, B(0, \delta)^{(n)} \subset B(0, r)^{(n)}$ and a unique holomorphism $\phi(z_1, \cdots, z_{n-1})$ such that $\phi = z_{0n}$.*

Proof. Suppose f is regular of order in z_n of order 1, then we can uniquely write

$$f = u(z_n - \phi(z_1, z_{n-1}))$$

u being a unit and ϕ a holomorphism the corollary falls. $\qquad\square$

2.5. **Oka-Cartan Theory.** In this section we shall explore one of the most fundamental concepts of complex analysis, namley the phenomenon of continuation. We follow Vitushkin[6], and shall start with some definitions and then survey results to try to get a general feel for the subject. The envelope of holomorphy $\tilde{D}$ of a domain or compact set D is the largest set to which all functions holomorphic on D extend holomorphically, and is in general a multisheeted *Riemann domain* over $\mathbb{C}^n$. A domain of holomorphy is it's own envelope, i.e $D = \tilde{D}$, and it is also often called holomorfically convex[17]. Closley related to the envelope concept is the notion of hulls [18] with respect to some class or other of functions - there are lot's of them- and two of the most prominent are the polynomial hull and the holomorphic hull. The polynomial hull of a set $D \subset \mathbb{C}^n$ is the set of all $z \in \mathbb{C}^n$; $\forall$ polynomials $P(\zeta)$; $|P(z)| \leq \sup_{\zeta \in D} |P(\zeta)|$. [19] Every smooth curve is holomorphically convex but the polynomial hull of the curve is in general non-trivial, for example if it is closed and has no selfintersection then the hull is either trivial or it is a one dimensional analytic set whose boundary coincides with the curve. We call a set analytic if it is described as $ker(f)$, $f : \mathbb{C}^n \to \mathbb{C}^m$, i.e given as the common

---

[16] Not discussed in this text, but states that $\nexists$ holomorphisms with isolated point singularities in several complex variables, opposite to the one variable case(See Chirka among others in [6], Gunning and Rossi[2], Range[1]).

[17] The reason for this similar to the reason that people talk about projective algebraic varieties-someone made a theorem. (Although this time it's not Chow!)

[18] Which we denote by a hat.

[19] It might be worth knowing that the union of three or less disjoint balls in $\mathbb{C}^n$ is polynomially convex.

zeros of a finite set of holomorphisms. So why are domains of holomorphy of such an interest? Well, in these domains we can solve lots of classical problems in complex analysis like finding integral representations, polynomial approximations, solving Cauchy-Riemann equations and solving the problem of division.[20] We give two examples namely polynomial polyhedra, which are defined by a system of inequalities $|P_i| \leq 1$, P_i being polynomials, and strictly pseudoconvex domains. We say that a set is strictly pseudoconvex if in some neighbourhood of each of it's boundary points is strictly convex for a suitable choice of coordinates. Suppose the hypersurface bounding a the domain in question is described by $\rho(\zeta, \bar{\zeta}) = 0$ then

Theorem 2.11 (Levi 1910). *In order for M with a C^2 boundary to be strictly pseudoconvex it is sufficient that $\partial\bar{\partial}\rho$ be positive definite.*[21] [22]

We also have

Theorem 2.12 (Oka's Theorem). *Every strictly pseudoconvex set is holomorphically convex and conversely any domain of holomorphy can be exhausted from the interior by strictly pseudoconvex sets.*

Boundary points of a domain of holomorphy are not equivalent. An important rôle is played by the Shilov boundary $\mathfrak{S}(D)$ which is the smallest closed subset of the boundary of D; $\forall f \in \mathcal{C}(\bar{D}) \cap O(D)$ we have $|f(z)| \leq sup_{\zeta \in \mathfrak{S}(D)}|f(\zeta)|$. For a ball the Shilov boundary coincides with the topological boundary whereas for the polydisc $B(0,1)^n$ it coincides with $(S_{\mathbb{C}}^o)^n$, i.e the n-dimensional torus. Actually the following holds;

Theorem 2.13 (Basener 1973). *Assume the boundary to be C^2,/ then $\mathfrak{S}(D) =$ closure of strictly pseudoconvex points.*

For domains of holomorphy a strong maximum principle holds.

Theorem 2.14. *Let D be a domain of holomorphy and f a nonconstant holomorphism continuous on $\bar{D}$ assuming a maximum on $z \in \bar{D}$. Then $z \in \mathfrak{S}(D)$.*

It is also known that

Theorem 2.15. *If D is a domain of holomorphy it falls that $H_k(D, \mathbb{C}) \cong \{0\}, k > n$. and actually if it is polynomially convex* [23] $H_n(D, \mathbb{C}) \cong \{0\}$

---

[20]Later we will talk of another kind of 'domains' where we still can solve our problems.

[21]Note the resemblence to a Kahler potential(See Nakahara[8] and Kobayashi[13]).

[22]i.e it all comes down to looking at the eigenvalues of the Hessian $(\frac{\partial^2 f}{\partial \zeta_\sigma \partial \zeta_\rho})$.

[23]Serre 1953, Andreotti and Narasimhan 1962

also follows. The Dolbeault cohomomology groups of a domain of holomorphy are trivial.

Theorem 2.16 (Oka, Cartan 1950). *The following is equivalent; 1) D is a domain of holomorphy. 2) The problem of division is solvable in D.* [24] *3) Each function holomorphic on a complex submanifold of D is the restriction of some function holomorphic throughout D.*

The first cousin problem is said to be solvable if it is possible to construct a meromorphism with given poles, the second if it is possible to construct a holomorphism with given zero's.

Theorem 2.17 (Oka 1937, 39, Serre 1953). *The first cousin problem is solvable on a domain of holomorphy, and if in addition $H^2(D, \mathbb{Z}) \cong \{0\}$ then the second cousin problem is solvable.*

We call the manifold X a Stein manifold if is 1) holomorphically convex, i.e if the holomorphically convex hull of each compact set in X is compact in X, and secondly 2) $\exists$ on X a finite family of holomorphic functions such that each point has a neighbourhood in which these functions seperate points. Each domain of holomorphy is an example of a Stein Manifold, and so are closed complex submanifolds of $\mathbb{C}^n$. On Stein manifolds just as on domains of holomorphy, some problems are solvable, namely; the division problem, the $\bar{\partial}$-equation, the first cousin problem is solvable and also the second provided the second integer cohomology is trivial.

Theorem 2.18 (Cartan 1953). *X is Stein iff $H^1(X, S) \cong \{0\}$. $H^1(X, S)$ denoting first cohomology group with coefficents in an arbitrary coherent analytic sheaf.*

Theorem 2.19 (Oka-Grauert Principle). *Let X be Stein. Then the second cousin problem has a holomorphic solution iff it has a continuous solution.* [25]

Excercise 2.5. *Derive a complex version of the Main theorem of integration calculus by using stokes formula on cochains of bidegree (1,1) one a one dimensional complex manifold. When calculating with this formula the reader will discover that it is not of much practical use - Can it be extended to be more useful? Discuss plausible applications of*

---

[24]This is a *global* result on D, the Weierstrass result is a *local* statement.

[25]The token principle emanates from the filosofy that seems to come along with Stein manifolds - Any holomorphic problem is a continuous problem and vice versa. For example deformations of holomorphic vector bundles on a Stein manifold bases can be looked upon either continuously or holomorphically, thus we may sometimes use holomorphisms to work on continuous deformations of bundles!

knowing $H_{\bar{\partial}}^{(p,q)}(\Sigma_g)$, Σ_g being a compact complex 1-dimensional manifold with genus g.

Excercise 2.6. *In texts in complex geometry authors often name a special form called the Kahler form Ω. Depending on author one writes*

$$\Omega = \frac{i}{2}h_{\sigma\rho}d\zeta_\sigma \wedge d\bar{\zeta}_\rho$$

or twice that [26], h being a hermitian metric on a complex manifold. If this form is closed the manifold is called Kahler. Kahler manifolds are a very restricted class of manifolds, yet by coincidence at least 3 of the canonical manifolds studied are Kahler. $\mathbb{C}P^n$ has a 'standard' metric often used, called the Fubini-Study metric, which satisfies that requirement among others. Find examples of other Kahler manifolds. [27]

2.6. Some Elementary Cohomology.

In this section we shall get aquainted with some algebraic topology and its applications to analysis. In one as in several variables, either real or complex, it is important to understand basic algebraic topology to understand the consequences it has for evaluation of integrals, solutions of partial differential equations, analytic continuation, integral representaions and functorial properties of various analytical objects like the Cauchy integral formula. It also intimatley related to the concept of a fibration, finite or infinite-dimensional(which is something that the present text will not go into), which is nowadays as fundamental to geometry, topology and analysis as tying your shoe laces before going outdoors.

Example 2.4. *The smooth kernal of the Laplacian Δ on a closed compact orientable manifold X has dimensionality $b^0 = dim_\mathbb{R}(H^0(X,\mathbb{R}))$.*

Example 2.5. *The value of an integral incerceling a pole is invariant under homotopies of the integration contour .*

Example 2.6. *The existance of a primitive to a meromorphism $f(\zeta)$ over $\mathbb{C}$ is determined by the topology of the Riemann surface that is the range of the meromorphism considered under all analytic continuations, called the the global analytic sheaf of the function (See Ahlfors[4]). If it has non-trivial topology (monodromy $\equiv$ isomorphism of homotopy groups is an often used word.) over the subset of $\mathbb{C}$ considered one will not find a primitive.*

---

[26]We fix it to be the latter for now.

[27]This by the way provides us with a good example of the fact that closed forms need not be exact. $H_{\bar{\partial}}^{(1,1)}(\mathbb{C}P^3,\mathbb{C}) \cong \mathbb{C}$ nontriviality actually generated by Ω.

Example 2.7. *To be able to use the general Martinelli-Bochner formula, which is an integral representation over arbitrary domains in $\mathbb{C}^n$, one has to first know what a cycle is.*

Let us get aquainted with the elementary calculus(We will unfortunately not go into homotopy in this text!). Assume M to be a smooth real manifold of dimension n and consider $\Lambda M \equiv \Gamma(M, \Lambda T^* M)$ [28]. We have the sequence

$$\{0\} \quad \overset{d}{\to} \quad \Lambda^0 M \quad \overset{d}{\to} \quad \Lambda^1 M \quad \overset{d}{\to} \quad \cdots \quad \overset{d}{\to} \quad \Lambda^n M \quad \overset{d}{\to} \quad 0$$

$$\| \mathbb{R}$$

$$C^\infty(M) \quad .$$

Pointwise on ΛM a smooth map $f : M_y \to X_x$, x a coordinate system on X,generates the map $f^* \omega = f^* \omega_\alpha dx_{\alpha 1} \wedge dx_{\alpha 2} \wedge \cdots \wedge dx_{\alpha p}$ $= \omega_\alpha f^* dx_{\alpha 1} \wedge dx_{\alpha 2} \wedge \cdots \wedge dx_{\alpha p} = \omega_\alpha \frac{\partial(x_{\alpha 1}, x_{\alpha 2}, \cdots, x_{\alpha p})}{\partial(y_{\beta 1}, y_{\beta 2}, \cdots, y_{\beta p})} dy_{\beta 1} \wedge dy_{\beta 2} \wedge \cdots \wedge dy_{\beta p}$. If f is a diffeomorphism this determinant is seen to be nonsingular and thus since by definition this map is seen to be a homomorphism of exterior algebras noting , f still a diffeomorphism, that *each* $\omega \in \Lambda X_x$ has a *unique* $\eta \in \Lambda M_y$ it falls that f^* is an algebraic isomorphism of exterior algebras ΛM under such circumstances.[29] But we also note that $[d, f^*] = 0$ since it can not matter in which order we take the diffrential or change the coordinates on M for example. Thus the diagram

$$C^\infty(X)$$

$$\| \mathbb{R}$$

$$\{0\} \quad \overset{d}{\to} \quad \Lambda^0 X \quad \overset{d}{\to} \quad \Lambda^1 X \quad \overset{d}{\to} \quad \cdots \quad \overset{d}{\to} \quad \Lambda^n X \quad \overset{d}{\to} \quad 0$$

$$f^* \downarrow \qquad f^* \downarrow \qquad \qquad f^* \downarrow$$

$$\{0\} \quad \overset{d}{\to} \quad \Lambda^0 M \quad \overset{d}{\to} \quad \Lambda^1 M \quad \overset{d}{\to} \quad \cdots \quad \overset{d}{\to} \quad \Lambda^n M \quad \overset{d}{\to} \quad 0$$

$$\| \mathbb{R}$$

$$C^\infty(M)$$

---

[28]See previous sections.

[29]This f^* has a fancy name by the way-It's called the pullback of f for obvious reasons.

is seen to commute and so $\frac{ker(d|_{C^p(M,\mathbb{R})})}{Ran(d|_{C^{p-1}(M,\mathbb{R})})} \equiv \frac{Z^p(M,\mathbb{R})}{B^p(M,\mathbb{R})} \equiv H^p(M,\mathbb{R})$[30] has to be invariant under diffeomorphisms of M, i.e is functorial. Hence we have something which measures topological [31] inequivalence. If the M is an open convex subset of $\mathbb{R}^n$ we can always pull it back to a point, hence in those cases $H^p(M,\mathbb{R}) \cong \{0\}$.

Using the same line of reasoning with the additional requirement that f be a biholomorphism(i.e a holomorphic diffeomorphism) one can prove that

$$
\begin{array}{ccccccccc}
C^\infty(X) & & \ncong & & O(X) & & & & \\
\bar{\partial} & \| & \bar{\partial} & & & & \bar{\partial} & & \bar{\partial} \\
\{0\} & \to & \Lambda^{(0,q)}X & \to & \Lambda^{(1,q)}X & \to & \cdots & \to & \Lambda^{(n,q)}X & \to & 0 \\
& & f^* \downarrow & & f^* \downarrow & & & & f^* \downarrow & & \\
\bar{\partial} & & \bar{\partial} & & & & \bar{\partial} & & \bar{\partial} \\
\{0\} & \to & \Lambda^{(0,q)}M & \to & \Lambda^{(1,q)}M & \to & \cdots & \to & \Lambda^{(n,q)}M & \to & 0 \\
& & \| & & & & & & & & \\
& & C^\infty(M) & & \ncong & & O(M) & & & &
\end{array}
$$

q fixed,$0 \leq q \leq n$, commutes. Consequently one defines $H_{\bar{\partial}}^{(p,q)}(M,\mathbb{C}) = \frac{ker(\bar{\partial}|_{\Lambda^{(p,q)}(M,\mathbb{C})})}{Ran(\bar{\partial}|_{\Lambda^{(p,q-1)}(M,\mathbb{C})})}$ on complex spaces.

Excercise 2.7. *Prove that the above diagram commutes, i.e that Dolbeault Cohomology groups behave functorially under biholomorphisms.*

We have now briefly mentioned the basic facts about the two most common cohomology theories, but there are more. However this text cannot address the needs of someone looking for a textbook in algebraic topology so we will have to omit these and instead refer to the topology course which is give in parallel with this course. But before closing we will mention a couple of useful theorems.

Theorem 2.20. *Two homotopically equivalent manifolds have isomorphic cohomology groups.*

---

[30]The relation $\omega \sim \eta$ is said to hold if $\omega = \eta \bmod B^p(M,\mathbb{R})$ under addition, and it is it that defines the quotient. Two forms that are equivalent in this sense are called cohomologous.

[31]Topological invariance is invariance under homeomorphisms. As shown by Milnor in the 50's in dimension 7 and Donaldson in dimension 4 (1984) there are toplogically equivalent manifolds which are not diffeomorphic. However every diffeomorphism is trivially seen to be a homemorphism so this is why we restrict our attention to diffeomorphisms in this discussion(See Novikov[17]).

Proof. See Novikov, Fomenko and Dubrovin, Modern Geometry III [7], theorems 1.3, 1.5 and lemma 1.4. □

Lemma 2.6 (Poincare's lemma). *Let M be diffeomorphic to the unit ball in $\mathbb{R}^n$. Then $d\omega = 0$ IFF $\omega = d\eta$, $\omega \in \Lambda^{(p+1)}M$, $\eta \in \Lambda^p M$, $p \geq 0$.*

Proof. Note that the origin is a deformation retract of the unit ball. Thus all cohomology groups trivialize in view of the previous theorem. By invariance of cohomology groups under diffeomorphisms the assertion now falls. □

Theorem 2.21 (Poincare duality). *Let M be compact, closed[32] and orientable. Then $H^{n-q}(M, \mathbb{F}) \cong H^q(M, \mathbb{F})$, $\mathbb{F}$ being the complex or real field.*

Theorem 2.22 (Non-abstract deRham Theorem). *Let M be compact closed. Then $H_p(M, \mathbb{C}) \cong H^p(M, \mathbb{C})$, being finite dimensional and dual to each other.*

Theorem 2.23 (Hodge's Theorem). *Let M be compact Kahler. Then*

$$H^p(M, \mathbb{C}) \cong \bigoplus_{p=r+s} H_{\bar{\partial}}^{(r,s)}(M, \mathbb{C})$$

Lemma 2.7 (Dolbeaults's lemma). *Let M be Biholomorphic to the unit ball in $\mathbb{C}^n$. Then $\bar{\partial}\omega = 0$ IFF $\omega = \bar{\partial}\eta$, $\omega \in \Lambda^{(p,q+1)}M$, $\eta \in \Lambda^{(p,q)}M$, $p \geq 0, q \geq 0$.*

The proofs of the four previous theorems would necessitate a far too deep plunge into various subjects and are therefore omitted.

Excercise 2.8. *Calculate $H^1(U(1), \mathbb{C})$ via Poincare duality.*

Excercise 2.9. *Prove $H_{\bar{\partial}}^{(0,0)}(M, \mathbb{C}) \cong O(M)$. Hodge's theorem should then imply something about the number of holomorphisms on a compact kahler manifold. The relation derived actually holds on any compact complex manifold.*

Excercise 2.10. *Digress on the relation between $H^1(U(1), \mathbb{Z}) \cong H^1(S_{\mathbb{R}}^1, \mathbb{Z}) \cong \mathbb{Z}$ the Cauchy integral formula*

$$2\pi i \; n(\gamma, z) \; Res\{f, z\} = \int_\gamma f(\zeta), \; f \in \mathcal{M}(\Omega)$$

with a pole at z and holomorphic otherwise.

---

[32]i.e without boundary.

Excercise 2.11. *The function $f(\zeta) = \frac{1}{\zeta}$ is holomorphic in $\mathbb{C} \sim \{0\}$. Yet we can't for arbitrary paths γ in the mentioned set use the main theorem of integral calculus. Why is that and what is the relevance of the homology class* [33] *of the path γ around the origin?*

Excercise 2.12. *Assume that for $\mathcal{B} : \mathbb{C}^2 \to \mathbb{C}$ defined pointwise by*

$$\mathcal{B}(z, \zeta) \equiv \int_0^1 t^{z-1}(1 - t)^{\zeta-1}dt$$

we knew $\mathcal{B}(z, \zeta) = \frac{\Gamma(z)\Gamma(\zeta)}{\Gamma(z+\zeta)}$ on $B((1, 1 + i/\sqrt{2}), 1/2) \subset \mathbb{C}^2$. Extend this to hold on a larger subset of $\mathbb{C}^2$. [34]

---

[33]Often called index.

[34]Answer: At least $\mathbb{C}_+^2$.

3. Quaternionic, Octonionic and Clifford Analysis

In this section we shall outline the elementary aspects of quaternionic, octonionic and Clifford analysis. The main goal will be to get a holomorphic mathematical apparatus that works in some (analytical) sense on these spaces.

3.1. Basic Definitions of Hyperanalysis. We begin by noting that the operation of differentiation is well defined(as the usual limit) for real partial differentiations with noncommutative coefficients and that we can split the exterior derivative/use coordinates in a manner similar to the $\partial, \bar{\partial}, z, \bar{z}$ of complex analysis in these spaces. Let us define such coordinates for the quaternionic case first;

Definition 3.1. *We define quaternionic coordinates by*
$$\zeta = t + ix + jy + kz, \zeta^i = t + ix - jy - kz,$$
$$\zeta^j = t - ix + jy - kz, \zeta^k = t - ix - jy + kz,$$
$$(t, x, y, z) \in \mathbb{R}^4$$
and call ζ the holomorphic coordinate.

We also have a duality, often misleadingly confused to be a real inner product duality,[35]

Definition 3.2. $\zeta \mapsto \bar{\zeta} = t - ix - jy - kz$

and differentiations

Definition 3.3.
$$\partial_\zeta = \frac{\partial}{\partial\zeta} = \tfrac{1}{4}(\partial_t - i\partial_x - j\partial_y - k\partial_z),$$
$$\partial_{\zeta^i} = \frac{\partial}{\partial\zeta} = \tfrac{1}{4}(\partial_t - i\partial_x + j\partial_y + k\partial_z),$$
$$\partial_{\zeta^j} = \frac{\partial}{\partial\zeta} = \tfrac{1}{4}(\partial_t + i\partial_x - j\partial_y + k\partial_z),$$
$$\partial_{\zeta^k} = \frac{\partial}{\partial\zeta} = \tfrac{1}{4}(\partial_t + i\partial_x + j\partial_y - k\partial_z),$$

with domain P^1. We interrupt to define P^1 and some other objects that we will need in the following.

Definition 3.4. $\mathcal{O}_\mathbb{R}(\Omega, \mathbb{M})$ *is the algebra of real analytic $\mathbb{M}$-valued functions over Ω, $\mathbb{M}$ a field, algebra or similar object. P^n is the space of $\mathbb{M}$ valued polynomials i.e $\mathcal{O}_\mathbb{R}(\Omega, \mathbb{M}) \subset P^\infty = \cup_{n \in \mathbb{N}} P^n$.*

It is then a matter of simple calculus left to the reader to prove one of the things we need;

---

[35]This derives it's name by the fact that it is the projection of a duality defined on the entire Clifford algebra onto the lefthanded part. It is also a duality directly in the 'hyper' sense, just like on $\mathbb{C}$.

Lemma 3.1. *The above quaternionic coordinates are mutually orthogonal in the sense of differentiations, i.e*

$$\partial_\zeta \zeta^i = \partial_\zeta \zeta^j = \partial_\zeta \zeta^k = 0$$
$$\partial_{\zeta^i} \zeta = \partial_{\zeta^i} \zeta^j = \partial_{\zeta^i} \zeta^k = 0$$
$$\partial_{\zeta^j} \zeta^i = \partial_{\zeta^j} \zeta = \partial_{\zeta^j} \zeta^k = 0$$
$$\partial_{\zeta^k} \zeta^i = \partial_{\zeta^k} \zeta^j = \partial_{\zeta^k} \zeta = 0$$

Let us list some properties that can easily be checked. We list them as a lemma whose proof we leave to the reader;

Lemma 3.2.

$$\overline{\zeta \eta} = \bar{\eta} \bar{\zeta},$$
$$(\zeta \eta)^i = \zeta^i \eta^i, \cdots, (\zeta \eta)^k = \zeta^k \eta^k,$$
$$\zeta i = i \zeta^i, \cdots, \zeta k = \zeta^k,$$
$$i^{-1} \zeta i = \zeta^i, \cdots, k^{-1} \zeta k = \zeta^k,$$
$$\mathfrak{Re}\zeta = \tfrac{1}{2}(\zeta + \bar{\zeta}) = \tfrac{1}{4}(\zeta + \zeta^i + \zeta^j + \zeta^k)$$
$$\mathfrak{Im}\zeta = \mathcal{S}\zeta = \tfrac{1}{2}(\zeta - \bar{\zeta})$$
$$\mathfrak{Im}^i\zeta = \mathcal{S}^i\zeta = \tfrac{1}{4i}(\zeta + \zeta^i - \zeta^j - \zeta^k),$$
$$\mathfrak{Im}^j\zeta = \mathcal{S}^j\zeta = \tfrac{1}{4j}(\zeta - \zeta^i + \zeta^j - \zeta^k),$$
$$\mathfrak{Im}^k\zeta = \mathcal{S}^k\zeta = \tfrac{1}{4k}(\zeta - \zeta^i - \zeta^j + \zeta^k),$$

with $\mathcal{S}$ denoting either the spatial part [36], also called the imaginary cart, defined by $\mathcal{S}(\zeta) = \zeta - \mathfrak{Re}\zeta$. Im^l denotes the l:th spatial part, by convention taken to be real. There is also a real inner product $< \zeta, \eta >= \mathfrak{Re}\bar{\zeta}\eta = tt' + xx' + yy' + zz'$, $\zeta = t + ix + jy + kz$, $\eta = t' + ix' + jy' + kz'$, and a inner hyperproduct $< \zeta, \eta >= \bar{\zeta}\eta$.

We can also add the following easy lemma, again with proof left to the reader.

Lemma 3.3. *Set $\alpha = ix + jy + kz$ to be an 'angular' coordinate, and define $\zeta = t + \alpha$. Then, setting $\eta = e^\zeta$,*

$$e^\alpha = \cos(|\alpha|) + \tfrac{\alpha}{|\alpha|} \sin(|\alpha|),$$
$$ln\ \eta = ln|\eta| + arg\eta,$$
$$\arg\eta = \alpha + \tfrac{2\pi\alpha n}{|\alpha|}, n \in \mathbb{Z}.$$

---

[36]This nomenclature derives it's existance from physical considerations, which in string theory at times produce the convention of having circular coordinates on the world sheet. It would of course be just as justified to say angular part. Either way the real coordinate corresponds to either time or the logarithm of the modulus of a p-brane hypernumber from these p-brane considerations, depending on conventions, special case and coordinate choice. Notice that there goes a convention of left moving is holomorphic on a (hyper)p-brane along with this.

It is not uncommon to choose a branch of ln to make it singlevalued in hyperanalysis, with complex analysis as the canonical example.

We would like to define the concept of a hyperbolic quaternion for later purpose

Definition 3.5. *Upon continuation of the [37] real coordinates (x, y, z) to imaginary values in some representation of the quaternions one obtains hyperquaternions or hyperbolic quaternions $\mathbb{H}_{Hyp}$. In particular the hyperbolic quaternions have a Minkowski metric under the duality defined above.*

To do analysis in a satisfactory manner we have to have invertible objects at times, the main idea we shall use to do this in the use of hyperbolic Clifford algebras is continiuing the generators of the algebra corresponding to negative signs in the quadratic form that determines the algebra [38]-conversely we continue back to Euclidean signature if we are given an algebra with mixed signature so that we can do our calculations. Continuing back and forth like this we avoid trouble since we are only interested in having Cl^1 invertible for our purposes.

It should be pointed out that the above prescriptions also hold for the Clifford algebras Cl(1,3) and Cl(4,0) to get required orthogonality etc(then taking into account the effect of dualities. This affects the signs in various hyperreal Cauchy-Riemann operators.) In those cases the above coordinates are called lightcone coordinates[39].

For the general case including other hyperspaces one defines coordinates similar to the above with the crucial property contained in the lemma above. We presently have all the cases we will be needing in this thesis. We recommend the reader to read Bourbakis algebra for further elementary material on quaternions and hypercomplex algebra.

In the following it is understood that $\bar{n}$ means the multiindex with unit entries at the non-holomorphic coordinates and nill at the holomorphic, similarly n means the multiindex with unit entries at the holomorphic and nill at the antiholomorphic.

We shall denote a general hyperreal space $\mathbb{M}$, then meaning it over $\mathbb{R}$ and $\mathbb{M}^{\mathbb{C}}$ it's complexification. The hyperquaternionic space $\mathbb{H}^{\mathbb{C}}$ and the space $Cl(4, 0)^{\mathbb{C}}$ are common in physics ever since the days of Hamilton, Cayley, and Clifford. To Hamilton they were a necessary consequence of the rotations of 3-dimensional Euclidean space, and other

---

[37] The non-real coordinates of a quaternion are also called spatial coordinates while the real called the time coordinate. This seems to be more than an analogy at times-as already Pauli and Hamilton before him pointed out.

[38] Physicists call this process Wick rotation

[39] See Green,Schwarz, Witten-Superstring Theory I and II.

(meta)physical interpretations are known as well. In particular in the pre-war era it was among other things by use of quaternions that the Pauli exclusion principle came to daylight. A paper entiteled 'Uber das Pauliche eqvivalenzverbot', by Jordan and others, form the early 30:s bears witness to this. Similarly this reappears in spinor quantum electrodynamics and other gauge theory generalisations and actually amount to an economic way to do some angular momentum considerations in quantum field theory- simply embedding them into the Feynman rules. For example the gauge boson vertex

Example 3.1.

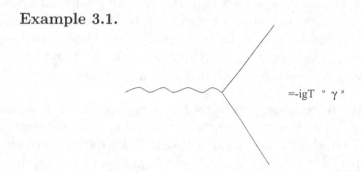

$$=\text{-igT}^{\alpha}\gamma^{\mu}$$

FIGURE 1. A Feynman rule that involves a hypernumber in $Cl(1,3) \subset Cl(4)^{\mathbb{C}}$. This hypernumber countes helicity states in order to conserve angular momentum in quantum theory. Note that the left-handed part of a complexified Clifford algebra lagrangian can and is commonly handeled by use of $\mathbb{H}^{\mathbb{C}}$. Schwinger in his days (approximately 1947-1948) invented the Schwinger rotation, writing $Cl(4)^{\mathbb{C}} = \gamma^0(\mathbb{H}^{\mathbb{C}} \oplus \bar{\mathbb{H}}^{\mathbb{C}})$, something that most physicists learn in prekindergarten as the 'Chiral' or 'Weyl' representaion of the Dirac algebra-The physicist name of $Cl(4)^{\mathbb{C}}$. The Dirac algebra also has convenient ways of implementing the operations C, P, T of quantum field theory. Experience from electroweak theory shows, however, that the left-handed or righ-handed objects of physics might differ at times, thus making $\mathbb{H}^{\mathbb{C}}$ a more fundamental object. It is unusual nowadays to see fermionic lagrangeans that are not broken up in lefthanded and right-handed piece in more practical use of the standard model physics.

with $T^{\alpha} \in \mathfrak{g}$ a lie algebra, g a real called charge, γ^{μ} genertors of a Clifford algebra and

$$\gamma^0 = \begin{pmatrix} 0 & 1 \\ 1 & 0 \end{pmatrix}$$

Finally use of hyperquaternions, or spinors as they are also called, is the simplest way to implement Wigner rotations, i.e the lift from $SO(1,3)$ to $SL(2,\mathbb{C})$ needed to implement Lorentz transformations on the Hilbert spaces of physics. Since $SO(3) \subset SO(1,3)$, $SU(2) \subset SL(2,\mathbb{C})$ and $SU(2)^{\mathbb{C}} = SL(2,\mathbb{C})$ this motivated the- somewhat abusive-name complex angular momentum among some particle physicists in 1960's. P. Deligne has further material on spin groups and complexification of varieties(of course a group can often be seen as a variety with some operations because of the nature of the conditions that define it) in his IAS lecture notes, in particular a good set of Dynkin diagrams and 'accidential' homo/isomorphisms in spin groups.

3.2. Some Algebraic Trivia of Hyperanalysis.
It is important to understand the elementary properties of hyperanalysis algebra in order to to be able to make calculations and to appreciate the differences, especially between complex and hypercomplex case, due to non-commutativity.

We would like to warn the reader-extreme care has to be taken when doing calculations, the differences are much more subtle than a simple transition from complex analysis and are reminicent of superalgebra with noncommutative coefficents.

Let us define the concept of a hypertensor.

Definition 3.6. $\zeta = x^A e_A$ *is the holomorphic coordinate for* $\mathbb{M}$ *a hyperspace (minimally) generated by* $\{e_A\}$, *x^A n real coordinates, n the cardinality of the set of generators, Thus* $\zeta \in \mathbb{M}^1 \equiv span_{\mathbb{R}}\{e_A\}$.

Definition 3.7. *A hypertensor T is a real multilinear map from a tensor product of modules V, V^*, V^* being a dual space defined by some duality, over $\mathbb{M}$ as follows; $T : \pi(\bigotimes V \bigotimes V^*) \mapsto \mathbb{M}$, π being some permutation of vector spaces in this tensor product. It is given in terms of hypercoordinates ζ as*

$$T = \pi[T^{\nu_1 \cdots \nu_m}_{\mu_1 \cdots \mu_n} d\zeta^{\mu_1} \cdots \otimes d\zeta^{\mu_n} \otimes \partial_{\nu_1} \cdots \otimes \partial_{\nu_m}]$$

π a permutaion of symbols, Einstein sum ranging over holomorphic as well as non-holomorphic indices. We often use the word tensor to mean hypertensor when there is no risk for confusion.

We have then a natural choice for a definition of antisymmetric hypertensor, a hyperform,

42

Definition 3.8. *A hyperform is an element in $\bigwedge V$, V a module or bundle module over $\mathbb{M}$. We often use the word form in a broader context to denote a $\bigwedge V$ section on hypermanifold X with $V \equiv T^*X$ the cotangent bundle over $\mathbb{M}$.*

Definition 3.9. *We define $\bigwedge(X, \mathbb{M}) \equiv \mathcal{O}_{\mathbb{R}}\Gamma(X, \bigwedge T^*X)$, i.e to the exterior cotangent bundle sections taking real analytic $\mathbb{M}$-valued function coefficents.*

This in turn makes it possible for us to define exterior differentiations on our spaces. We begin first with the split holomorphic/non-holomorphic of the exterior derivative.

Definition 3.10. *We define $\partial_L = \sum d\zeta^\mu \partial_\mu$, sum over holomorphic indices, $\partial_L^A = \sum d\zeta^\mu \partial_\mu$, sum over non-holomorphic indices. Similarly we define $\partial_R = \sum \partial_\mu d\zeta^\mu$, sum over holomorphic indices, $\partial_R^A = \sum \partial_\mu d\zeta^\mu$, sum over non-holomorphic indices. Both of these operators have $P^1 \bigwedge$, the exterior algebra of forms with first degree coefficients as domain. A stands in the above for antiholomorphic, which we take to mean non-holomorphic.*

We now use this to define extensions, for which we use the same symbols, to all of $\bigwedge(X, \mathbb{M})$ via the Leibniz rule.

Definition 3.11. *∂_L, ∂_L^A, ∂_R, ∂_R^A are defined on $\bigwedge(X, \mathbb{M})$ via the Leibniz rule on real factors.*

Lemma 3.4. *This definition determines the various exterior derivatives uniquely.*

Proof. Let δ denote this differentiation . Then $\delta\Pi f_i$ is claimed to be unique, f_i real analytic real factors. For f, g, h we have $L.S = \delta(fgh) = \delta(fg)h + fg\delta h = \delta fgh + f\delta gh + fg\delta h$, $R.S = \delta fgh + f\delta gh = \delta fgh + f\delta gh + fg\delta h$. But then the claim follows inductively for real factors. For the general case of non-real factors uniqueness of differentiation follows by uniqueness of commutation. $\square$

Example 3.2. *For the quaternionic case in a holomorphic/ anti-holomorphic split we have $\partial i\zeta = i\partial^i\zeta = 0$ according to the above definition and the elementary properties listed in previous sections, among them $\zeta i = i\zeta^i$.*

We now define another split for which we will have equal use-the split holomorphic dual holomorphic. It necessitates another definition of ∂_ζ, so one has to remember which definition one is using and which version of ∂_ζ that corresponds to it.

43

Definition 3.12. *Define* $\partial_\zeta = \frac{1}{2}(\partial_t - i\partial_x - j\partial_y - k\partial_z)$, *and use this definition whenever one is using the following split;* $\partial_L = \sum d\zeta^\mu \partial_\mu$, $\partial_L^d = \sum d\bar\zeta^\mu \bar\partial_\mu$, *summations running over holomorphic indices, and* $\partial_R = \sum \partial_\mu d\zeta^\mu$, $\partial_R^d = \sum \bar\partial_\mu d\bar\zeta^\mu$, *again sums over holomorphic indices, d as a superscript here stands for duality.*

Definition 3.13. ∂_L, ∂_L^d, ∂_R, ∂_R^d *are defined on* $\bigwedge(X, \mathbb{M})$ *via the Leibniz rule.*

We then have our first important theorem, namely that the above two really are splits of the exterior differentiation.

Theorem 3.1. *On* $\bigwedge(X, \mathbb{M})$ *we have holomorphic-antiholomorphic splits*

$$d = \partial_L + \partial_L^A, d = \partial_R + \partial_R^A$$

and holomorphic-dual holomorphic splits

$$d = \partial_L + \partial_R^d, d = \partial_R + \partial_L^d$$

which furthermore all satisfy exactness, i.e $\delta^2 = 0$, δ any of the operators appearing in the above split.

Proof.   (1) STATEMENT 1 The above operators are splits of d in the above sense.

Proof. We show this to hold for the quaternionic case for simplicity. Since the above opertors are all defined via Leibniz rules inductively we note that by calculations similar to

$(\partial_R + \partial_R^A)(\omega \wedge \eta)$
$= \partial_R \omega \wedge \eta + (-1)^p \omega \wedge \partial_R \eta + \partial_R^A \omega \wedge \eta + (-1)^p \omega \wedge \partial_R^A \eta$
$= ((\partial_R + \partial_R^A)\omega) \wedge \eta + (-1)^p \omega \wedge ((\partial_R + \partial_R^A)\eta)$

it suffices to show this to hold on $P^1 \bigwedge$, the exterior forms with first degree polynomial coefficients. But then we get for the holomorphic/ non-holomorphic case

$L.S = d = dx^\mu \partial_\mu,$
$R.S = \partial_R + \partial_R^A$
$= \partial_\zeta d\zeta + \partial_\zeta^i d\zeta^i + \partial_\zeta^k d\zeta^k + \partial_\zeta^k d\zeta^k$
$= \frac{1}{4}((\partial_t - i\partial_x - j\partial_y - k\partial_z)d(t + ix + jy + kz) + \cdots$
$+ (\partial_t + i\partial_x + j\partial_y - k\partial_z)d(t - ix - jy + kz))$
$= \mathfrak{Re}[(\partial_t - i\partial_x - j\partial_y - k\partial_z)d(t + ix + jy + kz)]$
$= \partial_t dt + \partial_x dx + \partial_y dy + \partial_z dz = L.S$

(The lefthanded case is treated similarly) and for the holomorphic/ dual holomorphic case

$$L.S = d = dx^\mu \partial_\mu,$$
$$R.S = \partial_R + \partial_L^d$$
$$= \partial_\zeta d\zeta + d\bar{\zeta} \bar{\partial}_\zeta$$
$$= \partial_\zeta d\zeta + \partial_\zeta d\zeta$$
$$= \mathfrak{Re}[(\partial_t - i\partial_x - j\partial_y - k\partial_z)d(t + ix + jy + kz)] = dx^\mu \partial_\mu$$

$\square$

(2) STATEMENT 2 Exactness.

Proof. Again since the above operators are all defined via Leibniz rules inductively we note that by

$$\delta^2 \omega \wedge \eta = \delta(\delta\omega \wedge \eta + (-1)^p \omega \wedge \delta\eta)$$
$$= (-1)^p \delta\omega \wedge \delta\eta + (-1)^{p+1} \delta\omega \wedge \delta\eta + \delta^2\omega \wedge \eta + \omega \wedge \delta^2\eta$$
$$= \delta^2\omega \wedge \eta + \omega \wedge \delta^2\eta$$
$$\omega \in \bigwedge^p(X, \mathbb{M}), \eta \in \bigwedge^q(X, \mathbb{M}), p \in \mathbb{N}.$$

it suffices to show this to hold on $P^1 \bigwedge$. But then $\delta^2 = 0$ holds trivially. $\square$

$\square$

We point out that when the choice of split is arbitray or a priori defined we use $d = \partial + \bar{\partial}$ to denote the above splits.

Finally we mention four other splits, namely $d = \partial_L + \partial_L^A$ $d = \partial_R + \partial_R^A, d = \partial_L + \partial_R^d, d = \partial_R + \partial_L^d$ defined as above on $P^1 \bigwedge$ but extended by the same definition as on P^1 rather than the Leibniz rule, e.g $\partial_L = d\zeta^\mu \partial_\mu$ with a sum over holomorphic indices. This later definition does not satisfy exactness for the different operators, but is useful in various solutions of $P.D.E$'s.

3.3. Explicit Examples of Simple Calculations in Hyperanalysis.
We start with some examples of calculations in hyperanalysis, and then we calculate the Cauchy-Fueter kernal as an exercise to get aquainted with how calculations usually proceed.

First of all we point out that some relations common to abelian exterior algebra generally do not hold, for example in general for X a hypermanifold, i.e a smooth manifold locally homeomorphic to $\mathbb{M}^n$, $n \in \mathbb{Z}_+$, with transition functions that are $\mathbb{M}$ hyperanalytic, $\mathbb{M}$ a hyperspace with generators e_a,

$$d\zeta \wedge d\zeta = \tfrac{1}{2}[e_a, e_b]dx^a \wedge dx^b \neq 0, \mathbb{M} = Span^{\mathbb{K}}\{e_a\}.$$
$$\omega \wedge \eta \neq (-1)^{pq}\eta \wedge \omega,$$
$$\omega \in \Gamma(X, \bigwedge^p T^*\mathbb{M} \otimes \mathbb{M}) \equiv \bigwedge^p(X, \mathbb{M}), \eta \in \bigwedge^q(X, \mathbb{M}).$$

45

in contrast to the real case. This does NOT lead to inexistence of hyper-cohomological theories [40] however, it only makes their representations more complicated.

Example 3.3. *In general we have*

$$\partial^2 \zeta^2 = \partial(d\zeta \wedge \zeta + \zeta \wedge d\zeta)$$
$$= d\zeta \wedge d\zeta - d\zeta \wedge d\zeta = 0$$

Example 3.4. *For the quaternionic case with* $\Phi \in \mathcal{O}_{\mathbb{R}}(\Omega, \mathbb{H})$, *coordinates as usual and* $\Omega \in \mathbb{H}$, *we have for the right exterior non-holomorphic derivative* $\bar{\partial} := \bar{\partial}_R$:

$$\bar{\partial}(\Phi(\zeta^i \zeta^j \zeta^k)^{-}1 d\zeta^i \wedge d\zeta^j \wedge d\zeta^k)$$
$$= (\bar{\partial}\Phi)(\zeta^k)^{-1}(\zeta^j)^{-1}(\zeta^i)^{-1} d\zeta^i \wedge d\zeta^j \wedge d\zeta^k$$
$$- \Phi(\zeta^k)^{-1} d\zeta^k (\zeta^k)^{-1} d(\zeta^j)^{-1}(\zeta^i)^{-1} d\zeta^i \wedge d\zeta^j \wedge d\zeta^k$$
$$- \Phi(\zeta^k)^{-1}(\zeta^j)^{-1} d\zeta^j (\zeta^j)^{-1}(\zeta^i)^{-1} d\zeta^i \wedge d\zeta^j \wedge d\zeta^k$$
$$- \Phi(\zeta^k)^{-1}(\zeta^j)^{-1}(\zeta^i)^{-1} d\zeta^i (\zeta^i)^{-1} d\zeta^i \wedge d\zeta^j \wedge d\zeta^k$$

Notice that there would have no difference between left and right diffrentiation in this example.

Finally we compute the Cauchy-Fueter Kernal. We have $G(\zeta, z) = \frac{-1}{2\pi^2} \frac{1}{|\zeta - z|^2}$, is a kernal for $\Box = \partial^\dagger \partial$, hence we have

$$\partial \frac{-1}{2\pi^2} \frac{1}{|\zeta - z|^2}$$
$$= \frac{1}{2\pi^2} \frac{1}{|\zeta - z|^2 (\zeta - z)} * d\bar{\zeta}$$

which we recognize as the Cauchy-Fueter kernal mentioned in Dunford's article.

3.4. **Integral Representations of Hyperanalysis I.** In this subsection we shall learn about the integral representations of functions in hyperanalysis. We are already aquainted with the complex case and would like to generalize this the general case considered in hyperanalysis. We cut stright through to the main point. In the following the splits of the exterior diffrentiation will be an arbitray choice from the previous section. Let Ω be compact to make the integrals well defined in the following, f continuously differentiable, and $\zeta = x^A e_A, z = x'^A e_A$, the usual holomorphic coordinates, x, x' real coordinates.

Example 3.5. *For* $Cl(42, 0)$ *the holomorphic coordinate is* $\zeta = x^A e_A \in Cl^1(42, 0)$, e_A *generating the algebra. One might also only take the lefthanded part* $x^A e_{A+}$, $e_{A+} = P_+ e_A$, P_+ *lefthanded Chirality projector.*

---

[40]The author has some interest in noncommutative topology, i.e. hypertopology, however he finds it unlikely that he will find space to touch that topic at any depth in this thesis.

Theorem 3.2 (Cauchy 1843, Fueter 1935). *We have*

$$f(z) = [< \partial\Omega, f\frac{1}{(\zeta-z)^{\bar{n}}}d\zeta^{\bar{n}} >$$
$$- < \Omega, (\partial+\bar{\partial})f\frac{1}{(\zeta-z)^{\bar{n}}}d\zeta^{\bar{n}} >]\frac{1}{\omega_R}, z \in \Omega \in C_{|n+\bar{n}|}(\mathbb{M}^1, \mathbb{Z}).$$

with $\Omega \subset \mathbb{M}^1$ corresponding to the generator of the unit element in the chain group $C_{|n+\bar{n}|}(\mathbb{M}^1, \mathbb{Z})$ over $\mathbb{Z}$ of the hyperspace $\mathbb{M}^1$ belonging to $\mathbb{M}$ and

$$\omega_R = < S_{\mathbb{R}}^{dim^{\mathbb{R}}(\mathbb{M}^1)}, \frac{1}{\zeta^{\bar{n}}}d\zeta^{\bar{n}} > .$$

Proof. We begin by noting that since $|\zeta^{\bar{\neg}}| = |\zeta|^{|\bar{\neg}|} \neq 0 \; \forall\zeta \neq 0$ we have that

$$f\frac{1}{(\zeta-z)^{\bar{n}}}d\zeta^{\bar{n}}$$

is smooth $\forall\zeta \neq z$. Hence using Stoke's theorem and taking out a disc of radius R around z, $D_{z,R}^{\mathbb{M}}$,we have

$$< \partial\Omega, f\frac{1}{(\zeta-z)^{\bar{n}}}d\zeta^{\bar{n}} >$$
$$= < \Omega \sim D_{z,R}^{\mathbb{M}^1}, \underbrace{(\partial+\bar{\partial})}_{=d} f\frac{1}{(\zeta-z)^{\bar{n}}}d\zeta^{\bar{n}} > + < \partial D_{z,R}^{\mathbb{M}^1}, f\frac{1}{(\zeta-z)^{\bar{n}}}d\zeta^{\bar{n}} >$$

$$= \{R \to 0\} = < \Omega, (\partial+\bar{\partial})f\frac{1}{(\zeta-z)^{\bar{n}}}d\zeta^{\bar{n}} > +f(z)\underbrace{< S_{\mathbb{R}}^{dim^{\mathbb{R}}(\mathbb{M}^1)}, \frac{1}{(\zeta-z)^{\bar{n}}}d\zeta^{\bar{n}} >}_{=\omega_R}$$

where we noted that in the second term only $f(z)$ depends on R, as is noticed by for example using polar coordinates, and took a limit to $R = 0$.

$\square$

Corollary 3.1 (The Quaternionic Integral Formula). *This is a second version of the Cauchy formula on the quaternions, intrinsically formulated as compared to the Cauchy-Fueter formula. Under the hypothesis of the previous theorem we have setting $\mathbb{H} := \mathbb{M}$*

$$f = [< \partial\Omega, f(\zeta^i\zeta^j\zeta^k)^{-1}d\zeta^i \wedge d\zeta^j \wedge d\zeta^k >$$
$$- < \Omega, (\partial+\bar{\partial})f(\zeta^i\zeta^j\zeta^k)^{-1}d\zeta^i \wedge d\zeta^j \wedge d\zeta^k >]\frac{1}{\omega_R}$$

Proof. $\mathbb{H} := \mathbb{M}$. Notice that $\mathbb{M}^1 = \mathbb{M}$ for this case. $\square$

Corollary 3.2 (The Dirac Algebra Integral Formula). *Under the hypothesis of the previous theorem we have setting $Cl(4,0) := \mathbb{M}$*

$$f = [< \partial\Omega, f(\zeta^i\zeta^j\zeta^k)^{-1}d\zeta^i \wedge d\zeta^j \wedge d\zeta^k >$$
$$- < \Omega, (\partial+\bar{\partial})f(\zeta^i\zeta^j\zeta^k)^{-1}d\zeta^i \wedge d\zeta^j \wedge d\zeta^k >]\frac{1}{\omega_R}, \Omega \subset \mathbb{M}^1.$$

Again there is a kin to this formula, this time called the CSIO formula which is formulated in a real manner opposite to this formula.

Proof. $Cl(4,0) := \mathbb{M}$. Notice that $\mathbb{M}^1 \neq \mathbb{M}$ for this case by the field property of the quaternions. $\square$

Corollary 3.3 (The Cauchy integral formula on Riemann Surfaces).

$$f(z) = \frac{1}{2\pi i}[\int_{\partial\Omega} \frac{\omega}{\zeta - z} - \int_{\Omega} \frac{\bar{\partial}\omega}{\zeta - z}] \; z \in \Omega, \; \omega \in \Lambda^{(1,0)}(\mathbb{C}_\zeta)$$

Proof. $\mathbb{C} := \mathbb{M}$ and noting that the abelian property of $\mathbb{C}$ implies that the exterior differential in the first expression lies in the kernel of $\bar{\partial}$ a simple calculation gives ω_R. Notice that the antiholomorphic version and dual holomorphic versions of the above integral formula are equivalent on $\mathbb{C}$, i.e in some sense antiholomorphic and non-holomorphic coincide for this particular case.

$\square$

Theorem 3.3 (Martinelli/Bochner 1943 for $\mathbb{C}^n$). *We have, setting* $d = dim^{\mathbb{R}}\{\mathbb{M}^1\}$, $n = dim^{\mathbb{M}^1}\{X\}$, $*$ *a Hodge star, an inner product* $<\zeta, z> = \sum \bar{\zeta}_\sigma z^\sigma$ *and* $\omega_N = \sum \zeta_\sigma d\zeta^\sigma$

$$f(z) = [< \partial\Omega, f\frac{1}{||(\zeta - z)||^{dn}} * \omega_N >$$
$$- < \Omega, (\partial + \bar{\partial})f\frac{1}{||\zeta - z||^{dn}} * \omega_N >]\frac{1}{\omega_R}, z \in \Omega \subset X.$$

with $\Omega \subset X$ contractible in the hyperspace X and

$$\omega_R = < S_{\mathbb{R}}^{dim^{\mathbb{R}}(\mathbb{M}^1)}, \frac{1}{||(\zeta - z)||^{dn}} * \omega_N > .$$

Proof. The proof closely follows the one-dimensional case. Noting again that

$$\frac{1}{||(\zeta - z)||^{dn}}$$

is smooth at all points except the origin we find via Stoke's theorem

$$< \partial\Omega, f\frac{1}{||(\zeta - z)||^{dn}} * \omega_N >$$
$$= < \Omega \sim D_{z,R}^{dn}, \underbrace{\partial + \bar{\partial}}_{=d} f\frac{1}{||(\zeta - z)||^{dn}} * \omega_N > + < \partial D_{z,R}^{dn}, f\frac{1}{||(\zeta - z)||^{dn}} * \omega_N >$$
$$= \{R \to 0\} = < \Omega, (\partial + \bar{\partial})f\frac{1}{||(\zeta - z)||^{dn}} * \omega_N > + f(z)\omega_R$$

again noting that only $f(\zeta)$ depends on R in the last expression.

$\square$

Theorem 3.4. *Assume that a smooth* $f(\zeta)\frac{1}{||\zeta-z||^{dn}}*\omega_N$ *has a integrable extension from a contractible punctured neighbourhood* $N(z)\sim z$ *of* z *to all of* $\Omega\sim\{z\}\subset X$ *called* $ext^*f(\zeta)\frac{1}{||\zeta-z||^{dn}}*\omega_N$ [41]. *Then under the notation of the previous theorem but with* X *a hypermanifold* [42], Ω *arbitrary in the chain group* $C_{dn}(X,\mathbb{Z})$,

$$f(z) = [< \partial\Omega_{z,R}, f\frac{1}{||(\zeta-z))||^{dn}}*\omega_N >$$
$$- <\Omega_{z,R},(\partial+\bar{\partial})f\frac{1}{||\zeta-z||^{dn}}*\omega_N> +discr.]\frac{1}{\omega_R}, z\in\Omega_R\subset X.$$

where $\Omega_{z,R}$ *is an arbitrary contractible set containing* z *and included in* Ω *and the discrepancy discr. is*

$$discr. =< \partial\Omega\sim\Omega_{z,R}, ext^*\frac{f(\zeta)}{||\zeta-z||^{dn}}*\omega_N >$$

Proof. Since $\Omega=\Omega\sim\Omega_{z,R}\cup\Omega_{z,R}$ is a partition with nill intersection the theorem follows by additivity of integrals, $\int_{\Omega\sim\Omega_{z,R}}+\int_{\Omega_{z,R}}=\int_\Omega$, under such conditions.

$\square$

We remark that

(1) The discrepancy discr. might be contained in the generalized period matrix [43] of X in the case of an f that gives a cocycle $ext^*\frac{f(\zeta)}{||\zeta-z||^{dn}}*\omega_N$.

(2) Should the reader have a form that is a superposition of the cocycles that generate his/hers gen. period matrix s/he only needs project his/hers form on the basis used using the inner product $<\omega,\eta>=\int_X\omega\wedge*\eta$.

(3) The contents of this subsection hold also for $f\in C^1(\Omega)$ by the last remarks of the previous section.

Finally we remind the reader that for integrals over hyperbolic spheres [44] $|\zeta|^2=\bar{\zeta}\zeta=m^2, \zeta\in\mathbb{M}, m\in\mathbb{R}$(also for natural reasons called hyperboloids) this gives for integrals of the above type with vanishing volume

---

[41]In terms of contravariant functors induced by transition functions on exterior cotangent bundles etc., or that it is smooth on $\Omega\sim z$ and only takes the form according to above on $N(z)\sim z$.

[42]A definition of this concept is in the hypergeometry section. For now the reader can think of a space that locally looks like $\mathbb{M}^1$

[43]Which is, a little bit more precisely but unconventionally put, rather an array of matrices.

[44]We remind the reader that these integrals are defined via continuation. Also the domains then considered are upon continuation not compact, so one uses σ-finiteness of the spaces often considered and/or exhausts Ω from the interior.

term an invariance under the choice of such spheres. This is the related to the famous Pauli smearing of the light cone when it comes up in physics and has been previously observed among other things in integral representations related to twistor theory.

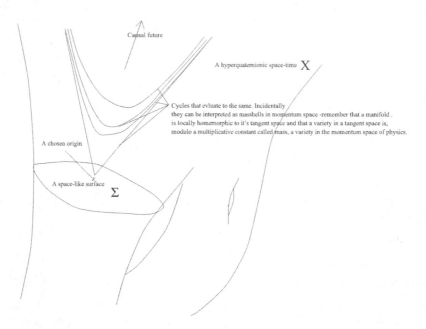

FIGURE 2. Cycles that evaluate to the same integral. It happens that one averages over a wide variety of integrals over such cycles-Pauli smearing. Fixing to one of these cycles would require gauge fixing of some kind of the group of bihyperholomorphisms. It is interesting to note that in some interesting cases the group of bihyperholomorphisms $Biholo^{\mathbb{M}}(X)$ is a subgroup of the conformal group $Conf(X)$, X a manifold.

3.5. The Riemann Mapping Theorem.

In this subsection we shall develop the 'hyper' version of the famous assertion of Riemann that essentially reduces the question of biholomorphic equivalence in the plane to topological equivalence. The percipient reader will probably see shadows of the first example of exotic structures relevant to hyperanalysis. This manifests itself as follows; We shall be able to prove that to each analytic diffeomorphism there is bihyperanalytic map and conversely. Since diffeomorphisms differ between topological manifolds with different structures this means that exotic structure in some sense, for cases where non-real methods can be applied, emanate

50

from the non-real strucure of some spaces. Actually the method of constructing exotics from non-real elements is the oldest known way of showing the existance of exotics, already John Milnor did so in the late 50's. Freedman/ Donaldson were then able to show amazing results in dimension 4, among other things the existence of an infinite number of structures in $\mathbb{R}^4$. Odd as it is all exotic structures(at least the ones I know of) seem to be generated by non-real considerations or things closely related. On the other hand concerning dimensions recognized as equipped with 'hyper' structure $\{2, 4, 8, 16, \cdots\}$ one easily sees that there are no conformal equivalences between smoothly equal sets in general for dimension 3 and greater; this would trivialize the Weyl tensor generically locally, which is not the case. We begin with the principal theorem in the complex case(where manifolds have no exotic structures but a lot of complex structures.);

Theorem 3.5. *Let Ω and Ω' be two diffeomorphic bounded subsets of $\mathbb{C}$, then they are biholomorphic[45].*

Proof. A long historical development initiated by Riemann, and followed up by Caratheodory, Montel and many others. The reader is referred to either L. Ahlfors excellent book in complex analysis or R.Remmert's Classical Topics in Complex Function Theory, equally excellent. $\qquad\square$

The main point of the above theorem is that it tells us that in some sense complex analyticity is a very natural concept in two real dimensions.

In several variables the situation changes considerably. This has been misunderstood to imply that hypergeometries are extremely rigid and unusual as compared to smooth objects of the same dimension. It is true that things are more rigid, but this seems to be so mainly in the case of several variables and multidimensional hypermanifolds (Conformal manifolds, which are linked to hypermanifolds of the kind we are considering are not uncommon e.g among the 4-manifolds and are just one example).

However there are approximative statements in that direction:

Theorem 3.6. *Let Ω and Ω' be topologically equivalent in $\mathbb{C}^n$. Then there is biholomorphism of Ω at most up to a distance $\epsilon > 0$ from which the boundaries of $f(\Omega)$ and Ω' differ when closest, i.e. $\inf_{x \in \partial f(\Omega),\ y \in \partial \Omega'} \|x - y\|_{\mathbb{E}^{2n} \cong \mathbb{C}^n} \leq \epsilon$.*

---

[45]We assume it understood that we mean bihyperholomorphic.

We now prove a theorem similar in nature but of weaker content for the general hyperanalytic case.

Theorem 3.7. *To each hyperanalytic morphism $f : \Omega \rightarrow \Omega'$ there corresponds a unique real analytic morphism and conversely in any hyperfield/algebra.*

Proof. Setting $\zeta = \sum x_a e^a$, $[x_a]$ local real coordinates on Ω and using various hyperreflections ζ^i we have for $f \in \mathcal{O}_{\mathbb{R}}(\Omega)$ by expressing the real coordinates in hyperreal coordinates

$$f = \sum a_\alpha x^\alpha = \sum a_\alpha (\sum b_j \zeta^j)^\alpha$$

normal convergence of the former of the former on Ω proves ditto for the latter. The converse is trivial.

$\square$

It might be worth while noting that the case $\mathbb{C}$ is of such a nature that the ring $\mathbb{C}[z]$ does not generate $\bar{z}$ 'naturally', as opposed to $\mathbb{H}, \cdots$. This means that a hyperanalytic theory is essentially a real analytic theory for the noncommutative cases, in congruence with results of Lie analysis, this in order to satisfy compatibility with the multiplicative algebraic structure. This does not, however, imply non-existance of hypertopological theories (the algebraic structure usually associated with sheaves of forms, hyperholomorphic structure sheaves and ditto is addition, which is *abelian*.).

4. NON-REAL GEOMETRY, NONCOMMUTATIVE GEOMETRY AND HYPERGEOMETRY

4.1. **Real geometry.** In this section we shall define some concepts of real geometry in a general enough manner for the uses of this thesis. The approach is hopefully short and powerful.

Definition 4.1. *A real topological manifold is a Hausdorff paracompact* [46] *topological space locally homeomorphic to* $\mathbb{R}^n, n \in \mathbb{Z}$ *with the Euclidean metric topology. The structural category of such manifolds is the called the category of topological manifolds and the morphisms taken to be the continuous maps. Structural isomorphism is then homeomorphism.*

Definition 4.2. *A differentiable manifold* $\check{X}$ *of class* $C^k, k \in \mathbb{Z}$ *is a topological manifold together with an open cover* $\{U_\alpha\}_{\alpha \in I}$, I *an index set, of X. Denoting local homemorphisms* $\phi_\alpha : U_\alpha \to \phi(U_\alpha) \subset \mathbb{E}^n$ *one requires the compositions* $t_{\alpha\beta} = \phi_\alpha \phi_\beta^{-1}$ *to be of class* C^k *for* $\forall \alpha, \beta; U_\alpha \cap U_\beta \neq \emptyset$. *The structural category of these manifolds is then called the class of* C^k*-differntiable manifolds. Structural morphisms of this structural category are the maps of class* C^k *and the isomorphisms the* C^k*-class diffeomorphisms.*

For metrics over modules(over rings) or vector spaces the reader is referred to Bourbaki's algebra or Analysis, Manifolds and Physics.

Definition 4.3. *A graded (left) differentiation* ∂ *over a field, ring* $\mathbb{K}$ *etc is a* $\mathbb{K}$*-linear map with a Leibniz rule, i.e* $\partial fg = (\partial f)(g) + (-1)^p f(\partial g), f, g$ *being in a (left) graded module over* $\mathbb{K}$ *for this differentiation, where f is assumed to have been assigned a integer* $n \in \mathbb{Z}$ *called the degree (order) of f. The set of graded derivations over* $\mathbb{K}$ *is denoted* $\mathfrak{der}_{\mathbb{Z}_2}^{\mathbb{K}}$. [47]

Definition 4.4. *The derivations over a point over* $\mathbb{K}$ *are the restriction of* $\mathfrak{der}_{\mathbb{Z}_2}^{\mathbb{K}}$ *to a assumed (left) submodule of nill degree and denoted by* $\mathfrak{der}^{\mathbb{K}}$.

Example 4.1. $\mathfrak{der}^{\mathbb{K}} \equiv \mathfrak{g}$, $\mathfrak{g}$ *being a Lie algebra over* $\mathbb{K}$*, then* $\partial := ad_{e_a}, a \in \{1, \cdots, dim^{\mathbb{K}}(\mathfrak{g})\}$ *spanning the Lie algebra is an example of a derivation over* $\mathbb{K}$.

---

[46]This seems slightly restrictive, however a famous theorem of Arthur Stone states that a metric (topological, topology induced by the metric) space is necessaraly paracompact. We hall mostly be interested in metric spaces in this thesis.

[47]For further information on supermathematics I would like to recommend Y.Manin's 'Complex Geometry and Gauge Field Theory' or Dan Freed's lecture notes on ditto subject.

Example 4.2. *The BRST operator s acting on a fermionic ghost superalgebra is an example of a graded derivation acting on a graded module. The degree of an element ω in this algebra is then p=f+g , with p the fermion number of ω and g the ghost number of ω.*

Definition 4.5. *A (graded, super)tangent space $TX^{\mathbb{K}}$,($TX^{\mathbb{K}}_{\mathbb{Z}_2}$), over $\mathbb{K}$ over a point belonging to a (smooth) (graded, super) manifold X is the set of differentiations over K acting on the differentiable K-valued (graded, super) functions over this point, i.e $TX^{\mathbb{K}} \equiv \mathfrak{der}^{\mathbb{K}}(X)$ or $TX^{\mathbb{K}}_{\mathbb{Z}_2} \equiv \mathfrak{der}^{\mathbb{K}}_{\mathbb{Z}_2}(X)$.*

Definition 4.6. *A cotangent (graded, super) space is the dual of a (graded, super)tangent space.*

Definition 4.7. *A Riemannian manifold is a manifold with a metric(in the sense of an non-degenerate positive inner product, see above) defined over the tangent space of each point. Such a manifold is also called proper Riemannian.*

Example 4.3. *Let X be a semisimple Lie group. Then X is a manifold and the Killing form $h_{ab} = -Tr(ad_{e_a}ad_{e_b})$ defines a non-degenerate positive definite metric h over each point.*

Example 4.4. *Let X be an infinite dimensional hermitian manifold, say for example an assumed direct sum space of fermionic states times gauge boson states on a fixed space-time background. Let A be this space of gauge boson states, G the gauge group and A/G a gauge slice. Let f be a map between X and $\mathbb{C}$. Let us make X Banach for the sake of the argument. Upon defining the Frechet differntial Df of a map f as a linear endomorphism of tangent spaces that satisfies*

$$f(x + h) - f(x) = Df(h) + o(|h|)$$

for an arbitrary fixed $x \in X$ we see, locally splitting $D = \delta + s + s_{\bigwedge T^*A/G}$, δ being an extension of the exterior derivative acting on the fermionic exterior cotangent space (by acting trivially on the rest and as usual on the fermionic cotangent space), s being the BRST operator and the last operator acting on the exterior cotangent gauge slice that the superalgebra can be interpreted as a gigantic exterior algebra above a point in X. The above operator split handles then much the same as the Dolbeault $\partial, \bar{\partial}$-operator split of complex analysis, in particular $0 = s^2 = \delta^2 = [s, \delta]_+$. The above is an important example and , upon taking the sum TX with the dual bundle one obtaines a bundle over X with fibre being the candidate for generalization of a symplectic manifold in finite dimensions. For the moment we suffice to note that a direct *sum of the laplacian and square Dirac operator would be a candidate metric on X in the obvious way by (not necessarily positive) hermicity.*

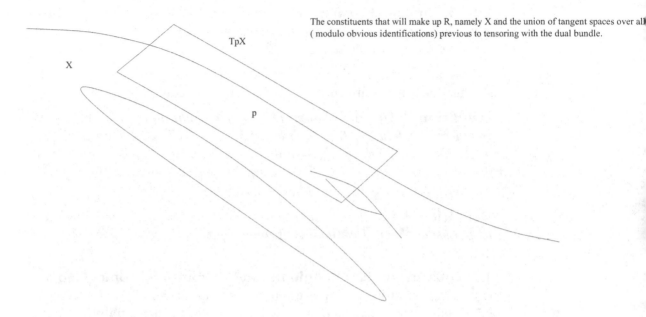

The constituents that will make up R, namely X and the union of tangent spaces over all (modulo obvious identifications) previous to tensoring with the dual bundle.

FIGURE 3. The manifold X together with a tangent space over a point p. Such tangent spaces and their duals are used to construct the double dual bundle $R = TX \oplus T^*X$. The relevance to physics is that physical Lagrangians are usually exterior powers of objects in R. If we change X to be only space-time itself R has a canonical generalization of the exterior derivative called the Dirac operator defined on it. Conversely we can define similar operators in the infinite-dimensional case-Linked to the Schwinger functional diffrentiation representation of the canonical Fock space formalism of quantum field theory.

Definition 4.8. *A pseudoriemannian (hermitian) manifold of signature (p,s) is a manifold where the metric h, viewed as a matrix, over each point can be set(by a change of basis and/or coordinates) (spectrally) into the form $h = \bigoplus_{\alpha \in I_+} 1_\alpha - \bigoplus_{\alpha \in I_-} 1_\alpha$, I_+, I_- being index sets. p,s are then the cardinality of I_+, I_- respectively.*

55

4.2. **Complex Geometry.** In this subsection we shall define the basic concepts of complex geometry, which rest upon the foundations of several complex variables we previously have to some part encountered.

Definition 4.9. *A complex topological manifold X is a Hausdorff paracompact topological space locally homeomorphic to $\mathbb{C}^n$ in the topology given by the identification $\mathbb{C}^n \cong \mathbb{E}^{2n}$.*

Just as in the real case there is another type of manifolds with stronger requirements that we shall preferably use;

Definition 4.10. *A holomorphic complex manifold is a topological complex manifold together with an open cover $\{U_\alpha\}_{\alpha \in I}$, I an index set, of X. To each of these U_α one defines a local holomorphic homemorphism $\phi_\alpha : U_\alpha \to \phi(U_\alpha) \subset \mathbb{C}^n$ and requires the composition $t_{\alpha\beta} = \phi_\alpha \phi_\beta^{-1}$ $\forall \alpha, \beta; U_\alpha \cap U_\beta \neq \emptyset$ to be a biholomorphism.*

Definition 4.11. *A pseudohermitian complex manifold is a manifold with a pseudohermitian metric of some signature according to definition 4.8.*

4.3. **Quaternionic, Octonionic and Hyperoctonionic Geometry.** The definitions of quaternionic, octonionic and hyperoctonionic manifolds closely parallells the definitions of complex analytic (holomorphic) [48] manifolds. For reasons of ease we shall choose an analytic prescription.

Definition 4.12. *A map shall be called $\mathbb{K}$-analytic if it's local power series expansion only involves powers of the $\mathbb{K}$-holomorphic coordinate and constant coefficients in $\mathbb{K}$.*

Definitions of quaternionic, octonionic and hyperoctonionic manifolds now follow by more or less replacing the word holomorphic with $\mathbb{K}$-analytic in the previous subsection. Never the less we might want to also have a holomorphic/ non-holomorphic prescription at times, this then is defined by annihilation by ∂ or $\bar{\partial}$ in such cases. We define analytic manifolds for later use;

Definition 4.13. *A topological hypermanifold is a Hausdorff topological space locally homeomorphic to $\mathbb{M}$ a hyperspace.*

Definition 4.14. *A analytic hypermanifold X is a paracompact topological hypermanifold together with an open cover $\{U_\alpha\}$; $X \subset \cup U_\alpha$. To each of these U_α one associates a homeomorphism $\phi_\alpha : U_\alpha \mapsto \phi_\alpha(U_\alpha)$ and requires the composition $\phi_\alpha \phi_\beta^{-1}$ to be hyperanalytic.*

---

[48]This is essentially the same for the complex case by the theorems of pluricomplex analysis and the local character of the definitions involved.

We will some time abuse language and call the part looking locally as $\mathbb{M}^1$ a hypermanifold. Indeed we shall at times identify space-time with $\mathbb{M}^1$ turning it into a hypermanifold in this latter abusive sense in a development bearing resemblence to supergeometry.

4.4. Noncommutative geometry.

Noncommutative geometry can be interpreted in two ways a priori distinct; In the way mentioned in this thesis and in Alain Connes way. Indeed the word hypergeometry was first invented to avoid confusion with noncommutative geometry. For Alain Connes treatment we refer the reader to his book Noncommutative Geometry, which however has more the character of a review than a traditional book of mathematics. Hypergeometry was in part developed to provide the gap between ordinary differential geometry and what Alain Connes calls 'noncommutative geometry'. In physics noncommutative geometry in the sense of Connes is to some extent already implicit in the Heisenberg picture, e.g. in this vein an analysis by N.Seiberg an E. Witten completed by the summer 1999 shows that noncommutative Yang-Mills theory is equivalent to usual Yang-Mills theory with higher dimensional correction terms included for a number of scenarios. This does not, however, diminish the interesting perspective allowd by noncommutative algebraic geometry(See below for deformations of isospectral sets).

4.5. (Spectral) Hypergeometry.

Spectral hypergeometry will be what we are doing in this thesis in the mathematical part. Spectral means that we are looking at individual objects in hypercat egories rather that the entire category itself, or that we are looking at a (prime) ideal in an algebra when taking an algebraic perspective. This procedure, beginning with a local study and then do a global study in a hypercategory is in congruence with mathematical convention and tradition.

4.5.1. *Physical Interconnection and Relevance.*

The use of the word 'spectral' is by us thought to approximately coincide with the 'spectral' of noncommuative geometry, although this is but a conjecture at this point. In the physical parts we see more that endorses this suspicion as determinant homomorphism of a C*-algebra generating isospectral sets of partial differential operator, which in the canonical case can be viewed as smooth varieties with complex analytic complexification. This brings a nice perspective of noncommutative varieties and their determinant morphisms as isospectral level sets, the isospectral sets correspnding to 'space'(the 3 in a D3-brane) in space-time, with the Morse theory of these isospectral sets giving fluctuations of spatial topology, and restrainted cobordisms to 4 corresponding to a type of D-brane Feynman diagrams. This has similar higher and lower dimensional analogues and leads e.g. to phenomena like diconnected spaces connecting and disconnecting at different time instances. A direct summand of N such space-times, identifed with D3-branes, taking string geometry tensor coefficents in various representations of gauge groups such as $U(N_c)$, is then supposed to correspond to our world, N being the number of particles, N_c some arbitrary number of colors. The different stringy space-times in the product are supposed to be roughly independent due to reduction of parallel transport/holonomy and other physical effects. The appropriate mathematization of this thought, however, cannot be seriously or meticulously undertaken in this thesis, but is physically addressed in the physical part.

4.6. The Geometry of Fibre Bundles and Sheaves. In this section we shall define the notions of bundles and sheaves. For smooth bundles we refer the reader to Choq. et al. Analysis, Manifolds and Physics while for sheaves we refer to either L.Ahlfors, Complex Analysis, or L. Hormander, An introduction to Complex Analysis in Several Variables. Since we will not investigate hyperbundles (i.e with hyperstructure respecting transiton functions) we will not need more definitions than the ones used in those books.

4.7. Geometry of Projective (Hyper)Spaces.

Definition 4.15. *Let $\mathbb{M}$ be a module (in the group theoretic sense, i.e a domain for left/right action of a representation of a group-for example a group itself is a module for the same group.) for a group G. We call the space $\mathbb{M}^{n+1}/G, n \in \mathbb{Z}$, equivalent elements being elements that are in the same orbit, the projective space of order n of $\mathbb{M}$ over G and denote it by $G\mathbb{P}^n(\mathbb{M})$.*

It is common to avoid pathology and achieve clearness by omission of the origin in $\mathbb{M}$, when such an origin is well defined, previous to taking the quotient-especially on vector space dittos.

Example 4.5. *Some examples occur so often that we shall use a shorthand notation. Set $\mathbb{M} = \{\mathbb{R}, \mathbb{C}, \mathbb{H}, \mathbb{O}, Cl(p,s), \cdots, \}$. Then, upon omission of the origin and taking respective quotients of the multiplicative left actions, we get*

$$\mathbb{R}P^n = (\mathbb{R}^n \sim \{0\})/\mathbb{R},$$
$$\mathbb{C}P^n = (\mathbb{C}^n \sim \{0\})/\mathbb{C},$$
$$\mathbb{H}P^n = (\mathbb{H}^n \sim \{0\})/\mathbb{H},$$
$$\mathbb{O}P^n = (\mathbb{O}^n \sim \{0\})/\mathbb{O},$$
$$ClP^n(p,s) = (Cl(p,s)^n \sim \{0\})/Cl(p,s),$$

All of these projective spaces except the real ones are called hyperprojective spaces.

Example 4.6. *Grassmannians and homogenous spaces. Note, as an example of the latter, that topologically $S^3 \cong SU(2)$ and we have a group isomorphism $SU(2)/SO(3) = \mathbb{Z}_2$, actually since $SO(3)$ follows from identification of opposite points on S^3 we have a homemorphism $\mathbb{R}\mathbb{P}^3 \cong SO(3)$.*

Example 4.7. *This example is from gauge field theory and quantum geometry in theoretical physics. In theoretical physics the experimental predictions emanate from an inner product on a Hilbert space that gives upon taking the square absolute value the probablity(density) for something to occur. Naturally the group of isometries of such a Hilbert space, often constructed as the tensor product of Hilbert spaces, leaves the physical predictions, emanating as they are from the inner product, invariant. Also, the eigenvalue corresponding to a specific state vector is invariant under assumption of commutativity with the Hamiltonian[49]-these two facts lead to a projectivity in physics in a sense that we shall try to explain below by considering some cases. Let us investigate the space of physically different states for some cases. Let us first investigate a spin-1/2 system-we would like to construct a state moduli consisting of physically inequivalent states $Mod(\mathcal{H})$ of the corresponding Hilbert space $\mathcal{H}$. We know that it is described by the Hilbert space $\mathbb{C}^2$ with the inner product $< z, \zeta >= \sum \bar{z}_a \zeta^a$, $z, \zeta \in \mathbb{C}^2$ with canonical complex Euclidean coordinates. It is obvious that any states differing by a complex scalar will upon renormaliztion represent the same state, for example the pure eigenvectors represent physically the same thing under a multiplication by a complex scalar. Hence we quotient out the complex rays through the origin in $\mathbb{C}^2$ and obtain*

$$Mod(\mathcal{H}) = \mathbb{C}\mathbb{P}^1$$

---

[49]Or equivalent statements in terms of generalized BRST cohomology/ Charge or current operators/ Baitlin-Vilikovsky formalism or functional supercohomology.

Now for the generic situation of theoretical physics(for example quantum electrodynamics) we also have to take into account that we have an antifermion, thus $\mathcal{H} = \mathbb{C}^4$ and we have a helicity state moduli

$$Mod(\mathcal{H}) = \mathbb{CP}^3$$

-the famous $\mathbb{CP}^3$ of Penrose. In much the same manner one quotients out bigger groups-for example global/local gauge groups like

$$SL(2, \mathbb{C}), U(1), SU(2), SU(3), Diff(X), Conf(X), Biholo(X)$$

, X a manifold,[51]*. The most common examples of such non-trivial examples of state modulis of quantum states are the gauge modulis of Q.F.T and the moduli space of a fixed Riemann surface of some genus and punctures in string theory- general quantum (hyper) geometry generalizes this and offers more interesting examples with a wide variety of phenomenology.*

We would like to make sense of the majority of these projective spaces as manifolds. Here is a first theorem in this direction;

Theorem 4.1. *A (hyper)projective space of real, complex, quaternionic, octonionic type has the corresponding (hyper)structure and can thus be viewed as a (hyper)manifold of the corresponding type.*

Proof. Proof proceeds by explicit construction. Define an open cover of $G\mathbb{P}^n(\mathbb{M})$ by the sets $\{U_\alpha\}_{\alpha \in I}$, $I = \{0, \cdots, n\}(n \in \mathbb{Z})$ an index set, $U_\alpha = \{z = (z_0, \cdots, c_n) \in \mathbb{M}^{n+1}; z_\alpha \neq 0\}$. Calling these z homogenous coordinates we now define inhomogenous coordinates $\zeta^{(\alpha)}$ on each U^α by $\zeta^{(\alpha)} = z_\alpha^{-1}(z_0, \cdots, \widehat{z_\alpha}, \cdots, z_n)$, hat meaning omission. It is obvious that this division is well-defined, since $x^{-1} = \bar{x}/||x||^2$ and $||x|| \neq 0 \; \forall x \in A; x \neq 0$, A being some arbitrary choice in the family $\{\mathbb{R}, \mathbb{C}, \mathbb{H}, \mathbb{O}\}$. Transition functions are now seen to be

$$t_{\alpha\beta} = (\zeta_\alpha^{(\beta)})^{-1}(\zeta_0^{(\beta)}, \cdots, \zeta_{\alpha-1}^{(\beta)}, 1, \zeta_{\alpha+1}^{(\beta)}, \cdots, \zeta_n^{(\beta)})$$

clearly being (hyper)holomorphic on their domain.

$\square$

---

[50]It is by the way interesting to note that the above corresponds to either forward I^+ or backward I^- causal infinity in space-time. One way of obtaining the second from the first is to to do a complex conjugation on the relevant Hilbert space which goes down to ditto on the state moduli.

[51]In physics we call a transformation global if it is made by a constant section of a principal bundle on a physical Hilbert space. If this section is not constant we call the transformation local.

5. Projective Holomorphic Metrics in Hypermathematics

In this section we shall briefly get aquainted with the concept of a projective holomorphic metric, or a $P\mathbb{M}$-metric.

It should at once be said that a projective holomorphic metric is not a metric, at least not in the usual sense.

Definition 5.1. *A projective holomorphic metric is a map* $< .,. >:$ $\mathbb{V} \times \mathbb{V} \to \mathbb{M}$, $\mathbb{V}$ *a module over* $\mathbb{M}$;

(1) $< a\phi, \psi >= \frac{1}{a} < \phi, \psi >$, $\psi \in V, \phi \in V, a \in \mathbb{M}$.

(2) $< \phi, \psi >=< \phi, \psi > a$.

Upon integration a projective holomorphic metric induces a metric on functions spaces over Hypermanifolds X_g by either one of the formulas $< \psi(\zeta), \phi(\zeta) >= \int \psi(\zeta)^{-1} \phi(\zeta) \frac{d\zeta}{\zeta}$ or $< \psi(\zeta), \phi(\zeta) >= \int \psi(\zeta)^{-1} \phi(\zeta) \frac{d\zeta^{\bar{n}}}{\zeta^n}$ over appropriate contours encircling a point and some fixed convention concerning the ordering. Just like ordinary metrics it seems that projective holomorphic metrics induce ON-expansions. Please notice that the holomorphic norm $||.||^2 =< .,. >$ need not be real valued.

The following example will hopefully explain some of the inspiration for these ideas and names:

Example 5.1. *Let* f *be a holomorphic function on a neighbourhood of the origin in* $\mathbb{C}$. *Then it can be expanded in a holomorphic basis according to*

$$f = \sum_{n \in \mathbb{N}} < z^n, f > z^n = \sum_{n \in \mathbb{N}} \frac{1}{2\pi i} (\int_\gamma \frac{1}{z^{n+1}} f(z)dz))z^n.$$

Example 5.2. *To obtain a duality on a complex number of unit modulus* $e^{i\theta}, \theta \in \mathbb{R}$, *one can either use the common duality:* $\overline{e^{i\theta}} = e^{-i\theta}$ *or the holomorphic duality* $(e^{i\theta})^{-1} = e^{-i\theta}$.

The name holomorphic duality is inherited because of the natural interpretation of reflecting something into infinity as the duality. Hence one can get a righthanded phase by reflecting at the boundary of the unit circle to get something with a lefthanded phase, a *projective* property in some sense-this also reflects in the Euclidean norm changing to the inverse by the duality, also the duality corresponds to changing the origin 0 to infinity in $P^1\mathbb{M}$-another reason for the name projective holomorphic metric.

We will unfortunately not have much more space to go into the interesting properties of these metrics, which seem to be linked to topics like the Radon-Penrose transform and residue pairings and be the natural ancestors from which some twistor theory transforms drive their existance. Perhaps the most joyful thing to sum up about these "metrics" is that they give us the virtues of a metric without using any non-holomorphic components.

6. Moduli Spaces of Hypermanifolds

We shall briefly in this section mention the concept of a moduli space of a $\mathbb{M}^1$-hypermanifold, i.e a $\mathbb{M}$-hypermanifold where we are only looking at the $\mathbb{M}^1$ part locally. Let X_g be a fixed such manifold with hypertopology g. Let $\mathcal{M}_g$ be a contractible ball in the space of metrics of the manifold, and let $Biholo_{\mathbb{M}}(X_g)$ be the space of biholomorphisms[52]. We then define the hypermoduli space $Mod_{\mathbb{M}}(X_g)$ as

$$Mod_{\mathbb{M}}//(X_g) = \mathcal{M}_g/Biholo_{\mathbb{M}}(X_g)$$

and Teichmuller space as $Teich_{\mathbb{M}}(X_g) = \mathcal{M}_g/Biholo_{\mathbb{M}}^0(X_g)$. Teichmuller space is a covering space of the moduli and is related to the moduli as an often discrete fibration over the moduli. This fibration is pathological at branch points that have to be omitted. The road that we shall take will go through a mix of elliptic operator theory, analytical index theory and a generalization of a trick from string theory. We shall only be concerned with versions and kins of the compact case for simplicity.

Let us denote the space of hyperholomorphic tensors with n contravariant and m covariant indices over X_g, $T(n,m)(X_g) = T(n,m)^-(X_g)$, and the corresponding hypermoduli space

$$Mod^{\mathbb{M}}(X_g)// = T(n,m)(X_g)/Biholo^{\mathbb{M}}(X_g)$$

and similarly

$$Teich_{\mathbb{M},\,(n,m)}(X_g) = T(n,m)(X_g)/Biholo^{\mathbb{M}}(X_g)$$

We shall in the following interchangebly use $\displaystyle{\not{\partial}}\_{}$[53] and ∇. The first thing we would like to prove is

Theorem 6.1.

$$Teich_{(n,m)}^{\mathbb{M}}(X_g) = ker(\nabla^*\nabla|_{(n,m)})$$

Proof. The proof rests on two lemmas and a corollary.

Lemma 6.1.

$$T^*Biholo^{\mathbb{M}*} \subset Ran\nabla \oplus Ran\nabla^*$$

---

[52]It is common to supress hyper to make language easier on us.

[53]Please do not confuse this with the lefthanded part of the Dirac operator. The present object means the lefthanded part of the exterior derivative /connection on X_g. In hypergeometry lefthanded objects are holomorphic, just as in string theory. The present convention adhers to Polchinskis definitions in his books String Theory I and II, and is the generalization to heterotic p-branes, one of the incarnations that hypergeometries take.

Proof. Let $f : X_g \mapsto f(X_g)$ be a biholomorphism and T a tensor over X_g. Then f^*T is given in components by $(f^*T)^\nu_\mu = T^{\nu'}_{\mu'} \bar{\partial}_{-\mu} f^{\mu'}_- (\bar{\partial}_{-\mu} f)^{-1,\,\nu}_{\nu'} \, mod[\mathbb{M}, \mathbb{M}]$ with μ, ν and their primes hyperholomorphic multiindices. The commutator indicates that one in general has to differ between left and right tensor indices and that the above ordering only holds up to commutation. Letting $\delta_B T$ be the biholomorphic part of the variation we then get

$$\frac{\partial_B}{\delta f^\kappa_+} T^\nu_\mu = T^\nu_{\mu'} [\bar{\partial}_{-\mu_1} \delta^{\mu'_1}_\kappa \bar{\partial}_{-\mu_2} f^{\mu'_2} \cdots \bar{\partial}_{-\mu_m} f^{\mu'_m} (\bar{\partial}_{-\mu} f)^{-1,\,\nu}_{\nu'} + perm + \cdots$$
$$] \, mod[\mathbb{M}, \mathbb{M}] = \{ \text{ integration by parts } \}$$
$$= \bar{\partial}_{-\mu_1} T^\nu_{\mu'} \delta^{\mu'_1}_\kappa \bar{\partial}_{-\mu_2} f^{\mu'_2} \cdots \bar{\partial}_{-\mu_m} f^{\mu'_m} (\bar{\partial}_{-\mu} f)^{-1,\,\nu}_{\nu'} + perm + \cdots$$

Since the omitted part only, upon partial integration, includes more terms in ran $Ran\bar{\partial}_- \oplus Ran f^*_-$ we are done by noting that partial integration does not change the cohomology class.

$\square$

Lemma 6.2.

$$T^* Biholo^{\mathbb{M}*} \supset Ran\nabla \oplus Ran\nabla^*$$

Proof. Proof proceeds by contradiction. Assume $f : X_g \mapsto X_g$ is a diffeomorphism but not a biholomorphism. Then falls

$$f_* T$$
$$= f_* \sum_{\text{holomorphic indices}} a^{\nu_1,\cdots,\nu_n}_{\mu_1,\mu_2,\cdots,\mu_m} \pi^\nu_\mu (\partial_{\nu_1} \otimes \cdots \partial_{\nu_n} \otimes dx^{\mu_1} \otimes dx^{\mu_m})$$
$$= \sum_{\text{holomorphic indices}} a^{\nu_1,\cdots,\nu_n}_{\mu_1,\mu_2,\cdots,\mu_m} \pi^\nu_\mu ((\partial_{\nu_1} f^{\sigma_1}) \partial_{\sigma_1} \otimes \cdots (\partial_{\nu_n} f^{\sigma_n}) \partial_{\sigma_n} \otimes (\partial f)^{-1,\,\mu_1}_{\rho_1} dx^{\rho_1}$$
$$\otimes (\partial f)^{-1,\,\mu_1}_{\rho_m} dx^{\rho_m}) + f_* \sum_{\text{holomorphic indices}} a^{\nu_1,\cdots,\nu_n}_{\mu_1,\mu_2,\cdots,\mu_m} \pi^\nu_\mu (\partial_{\nu_1} \otimes \cdots \partial_{\nu_n} \otimes dx^{\mu_1} \otimes dx^{\mu_m})$$
$$+ \underbrace{\sum_{\text{non-holomorphic indices}} a^{\nu_1,\cdots,\nu_n}_{\mu_1,\mu_2,\cdots,\mu_m} \pi^\nu_\mu ((\partial_{\nu_1} f^{\sigma_1}) \partial_{\sigma_1} \otimes \cdots (\partial_{\nu_n} f^{\sigma_n}) \partial_{\sigma_n} \otimes (\partial f)^{-1,\,\mu_1}_{\rho_1} dx^{\rho_1}}_{\neq 0}$$
$$\otimes (\partial f)^{-1,\,\mu_1}_{\rho_m} dx^{\rho_m}) \, .$$

Since $\otimes$ is not singular we have $\delta T \notin Ran\nabla \oplus Ran\nabla^*$.

$\square$

Corollary 6.1. (1)

$$T^* Biholo^{\mathbb{M}*} = Ran\nabla \oplus Ran\nabla^*$$

(2)

$$Teich^{\mathbb{M}}_{(n,m)}(X_g) = ker(\nabla^* \nabla|_{(n,m)})$$

Proof. 1) is trivial in view of the preceding two lemmas and 2) falls from the orthogonal Hodge decomposition $T(n,m)^- = Ran\nabla \oplus Ran\nabla^* \oplus ker(\nabla^* \nabla)$.

$\square$

$\square$

64

Thus we have reduced the problem of calculating the dimension of (n,m)-tensor Teichmuller space to calculating the index of the sequence $\mathfrak{S}$;

$$\{0\} \overset{i}{\mapsto} T(n,m)^- \overset{\nabla}{\mapsto} T(n,m+1)^- \overset{i}{\mapsto} \{0\}$$

and thus the Atiyah-Singer theorem gives on compact[54] real d-dimensional hypermanifolds of arbitrary hyperdimension, say h,

Corollary 6.2 (The Hyperreal Riemann-Roch Theorem.). *Let X_g a hypermanifold be compact, then*

$$dim^{\mathbb{M}}(Teich^{\mathbb{M}}_{(n,m)}(X_g))$$
$$= (-1)^{\frac{d(d+1)}{2}}\{Ch(T^-X^{\mathbb{M}\,n} \otimes T^{-*}X^{\mathbb{M}\,m})(1 - ch(T^*))\frac{Td(TX^{\mathbb{M}})}{e(TX)}\}[X_g]$$

is the dimension of the Teichmuller space of (n,m)-tensors on X_g.

Substracting p punctures from X_g we realize that the dimension increases with ph since the hyperdimension of the space of punctures is ph. Thus we have a new theorem

Theorem 6.2 (The Hyperreal Riemann-Roch Theorem.). *Let X_g a hypermanifold be compact up to p punctures and of hyperdimension h, then*

$$dim^{\mathbb{M}}(Teich^{\mathbb{M}}_{(n,m)}(X_g))$$
$$= (-1)^{\frac{d(d+1)}{2}}\{Ch(T^-X^{\mathbb{M}\,n} \otimes T^{-*}X^{\mathbb{M}\,m})(1 - ch(T^*))\frac{Td(TX^{\mathbb{M}})}{e(TX)}\}[X_g] + ph$$

is the dimension of the Teichmuller space of (n,m)-tensors on X_g.

To conclude we would like to give an example of an application of the above.

Example 6.1. *We shall calculate the dimension of the moduli space of metrics on a Riemann surface Σ_g, one of the classical moduli spaces. We take the more general road of deriving a classical Riemann-Roch theorem to do this. Restricting the above formulae to nill punctures and complex one-dimensional hypermanifolds we obtain*

$$dim^{\mathbb{C}}(Teich^{\mathbb{C}}_{(n,m)}(\Sigma_g))$$
$$= (-1)^{\frac{d(d+1)}{2}}\{Ch(T^-X^{\mathbb{M}\,n} \otimes T^{-*}X^{\mathbb{M}\,m})(1 - ch(T^*))\frac{Td(TX^{\mathbb{M}})}{e(TX)}\}[\Sigma_g]$$

We notice that in terms of Chern classes

---

[54]A compact manifold has no boundary and is therefore also called closed.

(1) $Ch(T^- X^{\mathbb{M}\ n} \otimes T^{-*} X^{\mathbb{M}\ m}) = (1+c_1)^n (1-c_1)^m = (1+(n-m)c_1)$.

(2) $1 - ch(T^*) = c_1$.

(3) $Td(TX) = 1 + \frac{1}{2}c_1$.

(4) $e(TM) = c_1$.

Hence not forgetting $(-1)^{\frac{d(d+1)}{2}}|_{d=2} = -1$ *we have*

$$dim^{\mathbb{C}}(Teich^{\mathbb{C}}_{(n,m)}(\Sigma_g))$$
$$= -(1 + (n-m)c_1)(1 + \frac{1}{2}c_1)[\Sigma_g]$$
$$= -((n-m) + \frac{1}{2})c_1[\Sigma_g] = (2(n-m) + 1)(g-1)$$

Thus we conclude that(setting $n - m = -2$ and taking into account that dimensionalities have to be either positive or nill) for the Teicmuller space of covariant two-tensors over a Riemann surface we have $dim^{\mathbb{C}}(Teich^{\mathbb{C}}(\Sigma_g)) = 3g - 3$ for $g \geq 1$.

7. Hypertopology

During a brief period when I was studying elementary several complex variables, complex geometry and topology I often wondered about the existance of some non-real generalization of ordinary real topology. The reader who has read the elementary cohomology section has already encountered such a candidate: $\bar{\partial}$-cohomology. However $\bar{\partial}$-cohomology deals with smooth differential forms and exterior algebras and only indirectly with complex spaces themselves.

Perhaps the easiest way to understand what we are looking for is to formulate it as a search for duals of $\bar{\partial}$-cohomology-to put it a bit more precisely we are searching for a duality between exact sequences of spaces and dittos of exterior forms on space.

We shall begin with a remarkable theorem, where we shall give the chain group over a space the structure of a Hibert space. Upon that we shall have the foundational material to define several objects and obtain our goal, our hypertopology.

Theorem 7.1. *Let C(X,K) be the homology ring of X over a field K. Then C(X,K) can, up to a completion under a norm, be made into a Hibert space for X equipped with a volume measure.*

Proof. We have for $\omega \in C(X,K)$, $\eta \in C(X,K)$ a product $\omega \wedge *\eta$ defined as follows; $\wedge$ is $\mathbb{K}$-bilinear defined on a homology basis generated by simplical complexes $< \mu_1, \cdots, \mu_p >, < \nu_1, \cdots, \nu_q >$, by $< \mu_1 > \wedge < \mu_2 >=< \mu_1, \mu_2 >, \cdots$ or in general manner $< \mu_1, \cdots, \mu_p > \wedge < \nu_1, \cdots, \nu_q >=< \mu_1, \cdots, \mu_p, \nu_1, \cdots, \nu_q >$. We then define $*$ as follows; Supressing brackets $* : C_q \mapsto C_d - q$; $*\mu_1 \wedge \mu_2 \wedge \cdots \mu_p =$

66

$sgn(\pi(\mu_1, \cdots, \mu_p))\mu_{p-1} \wedge \cdots \mu_n$. Then pick a volume measure ϵ_X[55]. Our inner productis then defined by

$$< \omega, \eta >= \int_{\bar{\omega} \wedge *\eta} \epsilon_X$$

the bar indicating field duality in $\mathbb{K}$. That C is a vector space over $\mathbb{K}$ is obvious, with elements consisting of chains modulo equivalence classes with respect to integration with measure ϵ_X, i.e to clarify $\omega = \eta$ iff $\int_{|\omega-\eta|^2} \epsilon_X = \int_{(\omega-\eta)\wedge*(\omega-\eta)} \epsilon_X = 0$. It then comes down to prove

(1) $< .,. >$ is a (graded) inner product.
(2) $||.|| = \sqrt{< .,. >}$ is complete.

Lemma 7.1. *Statement 1.*

Proof. We first prove a graded conjugation rule,

$$\overline{< \omega, \eta >} = (-1)^{(\deg(\eta)+deg(\omega))\deg(\omega)} < *\eta, *\omega >$$

, d the dimension of X. We can assume our volume measure to be real for simplicity(This need not be the case in general, but a simple renormalization of the homology basis resolves this issue.), then

$$< \omega, \eta >= \overline{\int_{\omega \wedge *\eta} \epsilon_X}$$
$$= \int_{\overline{\omega \wedge *\eta}} \epsilon_X = \int_{(-1)^{\deg_*(\eta)\deg(\omega)}*\bar{\eta}\wedge\bar{\omega}} \epsilon_X$$
$$= (-1)^{\deg_*(\eta)\deg(\omega)} \int_{*\bar{\eta}\wedge\bar{\omega}} \epsilon_X$$
$$= (-1)^{\deg_*(\eta)\deg(\omega)}(-1)^{deg_*(\omega)\deg(\omega)} \int_{*\bar{\eta}\wedge**\bar{\omega}} \epsilon_X$$
$$= (-1)^{(\deg_*(\eta)+deg_*(\omega))\deg(\omega)} \int_{*\bar{\eta}\wedge**\bar{\omega}} \epsilon_X$$
$$= (-1)^{(d-deg(\eta)+d-deg(\omega))deg(\omega)} < *\eta, *\omega >$$
$$= (-1)^{(deg(\eta)+deg(\omega))deg(\omega)} < *\eta, *\omega >$$

consisting a proof of a graded conjugation rule for the inner product. $< .,. >$ is obviously bilinear. Ass. $\omega \wedge *\omega \neq 0$ in measure, then $||\omega||^2 = 0$ by def., hence $\omega = 0 \Rightarrow ||\omega|| = 0$. Conversely $||\omega|| = 0 \Rightarrow ||\omega||^2 = 0$ implies ω to be nill by definition of nullity. Note that the Cauchy-Schwarz inequality holds for the above product from quite general considerations of linear algebra. Thus finally $||\omega + \eta||^2 = ||\omega||^2 + ||\eta||^2 + < \omega, \eta > + < \eta, \omega > \leq (||\omega|| + ||\eta||)^2$ by the Cauchy-Schwarz inequality implies $||\omega + \eta|| \leq ||\omega|| + ||\eta||$. $\square$

Lemma 7.2. *Statement 2*

---

[55]The author uses the notation ϵ for the volume form since it is, of course, the same thing as the Levi-Cevita tensor.

Proof. Defining $C(X, K)$ to be it's completion $\bar{C}$ under the norm above we are done. □

□

Let us recollect what we have proved; If we define the chain group to be the completion in the above sense under the above norm topology we have a Hilbert space. Note how the above parallels Hilbert spaces of say smooth forms, and that we are really abusing language when calling C^K a Hilbert ring, rather we are dealing with equivalence classes under a norm, in a way reminicent to abusing language when saying that L^2-spaces consist of Lebesgue square integrable functions.

We now make a slight detour and introduce the notation used in this 'hypertopology'. We shall introduce the concept needed heuristically; Perhaps the simplest way is to again think of dual spaces. So what we could perhaps do is think of the $\bar{\partial}$-cohomology and then try to think about how we might construct a dual. A natural guess in this context is to create 'holomorphic' vertices-hypervertices- as the simplices one is using. Just as in the case of hyperanalysis/geometry there are two different generalizations of $\bar{\partial}$-cohomology to the hypercomplex case; The dual holomorphic/non-real cohomology and the anti-holomorphic cohomology.

Example 7.1. *Dual non-real cohomology is generated by the $\bar{\partial}^i = d\bar{\zeta}^i \partial^i_{\bar{\zeta}}$, i holomorphic or non-holomorphic, operators on a left module(Remember that we fix things a priori to mean either left or right differentiation. That means that we have a left theory and a right theory. It is not a priori trivial that these two theories coincide in terms of information-indeed it might not be so.), usually $\bigwedge^{(p_1, p_2, \ldots, p_n)}(X, \mathbb{K})$, p_i anti-holomorphic indices for $i \neq 1$. Thus to take a concrete example f a bihyperholomorphism would generate an exterior algebra isomorpism of the sequence below corresponding to dual non-real cohomology on X_g a hypermanifold X with hypertopology g [56];*

---

[56]This notation derives it raison d'étre from the fact that we are actually looking at differential operators acting on entire structural cathegories. We then envisage objects to be linear superpositions of objects in the category-when we have a structural category formed of the space of linear superpositions of objects with fixed structural topology we then get a diagonal action-a spectral action. This representation of the category is called the spectral representation. Structural categories are Hilbert spaces in the obvious way(direct sum of objects)- when we do quantum theory later on this will be most useful-it is for this reason that we want to make the reader accustomed to this way of thinking.

$$
\begin{array}{ccc}
C^\infty(f(X_g),\mathbb{K}) & \not\cong & O(f(X_g))
\end{array}
$$

$$
\begin{array}{ccccccccc}
 & & \parallel\!\mathbb{R} & & & & & & \\
\bar\partial & & & & \bar\partial & & \bar\partial & & \bar\partial \\
\{0\} & \to & \Lambda^{(p_1,p_2,\cdots,p_n)}(f(X_g),\mathbb{K}) & \to & \Lambda^{(p_1+\bar1,p_2,\cdots,p_{n,})}(f(X_g),\mathbb{K}) & \to & \cdots & \to & \\
 & & f^*\downarrow & & f^*\downarrow & & & & f^*\downarrow \\
\bar\partial & & & & \bar\partial & & \bar\partial & & \bar\partial \\
\{0\} & \to & \Lambda^{(p_1,p_2,\cdots,p_n)}(X_g,\mathbb{K}) & \to & \Lambda^{(p_1+\bar1,p_2,\cdots,p_{n,})}(X_g,\mathbb{K}), & \to & \cdots & \to &
\end{array}
$$

$$
\begin{array}{ccc}
 & \parallel\!\mathbb{R} & \\
C^\infty(X_g,\mathbb{K}) & \not\cong & O(X_g)
\end{array}
$$

There is another cohomological theory, equally, or perhaps even of greater interest, namely the one for the ∂ operator acting on $\bigwedge^{(p_1,p_2,\ldots,p_n)}(X_g,\mathbb{K})$. Thus we again have that $f \in Biholo_\mathbb{K}(X_g)$ would generate an algebra isomorphism for the below sequence corresponding to hyperholomorphic cohomology

$$
\begin{array}{ccc}
C^\infty(f(X_g),\mathbb{K}) & \not\cong & O(f(X_g))
\end{array}
$$

$$
\begin{array}{ccccccccc}
 & & \parallel\!\mathbb{R} & & & & & & \\
\partial & & & & \partial & & \partial & & \partial \\
\{0\} & \to & \Lambda^{(p_1,p_2,\cdots,p_n)}(f(X_g),\mathbb{K}) & \to & \Lambda^{(p_1+1,p_2,\cdots,p_{n,})}(f(X_g),\mathbb{K}) & \to & \cdots & \to & \\
 & & f^*\downarrow & & f^*\downarrow & & & & f^*\downarrow \\
\partial & & & & \partial & & \partial & & \partial \\
\{0\} & \to & \Lambda^{(p_1,p_2,\cdots,p_n)}(X_g,\mathbb{K}) & \to & \Lambda^{(p_1+1,p_2,\cdots,p_{n,})}(X_g,\mathbb{K}), & \to & \cdots & \to & \Lambda
\end{array}
$$

$$
\begin{array}{ccc}
 & \parallel\!\mathbb{R} & \\
C^\infty(X_g,\mathbb{K}) & \not\cong & O(X_g)
\end{array}
$$

One can of course obtain a totally holomorphic theory by only considering holomorphic exterior algebra modules. This corresponds to the dual of the right lower diagonal of the Hodge diamond for the complex case. The general hyperdiamond is a multidimensional lattice of (co)homology groups/ hyperhodge numbers. Finally we would like to point out that one can always consider the usual antiholomorphic, not dual, $\bar\partial$-operator theory, $\bar\partial = \sum_{a\ not\ holomorphic} d\zeta^a \partial_a$.

In much the same manner we have a dual construction of hyperhomology groups. It is this interesting object which shall to some extent elucidate the structure of complex spaces to us. Thus one forms a split of the boundary operator $\partial = \partial + \bar\partial$ into holomorphic and antiholomorphic part and obtains a homological theory by letting the holomorphic boundary operator act only on holomorphic indices and the antiholomorphic on antiholomorphic. Notice that for the case of $\mathbb{C}$ the antiholomorphic operator and dual holomorphic operators coincide. The new operators satisfy the usual graded Leibniz rules and are in particular exact as $0 = (\partial + \bar\partial)^2 = \partial^2 + \bar\partial^2 + \{\partial,\bar\partial\}$ shows (also, by the way, implying $0 = \{\partial,\bar\partial\}$).

Excercise 7.1. *Prove that ∂ satisfies a graded Leibniz rule.*

Let us take an example of the abstract constructions encountered so far.

Example 7.2. *We shall calculate the complex holomorphic hyperhomology groups of $\mathbb{C}P^1$. We have holomorphic/antiholomorphic generators(vertices) $\{\mu_1, \mu_2, \bar{\mu}_1, \bar{\mu}_2\}$ and a cell complex K generated by $\bigwedge\{\mu_1, \mu_2, \bar{\mu}_1, \bar{\mu}_2\}$ and some relations among them. We first calculate $H_{\partial(0,0)}(X, \mathbb{C})$. Thus we have to compute $ker(\partial) = Z_{\partial(0,0)}(\mathbb{C}P^1, \mathbb{C})$. By $0 = \partial(a\mu_1 + b\mu_2), \forall (a, b) \in \mathbb{C}^2$ maximally[57] we get $\mathbb{C}^2 = Z_{\partial(0,0)}(\mathbb{C}P^1, \mathbb{C})$. Thus $B_{\partial(0,0)}(\mathbb{C}P^1, \mathbb{C}) = Ran(\partial|_{\bigwedge^{(1,0)}}) = \partial(a\mu_1 \wedge \mu_2) = a(\mu_2 - \mu_1) \cong \mathbb{C}$ gives $H_{\partial(0,0)}(X, \mathbb{C}) = Z_{\partial(0,0)}(\mathbb{C}P^1, \mathbb{C})/B_{\partial(0,0)}(\mathbb{C}P^1, \mathbb{C}) = \mathbb{C}^2/\mathbb{C} = \mathbb{C}$. Next we note $Z_{\partial(1,0)}(\mathbb{C}P^1, \mathbb{C}) \cong \{0\}$ by $0 = \partial a\mu_1 \wedge \mu_2 = a(\mu_2 - \mu_1) \Rightarrow a = 0$, thus $H_{\partial(1,0)}(X, \mathbb{C}) \cong \{0\}$. Next we compute $Z_{\partial(1,1)}(\mathbb{C}P^1, \mathbb{C})$. We have $0 = \partial a\mu_1 \wedge \mu_2 \wedge \bar{\mu}_1 \wedge \bar{\mu}_2 = a(\mu_2 - \mu_1) \wedge \bar{\mu}_1 \wedge \bar{\mu}_2$. Interesting as it is the latter need not imply $a = 0$-this is realized as follows-the form spanning the volume element is equivalent to nill since $0 = (\partial + \bar{\partial})\mu_1 \wedge \mu_2 \wedge \bar{\mu}_1 \wedge \bar{\mu}_2 = (\mu_2 - \mu_1) \wedge \bar{\mu}_1 \wedge \bar{\mu}_2 + \mu_1 \wedge \mu_2 \wedge (\bar{\mu}_2 - \bar{\mu}_1)$ and module properties imply this upon identification of terms. The above gives that all of $C_{(1,1)}$ lies in the kernel of ∂ , hence $Z_{\partial(1,1)}(\mathbb{C}P^1, \mathbb{C}) \cong \mathbb{C}$. As for the relevant boundary group we notice $C_{(2,0)} \equiv \{0\}$, thus $B_{\partial(1,1)}(\mathbb{C}P^1, \mathbb{C}) \cong \{0\}$, and so $H_{\partial(0,0)}(X, \mathbb{C}) = \mathbb{C}/\{0\} \cong \mathbb{C}$ falls. The remaining group $H_{(0,1)}$ is now easily computed by noting that exterior multiplication with antiholomrphic vertices (anti)commutes with holomorphic exterior diffrentiation and that no special relations hold there, thus $H_{(0,1)} \cong \{0\}$. We can collect and display the hypertopological data as follows*

$$(H_{\partial(p,q)}(\mathbb{C}P^1, \mathbb{C})) = \begin{pmatrix} & \mathbb{C} & \\ \{0\} & & \{0\} \\ & \mathbb{C} & \end{pmatrix}$$

in analogy with the Hodge diamond. Actully, using deRham duality and the field duality we directly obtain as a consequence the result for $\bar{\partial}$-cohomology as follows;

$$(H_{\bar{\partial}}^{(p,q)}(\mathbb{C}P^1, \mathbb{C})) = \begin{pmatrix} & \mathbb{C} & \\ \{0\} & & \{0\} \\ & \mathbb{C} & \end{pmatrix}$$

Excercise 7.2. *Find the complex holomorphic hyperhomology groups for $\mathbb{C}P^n$.*

---

[57]∂ ignores antiholomorphic simplices.

We shall, just to show some non-trivial aspect of these theories, calculate the right quaternion holomorphic hyperhomology $H_{\partial_R,\ (p_1,p_2,p_3,p_4)}(\mathbb{H}P^1,\mathbb{H})$ of the first quaternionic projective space.

Example 7.3. *Set $\partial := \partial_R$. Since $Z_{\partial,\ 0}(\mathbb{H}P^1,\mathrm{M}) \cong \mathbb{H}$ follows from $\partial\mathbb{H} = 0$ and $\partial a\sigma^1 = 0, a \in \mathbb{H}$, σ_1 the generator of the holomorphic chain group(we have that ∂ ignores the non-holomorphic chains), we have $H_{\partial_R,\ 0}(\mathbb{H}P^1,\mathbb{H}) \cong \mathrm{M}$. Poincare duality gives then $H_{\partial_R,\ (1,1,1,1)}(\mathbb{H}P^1,\mathbb{H}) \cong \mathrm{M}$. For the other groups we note that we have an orthogonal decomposition of corresponding chain groups, hence after some non-trivial work one can prove $H_p(\mathbb{H}P^1,\mathbb{H}) \cong \bigoplus_{p=|\alpha|} H_{\partial_R,\ \alpha}(\mathbb{H}P^1,\mathbb{H})$, α a muliindex. Triviality of $H_p(\mathbb{H}P^1,\mathbb{H})$ $p \neq 0,\ 4$ now implies that the remaining hyperhomology groups are trivial.*

To sum up:

$$H_{\partial_R,\ 0}(\mathbb{H}P^1,\mathbb{H}) \cong \mathrm{M},$$
$$H_{\partial_R,\ (1,1,1,1)}(\mathbb{H}P^1,\mathbb{H}) \cong \mathrm{M},$$
$$H_{\partial_R,\ \alpha}(\mathbb{H}P^1,\mathbb{H}),\ |\alpha| \neq 0,\ 4.$$

8. Integral Representations of Hyperanalysis II

Now that we have gained a little bit of knowledge about the general structure of hypermathematics we shall revisit integral representations, in particular constructing some far-reaching but simple generalizations. Three formulas shall be presented

(1) A formal integral representation with holomorphic kernal of a holomorphic function on compact domains of $\mathbb{C}^n$. Notice that the Shilov boundary of collections of sets with the same dimension might be objects of different dimensions, hence this is highly non-trivial[58].

(2) An integral representation on product submanifolds $\mathbb{M} = I \times M \subset X_g$ of X_g a smooth manifold, $I \cong [0,1]$ for real cases.

(3) An integral representation on a family of conformal equivalences of submanifolds $M_t \subset \mathbb{M} = I \times M \subset X_g$, $t \in I$, X_g a smooth manifold.

Definition 8.1. *A formal integral representation in the present sense is a map from the homology Hilbert ring C times its dual to a field, algebra, ring etc of functions over a field, ring or other suitable algebraic structure $\mathbb{K}$, i.e pointwise a map $\mathcal{I} : C \times C^* \mapsto \mathbb{K}$.*

Then

Theorem 8.1. *Let Ω be arbitrary and compact in $\mathbb{C}^n$. Then $\exists$ a formal integral representation on a Ω with holomorphic kernal for holomorphic $f \in \mathcal{O}(\Omega)$. The formal integration domain is contained in the usual boundary of Ω.*

Proof. We can use cubes C_i as basis chains, then on C_i Cauchys polydisc theorem provides us with

$$f(z) = \frac{1}{(2\pi i)^{|n|}} \int_{\partial_+ C_i} \underbrace{d\zeta^n}_{\epsilon_X} \frac{f(\zeta)}{(\zeta - z)^n}, z \in C_i.$$

∂_+ being the holomorphic boundary operator and zero otherwise. Hence for $z \in \sum C_i$, $C_i \cap C_j = \emptyset, i \neq j$ we get

$$f(z) = \frac{1}{(2\pi i)^{|n|}} \int_{\sum \partial_+ C_i} \underbrace{d\zeta^n}_{\epsilon_X} \frac{f(\zeta)}{(\zeta - z)^n}, z \in \sum C_i.$$

thus pick a sequence of families $\{C_i\}^\sigma$ of families of such chains indexed by σ that exhausts Ω from the enterior with index sets I^σ, then

---

[58]Thanks goes to O. Stormark of the mathematics department of the Royal Institute of Technology, Stockholm, Sweden, for pointing this out to me.

by completeness we get, defing an extension $\partial_+\Omega$ by linearity our result by completeness of C, setting $\partial_+\Omega = lim\ \partial_+ \sum_\sigma C_i^\sigma = lim\ \sum_\sigma \partial_+ C_i^\sigma$. I now claim the existance of this limit.

Lemma 8.1. *Let the topology on C be the one inherited by it's Hilbert space structure. Then the above limit exists, i.e $lim\ \sum_\sigma \partial_+ C_i^\sigma\ \exists$.*

Proof. It suffices to note with μ the corresponding Lebesgue volume from ϵ_X that

$$
\begin{aligned}
&||\Omega - \sum_{I^\sigma} \partial_+ C_i^\sigma||^2 \\
&= \int_{(\Omega - \sum_{I^\sigma} \partial_+ C_i^\sigma)^*(\Omega - \sum_{I^\sigma} \partial_+ C_i^\sigma)} \epsilon_X \\
&= \mu(\Omega - \sum_{I^\sigma} \partial_+ C_i^\sigma)
\end{aligned}
$$

which by exhaustion from the interior will tend to nill as σ goes to infinity, since null sets are defined to be in the nill equivalence class of our Hilbert space. $\square$

$\square$

We would like to point out directly to the reader that s/he can in favorable cases use our hypertopology to compute the holomorphic pseudoboundary directly in an algebraic manner. In the pluricomplex case other, more function-theoretic, ways are known, indeed a combination seems to be the most effective way.

We now move on to the next theorem.

Definition 8.2. *A hyperhomotopy of (hyper)manifolds $\mathbb{M}$ is a structural isomorphism of a cylinder $M \times I$, M a hypermanifold, $I = D^1$ the appropriate hyperdisc.*

Example 8.1. *For the real case a hyperdisc is the interval, for complex case it is the unit disc in the complex plane, and for quaternionic case it is the real 4-dimensional closed ball of unit radius around the origin.*

Sometimes we abuse language and say homotopy for contraction, with contractible meaning homotopic to a point (See Novikov et al., Modern Geometry III).

Definition 8.3. *A hyperhomotopy is called centered at a point q of an embedding space if it can contract a manifold to the point q, with the origin in the corresponding homotopy disc corresponding to this point.*

Fix f to be real analytic or/and smooth depending on circumstances.

Theorem 8.2. *Let $\mathbb{M} \subset X_g$ be a smooth hyperhomotopy of closed commutative (pseudo-)Riemannian hypermanifolds centered at $q \in X_g$,*

73

and $g = diag(1, h_t^q)$ a metric on X with h_t^q a metric on M_t^q, t the corresponding value on the homotopy hyperdisc D. Then for $< M_{0+}^q, \frac{1}{\sqrt{h_t^q}} \epsilon_{M_{0+}^q} > \neq$ 0, with ∂ denoting the exterior differentiation operator holomorphically normal to the manifolds, i.e $\partial = dt \partial_t$ for the right choice and $\partial = \partial_t dt$ for the left,

$$f(0+, q) = [< \partial \mathbb{M}_q, f(t, p) \frac{1}{\sqrt{h_t^q(t,p)}} * dt >$$
$$- < \mathbb{M}_q, (\partial f(t,p)) \frac{1}{\sqrt{h_t^q(t,p)}} * dt >]_{<M \times \partial D_{0+}, \frac{1}{\sqrt{h_t^q}} * dt >}$$

Remark: Please note the parentheses at the last line around ∂f.

Proof. Let D_R be the hyperdisc of radius R.

$$L.S. = < \partial \mathbb{M}_q, f(t,p) \frac{1}{\sqrt{h_t^q(t,p)}} * dt >$$
$$= < M \times (I \sim D_R), d\, f(t,p) \frac{1}{\sqrt{h_t^q(t,p)}} * dt >$$
$$+ < \partial(M \times D_R), f(t,p) \frac{1}{\sqrt{h_t^q(t,p)}} * dt >$$
$$= < M \times (I \sim D_R), (\partial f(t,p)) \frac{1}{\sqrt{h_t^q(t,p)}} * dt >$$
$$+ < M \times \partial D_R, f(t,p) \frac{1}{\sqrt{h_t^q(t,p)}} * dt >$$
$$= \{R \mapsto 0\}$$
$$= < M, (\partial f(t,p)) \frac{1}{\sqrt{h_t^q(t,p)}} * dt >$$
$$+ f(0+, q) < M \times \partial D_{0+}, \frac{1}{\sqrt{h_t^q(t,p)}} * dt >$$

Where we in the second line noted Stoke's theorem, in the second that

(1) The quotient $\frac{1}{\sqrt{h_t^q(t,p)}} * dt$ is a constant differential and hence lies in the kernel of the exterior derivative.
(2) That the resulting expression contains the commutative Hodge-DeRham inner product and thus projects df onto its holomorphic part ∂f.

In the last line we noted that the only thing in the integrand of the last expression that depends on any coordinate is the function f itself. $\square$

Notice that we have all of a sudden a new tool that permits us to calculate and prove very general results. There is a possible generalization of this result which deals with the case where the deformation retract is not singular which we shall not go into in this thesis, but is virtually the same as above with the soul exception that q is then on a submanifold of X, of which some variations have been considered by Grothendieck and others as a consequence of some deep work in residue theory. The proof is essentially the same as above but with the origin of the hyperdisc replaced with this retract rather than the point above.

Excercise 8.1. *Prove this theorem.*

Of course there are many special cases of this theorem holding under weaker circumstances. The Cauchy integral formula, Martinelli-Bochner and the below theorem on conformal equivalences are examples.

Theorem 8.3. *Let $\mathbb{M} = M \times I$ be a family of conformal equivalences of smooth pseudo-Riemannian real manifolds parametrized by $t \in I = [0, 1]$, with metrics $h_t = e^{2\sigma} h_0$, h_0 the metric corresponding to $t = 0$. Then with d the dimension of M we have*

$$f(0+, q) = [< \partial \mathbb{M}_q, f(t, p) \tfrac{1}{e^{d\sigma}} * dt >$$
$$- < \mathbb{M}_q, (\partial f(t, p)) \tfrac{1}{e^{d\sigma}} * dt >]_{\frac{1}{<M \times \partial I_{0+}, *dt>}}$$

Proof. Repeating the argument from the last theorem and noting that the quotient

$$\frac{1}{e^{d\sigma}} * dt$$

lies in the kernel of the exterior derivative d since it does not depend on t and that any exterior differential other than dt would annihilate $*dt$ we are done. $\qquad \square$

We remark that the weaker

Corollary 8.1. *Under the assumptions of the previous theorem*

$$f(0+, q) = [< \partial \mathbb{M}_q, f(t, p) \tfrac{1}{e^{d\sigma}} * dt >$$
$$- < \mathbb{M}_q, (df(t, p)) \tfrac{1}{e^{d\sigma}} * dt >]_{\frac{1}{<M \times \partial I_{0+}, *dt>}}$$

holds.

Example 8.2. *We would like to show one small application of the previous theorem. Set $f = \partial_t u$ and everything to be smooth. Using $f * dt = *\partial u$ we easily see $df * dt = (-1)^s * *d * \partial u = (-1)^{s+1} * d^* du$, where we noted that $< M \times I, \bar{\partial}\omega >=< \underbrace{(\partial M)}_{=\emptyset} \times I, \omega >= 0$ for arbitrary smooth ω and closed M. Hence we can write for $u \in C(\mathbb{M}, \mathbb{R})$*

$$\partial_t u = [< \partial \mathbb{M}_q, \tfrac{1}{e^{d\sigma}} * \partial u >$$
$$- < \mathbb{M}_q, (-1)^{s+1} \tfrac{1}{e^{d\sigma}} * \Box u >]_{\frac{1}{<M \times \partial I_{0+}, *dt>}}$$

which upon integration yields for $(t, x) \in I \times M$

75

$$u(t,x) = u_{//ker(\partial)} + \int^t [< \partial \mathbb{M}_q, \frac{1}{e^{d\sigma}} * \partial u >$$
$$- < \mathbb{M}_q, \frac{(-1)^{s+1}}{e^{d\sigma}} * \Box u >]\frac{1}{<M \times \partial I_{0+}, *dt>}$$

9. Linear PDE and Congruences of Vector Fields.

Often we learn in analysis to look upon solutions of differential equations in one variable as flows of vector fields. In this section we shall similarly look upon the solutions of PDE's as flows of vector fields. We will return to this point when considering the evolution of D-branes in physics.

We begin as usual with a theorem.

Theorem 9.1. *Let M be a complex manifold and $\mathbb{M}$ the corresponding cylinder $M \times B^1_{\mathbb{C},\epsilon} \subset \mathbb{C}$, with this latter ball of radius $\epsilon \in \mathbb{R}_+$ small enough. Let the analytic PDE with analytic Cauchy data*

$$Pu = J,$$
$$u|_M,$$
$$\partial_t u|_M,$$
$$\cdots \partial_t^q u|_M.$$

$q = deg(P)$ set to be the degree of P as a linear differential operator in t over the ring of differential operators with constant coefficients on M, be given, in such a way that M is not charecteristic in the sense of Cauchy-Kovalevskaja w.r.t the operator P. Then the above equation has a real analytic solution that can be written

$$u = \sum_{IJ} a_{IJ} \lambda^I e^{\lambda O_J} u_I + P^{-1}J$$

on this cylinder, $u_I = \partial_t^q u|_M$ being the I:th Cauchy data with P understood to act 'projectively' on the quotient of it's domain and it's kernel and $\lambda \in B^1_{\mathbb{C},\epsilon}$ a complex transversal parameter. In case the coeffients are all holomorphic operators on M, the solution above is holomorphic.

Sketch of Proof: The proof presents quite an arduous task in it's thorough version. Since it is not hard, just technical, we only sketch it. Cauchy-Kovalevskaja's theorem gives existance of an analytic solution. As for the rest, since M is analytic it is smoothly embeddable in $\mathbb{R}^{4dim^{\mathbb{C}}(M)}$, extending by the nill extension from this cylinder to the total of this embedding space we get a distribution by using a partition of unity. Inverting P via canonical Fourier theory we obtain an inverse $P^{-1}J$. Solving the PDE taking the negative of this solution as Cauchy data with no source term we now have a new inverse $P^{-1}J$ that will not affect the boundary value data by addition with the former. Letting our operator act on the Fourier representation of the boundary value data, we see that we a have a polynomial in the Fourier space coordinates of the manifold $P(\partial_t) = f_a(\partial_t)^a = 0$ as the equation defining the solutions, with coefficents f_a being polynomials in the Fourier coordinates. Hence in the Fourier space of the 'homotopy' ball above the solutions are spectrally represented as lying as a discrete singular fibration over the Fourier space of the manifold, with possible branching points of lower dimension inducing singularities and conversely. Taking this into account when taking the transition back we obtain our solution in the general form above since this latter polynomial would split locally as $0 = \Pi(\zeta(\omega) - \partial_t)$ in the Fourier space at zero's of minimal mutipicity, ω coordinates in the Fourier space of the manifold, hence making explicit construction of the kernel possible as the direct sum of factors, with $\zeta(\omega)_i = \partial_t$ holding on each subkernel, thus we obtain the solution via exponentiation by linearity. When zero's are of higher order in the Fourier space of the ball a simple argument shows that the introducing a polynomial coefficient in front of the exponential corresponding to the appropriate factor of the same order as the order of the zero suffices to correct this. The proof consists then of tying together all the loose ends above, such as various existances, and then showing that the solution is analytic by an application of Payley-Weiner's theorem or some other device.

■

The interesting thing about this theorem is that it displays the solutions in a way that reveals some of the dynamics involved and in doing so connects to [generalized] Lie flows in a very direct way.

10. Modular Identification of Tangent Spaces.

It is well known that the exponential map over a proper Riemannian manifold X_g generates a local homeomorphism between the an open subset of the tangent and an open subset of the manifold. For physical reasons, linked to general quantum field theory, pseudodifferential operator theory and symplectic geometry, we wish to extend this. For compact homogenous spaces a similar construction is well known, we give here another construction similar in nature.

Theorem 10.1. *Let X_g be an orientable proper Riemannian manifold with trivial zeroth monodromy $b_0(X_g) = 1$, then*

$$T_p X_g / \sim \cong X_g$$

$\sim$ being the equivalence generated by having the same exponential action, with the quotient being regarded as a the metric topological space

77

induced by

$$d_{T_pX_g/\sim}(v, v') = d_{X_g}(exp(v), exp(v')), v, v' \in T_pX_g/ \sim$$

the latter being the usual geodesic distance between two points.

Notice that injectivity is trivial, likewise bicontinuity above. Hence surjectivity remains, which falls from work by Hopf, Rinow and Kobayashi and is outside of the scope of this thesis.

11. Twistor Solutions to Some Equations of M. Physics.

The following is an exerpt from a talk held as course work in a graduate course in several complex variable theory. It aims at giving a modest aquaintance with some perspectives in twistor geometry. The references cited at the end should be consulted for further information concerning these topics.

11.1. Some Facts of Twistor Geometry.

11.1.1. *The Ingredients.* The ingredients of twistor geometry(in this talk) are geometric to their nature. $\mathbb{T} \cong \mathbb{C}^4$ is called Twistor space and from it we obtain $\mathbb{F}$, our flag manifold

$$\mathbb{F} = \{S_i \subset \mathbb{T}; S_1 \subset S_2\}$$

S_i subspaces of dimension i. $\mathbb{M}$ is then defined to be the Grassmannian $G_{2,4}(\mathbb{C}) = G_2(\mathbb{C}^4)$ and called Minkowski space-time; it has dimension 4. Similarly $\mathbb{P}$ is defined to be the third projective space. By this one obtains a double fibration

$$\mathbb{P} \longleftarrow \mathbb{F} \longrightarrow \mathbb{M}$$

with projections $\mu : \mathbb{F} \to \mathbb{P}, \nu : \mathbb{F} \to \mathbb{M}$ and calles the correpondence $c = \mu\nu^{-1}$ the Klein correspondence. The image of a set under c is denoted by a hat $\hat{\cdot}$ and the image under c^{-1} is denoted by a tilde $\tilde{\cdot}$, similarly the inverse image under any projection by a prime $'$.

11.1.2. *Examples of Isomorphisms.* One of the main objectives of the twistor program has beeen to establish isomorphisms between various cohomology groups and spaces of solutions to partial differential equations. The isomorphism between these spaces is called the Penrose transform and denoted $\mathcal{P}$. In the following section we list some theorems exemplifying this.

Theorem 11.1. *Let U be in the topology of $\mathbb{M}$, and assume $n \geq 1$, then under condition of trivial fiberwise monodromy for $\mu : U' \mapsto \hat{U}$*

$$P : H^1(\hat{U}, \mathcal{O}(-n-2)) \mapsto \Gamma(U, \mathcal{Z}'_n)$$

is a natural isomorphism, $\hat{U} = c(U) = \mu\nu^{-1}$ the image of U under the Klein corrspondence c belonging to the double fibration above.

Proof is found in Ward and Wells 1990 for example.

Theorem 11.2 (Gindikin, Henkin 1978, Eastwood 1981.). *There is an isomorphism*

$$H^{(0,1)}(D_+, \mathcal{O}(-2)) \mapsto Sol(\square_{ij} F(Z) = 0, Z \in S_+)$$

D_+ a 1-linear concave domain of $\mathbb{C}P^3$, S_+, M_+ Stein,

$$\square_{ij} = \frac{\partial^2}{\partial u_i \partial v_j} - \frac{\partial^2}{\partial u_j \partial v_i}$$

$Z = (u, v) \in S_+$.

Notice that conditions of simple monodromy hide in the Stein property this time. There are also similar results for $\bar{\partial}$ equation with a source term which under a Penrose transform map to a inhomogenous Laplace equation on $\mathbb{M}$.

11.1.3. *Solutions in Terms of Contour Integrals.* The Penrose program originated as solutions to some PDE's via contour integrals. Say we want to solve $\nabla^{AA'} \phi_{A'\ldots C'} = 0$ for the massless free field case, i.e when $\nabla^{AA'} = \partial_a \sigma^{aAA'}$. Let f be holomorphic on a relevant subset of $\mathbb{T}$ and use $[\omega^A, \pi_{A'}]$ as a coordinate on the image of this subset as viewed in $\mathbb{F}$ mapped on $\mathbb{P}$. Let x be a point in $\mathbb{M} = \mathbb{M}^I$. Define $g(x, \pi) = f(ix^{AA'} \pi'_A, \pi'_A) = f(\omega^A(x), \pi_{A'})$. Then for n-helicity fields ϕ over space-time

$$\phi_{A'\ldots C'}(x) = \frac{1}{2\pi i} \int_\gamma \pi_{A'} \cdots \pi_{C'} g(x^d, \pi_{D'}) \Delta\pi$$

$\Delta\pi = \pi_{E'} d\pi^{E'}$, γ a closed contour in the projective line determined by the Klein correspondence $\hat{x}$ of a point x in complexified Minkowski space-time $\mathbb{M}$, is a solution. This can be explicitly checked by using $\nabla_{AA'} g = i\pi^{A'} \frac{\partial}{\partial \omega^A} f$ obtained by using the chain rule. Similarly for the negative helicity case $-n$

$$\phi_{A\ldots C} = \frac{1}{2\pi i} \int_\gamma \frac{\partial}{\partial \omega^A} \cdots \frac{\partial}{\partial \omega^C} g(x^d, \pi_{D'}) \Delta\pi$$

with f of homogenous degree n in ω.

79

References

[1] Yuri I.Manin, Gauge Field Theory and Complex Geometry, Springer, Grundlehren der mathematischen Wissenschaften 1997.

[2] Ward and Wells, Twistor Geometry and Field Theory, Cambridge University Press 1990.

[3] Mason, Tod, Tsou and Woodhouse, The Geometric Universe, Oxford University Press, 1998, containing S. Gindikin, Penrose Transform for Flag Domains.

[4] Chirka, Dolbeault, Khenkin, Vitushkin, Introduction to Complex Analysis, Springer, the Encyclopeadia of the Mathematical Sciences, among others containing a summary of results by Khenkin, Polyakov, Penrose and others concerning transforms on q-concave domains in $\mathbb{C}P^3$.

12. Physical Applications

In this section we shall discuss the physics of hypermathematics-It will turn out to embrace perhaps all of physical theory. Hypermathematics has been in the background of physical thinking, but most often attacked by real local methods, at least as a simplifyer—e.g. the theory of elliptic functions and spinning tops—,more or less since the day elementary classical mechanics with associated mechanics was born. It is no wonder that Leibniz is reported to have stated 'The complex numbers are the airy flight of God's spirit.'.

Equally old is the idea of using non-real methods to solve real problems, albeit seldom implemented in practice. Hitchin may attack integrable systems on line bundles on Riemann surfaces although not many physicists know sheaf theory, or realize what the Riemann surface might correspond to physically(e.g a generalized complexified time interval). Liouville's theorem not only means that an entire bounded function on the complex Argand plane is a constant but also means the conservation of volume in phase-space, this because Liouville was active in both areas. Legendre, Bessel, Gauss, Riemann etc were all active in complex function theory and mathematical physics-some of them founded important parts of these sciences-but they were also into mechanics and suchlike. This does not, however, imply that physicists in the practical subsciences of physics that follow them today think of physics in a non-real manner.

It is common knowledge the lefthanded side of the Dirac operator on Euclidean space-time is a quaternionic Cauchy-Riemann operator. Yet it used to be uncommon to investigate the spinor bundle(on which the operator acts) as a quaternion holomorphic bundle over a quaternionic base. This is very strange from a string theoretic perspective, as it is well known that supersymmetry arguments or the Einstein field equation naturally yield hyperkahler space-time in a number of situations. Thus everybody knows that an on-shell space-time satisfying some physicality conditions is hyperkahler but it is not common to treat it as such, e.g., a natural and intrinsic quaternionic 4-fold. Perhaps some of motivation for this is that one often seeks to look at off-shell space-times in physics, and that hypercomplex theories are often very restrainted, with quite small moduli. Another— very good—issue is that this might not simplify treatment, and that the effort learning such a theory might not be worth the gain.

More to the defence of the critics of such a programme; It must be said that it is not a priori trivial *which* hypercomplex theory to use, and such theories are not well developed presently-or at least not commonly in use. On the other hand *when* used such attempts often imply that different complex continuations, that make little mathmatical sense, have to be used in order to get the physical situation of Minkowskian space-time. Thus we see that the critique of such an extremist program is more than well motivated.

In the opinion of the present author this is best tackled in the obvious way— by using general methods in general circumstances and viewing hypercomplex theries as a mere bonus that can be used in the few cases where there is anything to gain by such a formulation. Since the hyperkahler case is very common as a suitable background this leads to a nice setting in which one can change picture at convinence. We shall see, later in the physical part, that we will be rewarded for our open minded perspective, and that this will unravel some interconnections between noncommutative geometry in various settings and string theory.

Let X be manifold and look on the Whitney sum fibration $TX \oplus T^*X$. Using a a basis $\{i(e^A - e^{A*}) = \gamma^a\}$ on the selfadjoint[59] part of the fiber, $\{e^A\}$ a basis of sections on T^*X, we see that, recognizing that the fiber over a point of our fibration is the classical phase-space, we have a natural way of constructing Clifford algbras. To put it simply, if we use the tangent of a manifold as a local chart over a contractible set(e.g matrix Lie groups when exponentiating and using coordinates on the tangent group manifold to describe elements in the group.) and then tensor with the dual to get the product of canonical coordinates times canonical momenta. Notice that a momentum then lives in the cotangent fiber and a canonical position in the tangent.

We can easily see that this is so on the basis of other considerations. Here is one example of two different such considerations.

(1) Notice that the kernal of the Fourier transform is e^{ipX}. If X was picked a point on the manifold there would be no inner product to pair it with the mometum p. Hence this expression is ill defined. But letting X be a tangent vector the above is well defined if we let p be a covector or a vector. The whole expression is a Lorentz scalar, so pX has to be invariant under the Lorentz group since the Lorentz group has continuous action and hence cannot map to pX to $pX + 2\pi n$ for maps close to the identity, n a non-zero integer. But setting the usual inner product and

---

[59]The real antiselfadjoint part corresponds to the usual representation of SO(n) upon projecting out the imaginary part, n the dimension of X, while the imaginary antiselfadjoint part to the spinor representation. See GSW II.

picking a one-dimensional case, we see that a Lorentz transformation that induces a contraction $x' = \sqrt{1 - v^2}x$ has to induce the transformation $p' = \frac{1}{\sqrt{1-v^2}}p$. Thus whatever p is it is in the dual of the space that X is in. But when doing automorphisms of the underlying base we know very well that tangent spaces will have a covariant functor induced on them, namely the Jacobian of the relevant map locally. Hence it is clear that X has to be a vector, otherwise it would not transform as it does under the Lorentz group.

(2) Elementary quantum mechanics teaches us that $p_\mu \psi = i\partial_\mu \psi$.[60]. This object is obviously a covector, just like $p\psi = id\psi$. Notice that the *operator* involved is a vector.

Please notice the existance of a symmetry. Although the above takes p to be a covector it does so because of *conventions*. We could equally well have picked things the other way around, indeed all of quantum theory has symmetry, which is takes guises such as supersymmetry. With the above conventions follow also the rule of having a *passive* Wigner rotations and maps.

12.1. Classical Physics in Terms of Symplectic and Complex Manifolds.

As mentioned at the start of the previous section we classical physics can be described in terms of symplectic manifolds. If this symplectic structure would be integrable, which is equivalent to vanishing of the Nijenhuis tensor field or that the holomorphic vector fields form a Lie algebra under the Lie bracket, the symplectic manifold in question is also a complex manifold.

12.2. Several Complex Variables in QFT.

More or less from the start in the early 30's of quantum theory the complex numbers have been associated with the quantum world. Indeed it is impossible to account for some of the dynamics of quantum mechanical systems, like way function interference, without the use of at least complex numbers up to isomorphism.

Perhaps the simplest and earliest examples of pluricomplex theory being used in practice is quantum field theory are the phenomena of Wick rotation and complex angular momentum. Wick rotation stems from various pathologies connected to integration of Green's function in Minkowskian space-time and the cure of Wick was to continue to Euclidean space-time, evaluate integrals, and then go back to non-Euclidean space-time, in effect implementing a little mini-renormalization

---

[60]This is the timelike convention.

already at the start of computations. This has led to some literature on several complex variables in quantum field theory to motivate this continuation of the amplitudes during the 50's and 70's. In a similar vein complexified treatment of the group $SL(2, \mathbb{C})$ grew under the same time under the perhaps slightly misleading concept of complex angular momentum in some parts of elementary particle physics, i.e a rotation of angle α around an axis corresponds to the boost of imaginary rapidity $i\alpha$ in the direction of this axis modulo a sign due to a coincidence of representation theory in the spin cover, $spin(n, m)$, of $SO(n, m)$ for the physical dimension and signature.

12.3. **Yang-Mill's Theory and Complex Gauge Theory.** Yang-Mills theory is what we use to describe three out of four positively known gauge fields. It's basic ingredients are sections of a spinor bundle called Dirac fields and connections on this bundle called Yang-Mills gauge boson fields. Thus the spinors ψ correspond to fermions and the connections A to bosons. A good introduction for the mathematician is Y.Manin's 'Complex Geometry an Gauge Field Theory'. A good introduction for the physicist is S.Weinberg's 'The Quantum Theory of Fields I,II III'.

Example 12.1. *Weak $SU(2)$ is a part of the $U(1) \otimes SU(2)$ electroweak theory. The spinor module is two dimensional and includes one fermion and it's neutrino for each generation. The connections are in a 3 dimensional structure group extended by $U(1)$, the dimension corresponding to the $3 + 1 = 4$ particles W^+, W^- ,Z_0 and the electromagnetic gauge boson A.*

12.4. **String Theory and D-branes.** String theory is perhaps the most accepted theory of quantum gravity and grand unification and has attracted the most attention hith herto. We cannot for lack of space go into the beautiful world of strings.

We shall however note this: A real two-dimensional one-dimensional hypermanifold is a world-sheet of a string. Left-moving is then holomorphic on this sheet, also called Riemann surface.

The natural generalization to more dimensions are hypermanifolds X_g of higher real dimension,(example: quaternionic manifolds) and the right-moving objects are then [classically] the dual manifolds $\bar{X}_g$. For this reason higher dimensional hypermanifolds will also be called D-branes.

12.5. Spinors and Twistor Theory.

By now it is hopefully clear that all of physics has spinors written all over it. The basic fundamental object, at our energies with our fundamental constants, and other circumstances we live under is after Wick rotation a holomorphic quaternionic one-manifold with an associated holomorphic one-bein. Let us us pick a complex symplectic representation for the holomorphic tangent space over a point. We can then write such a vector as $Z = z_0 + jz_1 \in \mathbb{H}$, $(z_0, z_1) \in \mathbb{C}^2$. Thus locally we have identified $\mathbb{C}^2 = \mathbb{H}$. Taking the unit sphere $S^3 \cong SU(2)$ in $\mathbb{H}$ and restricting to spatial directions $\{i, j, k\}$ we get a new sphere $S^2 \cong \mathbb{C}P^1$ which represents the complex rays through the origin, i.e $\mathbb{C}P^1 \cong S^3/U(1) \cong S^3/S^1 \cong (\mathbb{C}^2 \sim \{0\})/\mathbb{C}$. We also notice that our original $\mathbb{C}^2$ was the Hilbert space for a spin half system, so the first projective space could not only be interpreted as the directions from the origin of three space but also as the moduli of different helicity quantum states for such a particle. A similar construction gives $\mathbb{C}P^3$ when an antifermionic particle is included, i.e. upon taking the direct sum of the tangent with the dual, or as the perspective of hypergeometry would have by taking into account the antiholomorphic(or rightmoving/negative helicity- whichever the reader prefers.) partner.

Penroses program uses the insight that such ideas give to use these projective spaces as the space of inequivalent states together with double fibration techniques to solve the equations of physics. This is what we might call an approach that attacks twistor theory in inhomogeneous coordinates. In later years it seems that a return to homogenous coordinates has occured. The interested reader is referred to Ward and Wells, Twistor Geometry and Field Theory and Penrose and Rindler, Spinors and Space-Time I and II.

Thus twistor geometry is a particular case of hypergeometry. In the physics chapters we shall learn about quantum hypergeometry-that is what happens when we look at entire categories of manifolds that locally look like spinor/twistor spaces.

Beautiful as it is these manifolds are also what we just a section ago called D-branes.

12.6. Some Final Remarks Concerning Hypereal Methods and Physics.

Finally I would like to remark why we choose the selfadjoint part of $T_P X \oplus T_P^* X$, or in even greater generality why we have real spectra in physics.

Let us take a concrete example first and then generalize. We pick the lefthanded Dirac operator $\slashed{\partial}\,_-$, but this time with complexified coordinates. Let us forget about hypergeometry for a moment and only have a complexified space-time as a base. Notice that the corresponding span of matrices allowed is then $M(2,\mathbb{C})$, the set of 2×2 complex matrices. Then it is easy to see that $\frac{1}{i}\partial\!\!\!/\;_- = m\psi_-$ has a solution $\psi = e^{im\!\!/x}\;-$, x a coordinate on our complexified space-time. Setting $x = (x_0 + iy_0, x_1 + iy_1, x_2 + iy_2, x_3 + iy_3), (Re(x), Im(x)) \in \mathbb{R}^8_+$ we understand that we have different evolution in different directions, namely either periodic or exponentially decaying. This is realized as follows; for the complex time coordinate with othe others set to nill we have for real time $\psi_- = e^{imx_0}$, resonant[61], but for purely imaginary time we have e^{-my_0}, decaying. In the same manner we have evolution in purely spatial direction x_1 given by $e^{(i)^2 mx_1} = e^{-mx_1}$ for the real spatial coordinate x_1, fastly decaying (the i comes from the lefthanded Clifford algebra and not from the continuation) and fields in the imaginary spatial direction y_1, e^{-iy_1}, resonant. We also notice that the typical decay length of such a field is $\frac{1}{m}$, quite small with the Dirac mass spectra we have (For a neutron that is about a fifth of a Fermi, for a electron 0.02 Angstrom, less than fifty times the dimension of a small atom.).

So to sum it up imaginary space is resonant together with real time. That means that the holomorphic quadratic form x^2 can be very well approximated by $x^2 = x_0^2 - y_1^2 - y_2^2 - y_3^2$. Thus we have all of a sudden sketchedly and very naively justified two things

(1) That we have a Minkowski metric.
(2) That Wick rotation really does work and is ok even in the general global sense from a physical point of view. Problems of non-uniqueness still remain.

    The reader that is awear of historical developments might add from the point of view of hypergeometry/ twistor geometry might add an arbitrary choice of items to this list, e.g

(3) That what the particle physicists of the 60's called complex angular momentum occured in physics, i.e that we have a complexification $SU(2)^\mathbb{C} = SL(2,\mathbb{C})$. Thus rotations are the real parts of the complex angles that parametrize a trivial neighbourhood around the identity of a complex Lie group, which is precisely the automorphism group of the relevant tangent fiber. Notice that the antiholomorphic/righthanded partner has $\overline{SL(2,\mathbb{C})}$ that acts on it, so a direct sum bundle of the holomorphic and antiholomorphic partner would have automorphism group $\overline{SL(2,\mathbb{C})} \oplus SL(2,\mathbb{C})$.

_____

[61]A dimension where operator fields or similar objects have almost free evolution is here temporarily called resonant. Thus the imaginary time dimension is resonant in Minkowski space-time with the above conventions.

(4) A possible link to the success of holomorphic geometry and complexifed space-time in various parts of mathematical physics, e.g. twistor theory.

Of course it must be realized that we really have not accomplished much in terms of uniqueness matters etc, since the correct way of looking at our objects would be to to start with the complex thing and then obtain the real space as an approximation.

We must now scrutinize our arguments and find possible objections that one might have against them that might ruin them. Here are two objections we must deal with

(1) The above was an argument for free fields. What about interacting fields?
(2) What about $x \notin \mathbb{R}^8_+$? We do seem to have exploding solutions then.

Let us begin with the second as it seems to be the more dangerous objection. The answer is that we must be careful to evolve with time and not the other way around. So if we want to look at negative coordinates we must be careful to conjugate/ take negative mass, depending on which operation we choose to do, to get the backwards moving objects in different dimensions. Of course taking negative mass and negative coordinate at the same time means that the way function looks the same, so if we stick to only having our positive time coordinate this implies we only have positive mass by identifying the case with two minus signs with the former.

The first objection on the other hand is very easy to answer for a physicist. The pertubated solutions are so close to free fields asymptotically that we can without problems assume that they are free to run a qualitative argument. Indeed as any physicist who knows his path integrals knows the approximation of fields being almost free is quite ok for most cases and almost necessary in order to get anything out of his expressions, often using the Euler canonical product formula and the spectrum of the free d'Alembertian, essentially reducing space-time to a torus in the periodic dimensions. Much the same reasoning explains why we have real spectra in physics, namely the states corresponding to non-real eigenvalues of different operators are decaying.

13. Solving Some Equations of Mathematical Physics Using Hyperreal Methods.

Use ψ for a complex field satisfying the Dirac equation and Φ, u for complex and real fields satisfying only the D'Alembert-Laplace equation. Let Ω be a compact contractible subset of $\mathbb{E}^n$, $n \in \mathbb{Z}_+$, with C^1 rectifiable boundary.

13.1. Dimension 2 of Arbitrary Signature.

We shall in this section solve the Dirac and D'Alembert-Laplace equation for various cases, beginning with the Dirac equation and then taking the Laplace equation. Assume everything to be real analytic and differentiable enough throughout these sections, and then interpret the integral operators obtained to act on completions of spaces. For $\mathbb{E}^2$ we use $Cl_+(2,0) \cong \mathbb{C}$, and thus Cauchy-Fantappie's formula yields

$$\psi(z) = \frac{1}{2\pi i}[< \partial\Omega, \frac{\psi d\zeta}{\zeta - z} > - < \Omega, \frac{\bar{\partial}\psi d\zeta}{\zeta - z} >] \, , \, z \in \Omega.$$

which takes care of the Dirac equation for such cases, noting $\partial\!\!\!/_+ = 2\partial_{\bar{z}}$. Notice that compatibility requirements have to be satisfied among boundary value, source term, and the equation and that these cannot be arbitrarily prescribed. In particular for a massless ('Weyl') spinor in dimension 2 we see that it would satisfy CR conditions on the space-like hypersurface $\partial\Omega = \Sigma$. By real analyticity we now obtain the other cases $\mathbb{R}^{(1,1)}$, $\mathbb{R}^{(0,2)}$ via Wick rotation. This being understood we only deal with Euclidean cases from now on (What we are really doing is considering PDE's on real submanifolds of complex manifolds, hence this is justified.).

Substitution then gives

$$\partial_z\Phi = \frac{1}{2\pi i}[\langle \partial\Omega, \frac{\partial\Phi}{\zeta - z}\rangle - \langle \Omega, \frac{\bar{\partial}\partial\Phi}{\zeta - z}\rangle] \, z \in \Omega.$$

hence inverting ∂_z gives us

$$= \Phi_{//\bar{O}(\Omega)} + \frac{1}{2\pi i}\int^z [\langle \partial\Omega, \frac{\partial\Phi}{\zeta - z}\rangle - \langle \Omega, \frac{\bar{\partial}\partial\Phi}{\zeta - z}\rangle]$$

For for a real $\Phi = u$ we obtain

$u(z, \bar{z}) = \mathfrak{Re}\Phi = \frac{\Phi_+ + \Phi_-}{2}$
$= \mathfrak{Re} \; \Phi_{//\bar{O}(\Omega)}$
$+\mathfrak{Re} \; \frac{1}{2\pi i}\int^z[\langle \partial\Omega, \frac{\partial\Phi}{\zeta - z}\rangle - \langle \Omega, \frac{\bar{\partial}\partial\Phi}{\zeta - z}\rangle]$
$= \frac{1}{4\pi}\langle \partial\Omega, *d\ln|\zeta - z|^2\Phi\rangle + \frac{1}{4\pi}[\langle \partial\Omega, \ln|\zeta - z|^2 * d\Phi\rangle - \langle \Omega, \ln|\zeta - z|^2 \Box\Phi\rangle]$
$, \; z \in \Omega.$

where the last line shows how this relates to the purely real kernels (equivalence). This is not surprizing as $\square = 4\bar{\partial}^\dagger\bar{\partial}$, duality being the Hodge-DeRham dual, in our conventions[62], hence $2i\delta(\zeta - z) = \delta(x - y) = \square G(x,y) = -4\bar{*}\bar{\partial}(\bar{*}\bar{\partial}G(x,y))$ for the relevant elementary kernal G of the D'Alembert-Laplace operator. Notice that by the above $\slashed{\partial} = i\bar{*}\bar{\partial}$ generates another representation of the Dirac operator on the complex cohomology algebra of this case.

13.2. **Dimension 3 and More.** Next for the case $\mathbb{E}^n$, $n \geq 3$, we use
$$\square \underbrace{\frac{1}{k_n\|x - x'\|^{d-2}}}_{\mathcal{K}(x,x')} = \slashed{\partial}\ \slashed{\partial}\underbrace{\frac{1}{k_n\|x - x'\|^{d-2}}}_{:=G(x,x')} = \delta(x - x'),\ k_n \text{ some suitably}$$
defined constant. Hence we have a Green's function for the Dirac operator, and so

$$\psi(x) =< \partial\Omega, G(x,x')\psi(x') > - < \Omega, G(x,x')\slashed{\partial}\ \psi(x') >$$

For the Laplace operator one uses the canonical Newton formula

$$\Phi(x) =< \partial\Omega, \mathcal{K}(x,x')\Phi(x') > + < \partial\Omega, \mathcal{K}(x,x')*d\Phi(x') > + < \Omega, \mathcal{K}(x,x')d^*d\Phi(x') >$$

If one inforces C^2 requirements on the solutions obtained the above formulae are corresponding to overdetermined problems and cannot be applied. Other kernals have then got to be sought, corresponding to other problems like the Dirichlet or Neumann problem, varying from region to region. For the case of dimension 4 we are lucky, here the lefthanded Clifford algebra closes again, which can be used to some extent to solve partial differential equations. Let the following expressions be $\mathbb{H}$-analytic, and let the involved solutions satisfy the regularity condition
$$< \partial\Omega, \psi * d\zeta >= 0, z = t + ix + jy + kz$$
, a quaternionic version of the Morera criterion for holomorphicity. One can then write, supressing a normalization constant,

$$\psi(x) =< \partial\Omega, \psi\bar{*}\partial\frac{1}{\|\zeta\|^2} > + < \partial\Omega, \bar{*}\partial\psi\frac{1}{\|\zeta\|^2} > - < \Omega, *\square\psi\frac{1}{\|\zeta\|^2} >$$

with u the real part of this. It should be pointed out that, modulo a constant, the regular function satisfying the condition above is determined by it's real part on the boundary in a way canonical from elementary complex analysis on this same boundary. The solution given in this manner is unique.

---

[62]This differs by a factor of two from the more usual conventions, which are associated with different normalization of the Cauchy-Riemann operator.

The reader might wish to consider a symplectic representation of $\mathbb{H} \cong \mathbb{C} \times \mathbb{C}$ given by $\zeta = z_0 + kz_1$, $z_i = x_i + iy_i$. This thus essentially reduces matters to smooth analysis on $\mathbb{C}^2 \sim \{0\}$. But then the kernal we seek is on cohomological grounds, considering

$$H_{\bar{\partial}}^{(0,1)}(\underbrace{\mathbb{C}^2 \sim \{0\}}_{:=X}, \mathbb{C}) \cong H^1(X, \Gamma(X, \overset{n}{\bigwedge} T^*X)) \cong H^1(X, \mathcal{O})$$

, the latter isomorphisms holding in view of Dolbeault isomorphism induced by existance of a pseudoconvex Leray cover, the Martinelli-Bochner Kernal. An explicit check, computing $\bar{*}\partial \mathcal{K}(x, x')$, verifies this. Hence we have our symplectic representation with ψ taking values in a representation space $\mathbb{C}^2$ as follows

$$\psi(z) = \begin{pmatrix} \psi_1 \\ \psi_2 \end{pmatrix} = \frac{1}{(2\pi i)^2}[< \partial\Omega, \frac{\psi\omega'(\bar{\zeta}-\bar{z})\wedge\omega(\zeta)}{||\zeta-z||^4} > - < \Omega, \frac{\bar{\partial}\psi\omega'(\bar{\zeta}-\bar{z})\wedge\omega(\zeta)}{||\zeta-z||^4} >]$$

yielding a third solution to the Dirac equation. By the same token, taking quaternionic real parts, we have

$$u(z) = \Re\Phi = \Re \begin{pmatrix} \Phi_1 \\ \Phi_2 \end{pmatrix} = \begin{aligned} &\Re\frac{1}{(2\pi i)^2}[< \partial\Omega, \frac{\Phi\omega'(\bar{\zeta}-\bar{z})\wedge\omega(\zeta)}{||\zeta-z||^4} > + < \partial\Omega, \frac{\bar{*}\bar{\partial}\Phi}{||\zeta-z||^2} > \\ &+ < \Omega, \frac{\bar{*}\Box_{\bar{\partial}}\Phi\omega'(\bar{\zeta}-\bar{z})\wedge\omega(\zeta)}{||\zeta-z||^4} >] \end{aligned}$$

yielding another solution for the complex D'Alembert-Laplace equation $\Box_{\bar{\partial}}u = J$ with boundary value data.

13.3. **Summary.** Consider the lift $l : P_{\mathbb{R}} \to P_{\mathbb{M}}$ and conversely the projection $\pi : P_{\mathbb{M}} \to P_{\mathbb{R}}$, from the category of well posed problems of the above type(which is contained in the Hadamard category) for the real case and hypercomplex or complex case $\mathbb{M}$. The above then amounts to commutativity of the diagram

$$\begin{array}{ccc} P_{\mathbb{M}} & \longrightarrow & \mathcal{S}_{\mathbb{M}} \\ l \updownarrow \pi & & l \updownarrow \pi \\ P_{\mathbb{R}} & \longrightarrow & \mathcal{S}_{\mathbb{R}} \end{array}$$

with $\mathcal{S}$ indicating the various solution spaces. Thus nothing is gained by introducing hyperreal methods for the solvability of the partial differential equations considered and nothing is lost for the Cauchy problem.

For a closely related kin of the above D'Alembert-Laplace problems, namely the Dirichlet problem on Ω the above other formulations of solutions can however be more practical. From the real analysis perspective the consderations involved suffice to prove existance of the Poisson kernel, via arguments linked to Perron's principle. The above hypercomplex methods can be used to find such solutions in cases where the relevant kernal is hard to obtain via real methods. That this is so is because of the fact that the various purely hypercomplex/complex kernals considered are are reproducing[63] for dimension 2 and 4 respectively, in the sense that they create solutions to the partial differential equations involved from boundary value data that need not be bordant to a function satisfying the differential equation given.

Example 13.1. *We physicists often encounter the equation $\nabla_- \psi = J$ with some Cauchy data. Solve it in some sense for topologically trivial Ω in X_g. Notice that $\Sigma = \partial \Omega$ becomes space, and that the data prescribed corresonds to timelike evolution of spinors in normal direction to this surface at each point.*

Example 13.2. *"Solve", in formal sense, the Klein-Gordon equation $(\Box + m^2)\psi = J$ with appropriate Cauchy data and suitable requirements on toplogically trivial Ω on an arbitrary pseudo-Riemannian background in arbitrary dimension and signature. In particular if the equation is for a space-time way function, what does positive mass imply for the one-particle space-time itself?*

Example 13.3. *If for families of pseudo-riemannian manifolds with fixed space-time dimension and signature there would exist 3 and only three elliptic manifolds, how many tachyonic states would the corresonding field theory with lagrangean $\mathcal{L} = \bar{\psi}(i\slashed{D})^2 \psi$ have? (Hint: Wietz.- Bochner and a good conformal gauge slice.) Include the effect of degeneracies if possible.*

---

[63]The Martinelli-Bochner kernel should not be confused to be reproducing in dimension 2 complex for usual complex functions, although it is when considered to act on a tuple of two functions, reproducing property with respect to a hypercomplex Cauchy-Riemann operator.

References

[1] R.Michael Range, Holomorphic functions and integral representations in several complex variables, Springer GTM: ; 1986.

[2] Gunning and Rossi, Analytic functions of several complex variables, Prentice-Hall Series in Modern Analysis.; 1965.

[3] Reinhold, Remmert. Classical Topics in Complex Function Theory; 1998.

[4] L.V. Ahlfors, Complex Analysis.

[5] H.L. Royden, Real Analysis.

[6] E.M. Chirka, P.M. Dolbeault, G.M Khenkin and A.G Vitushkin, Introduction to Complex Analysis.

[7] Novikov,Dubrovin and Fomenko, Modern Geometry I,II,III.

[8] Nakahara, Geometry,Topology and Physics.

[9] L. Schwartz, Cours d'analyse.

[10] L. Hormander, An Introduction to Complex Analysis in Several Variables.

[11] Bell, Brylinski, Huckleberry, Narasimhan, Okonek, Schumcher, Van der Ven, Zucker, Complex Manifolds.

[12] W.Rudin, Function Theory in the unit ball of $\mathbb{C}^n$.

[13] S.Kobayashi, Hyperbolic Complex spaces.

[14] Eric D'Hoker, String Theory lecture notes from IAS.

[15] S.P Novikov, Topology I, Encyclopaedia of Mathematical Sciences, Springer.

[16] A.Sudbery, Quaternionic analysis, Math. Proc. Camb. Phil. Soc. (1979),85, page 199.

[17] Hugett, Mason,Tod, Tsou and Woodhouse, The Geometric Universe, Oxford.

[18] Feynman,Leighton and Sands, The Feynman Lectures on Physics

14. Appendix; Spin Cobordisms, The Singular Cauchy Problem, D-branes, Number Theory and Stochastic Flows.

14.1. Spin Cobordisms and The Singular Cauchy Problem. Let Σ be a chain, say a smooth rectifyable cycle, e.g. formed as a closed double in the sense of partial differential equations on manifolds with boundaries, and let Ω be a cobordism, i.e. $\partial\Omega = \Sigma$. Despite that we might consider Dirichlet problems we might consider other problems as well. In this subsection we shall only consider the analytic d'Alembert-Laplace problems of type

$$\displaystyle{\not D}^2\Phi = 0, \Phi \in C^2(N(\Sigma)), D_\mu = \frac{\partial}{\partial x^\mu}$$
$$\Phi|_\Sigma = B.V.1,$$
$$\partial_n\Phi|_\Sigma = B.V.2,$$

, $N(\Sigma)$ a small enough neighbourhood of Σ, n normal to Σ, or, if we force the spinor Φ to be real, problems of the type

92

$$\Box u = 0, u \in C^2(N(\Sigma))$$
$$u|_\Sigma = B.V.1,$$
$$\partial_n u|_\Sigma = B.V.2,$$

Just as remarked at the last section of the mathematical part, substituting the Dirichlet problem for the Cauchy problem usually means that one has an overdetermined system if one does not let go of twice differentiability throughout Ω. Because of growth conditions implicit with these equations much of the handy and usual spectral calculus associated with Dirichlet problems and associated physical entities, such as D-branes, collapses, and a— possibly extended—singularity usually arises in the middle of the cobordism. The mechanism giving both existance and singularity is easily recognized. The rough skeleton of an argument could proceed as follows; By elementary partial differential equations the analytic harmonic Dirichlet on Ω has a unique solution hence implying a 1-1 correspondence between relevant boundary values and harmonic bordisms, but by another classical theorem (Cauchy-Kovalevskaja) we know that that the infinitisimally transversal Cauchy also has a solution. As we know that a harmonic function is analytic, we can extend it to some envelope of harmonicity in Ω. If, however, we could always do this without obstructions then falls a contradiction, since the space of Cauchy data is strictly larger than the space of Dirchlet ditto. Thus in the space of Cauchy data $\mathcal{C} = \mathcal{D} \times \mathcal{C}_1, \mathcal{D}$ denoting analytic Dirichlet data and $\mathcal{C}_1$ the space of analytic normal derivatives, only a diagonal $\mathcal{D} \times \partial\mathcal{D} \subset \mathcal{D} \times \mathcal{C}_1$ can be non-singularly continued throughout the cobordism Ω. Thus the $C^2(\Omega)$ property combined with two boundary conditions makes the problem overdetermined—Hence one disposes of it , namely by requiring the P.D.E's to only hold infitisimally transversally to Σ. As the solution by necessity is analytic within the envelope of analyticity created by this function we will obtain a —possibly ramified, singular etc—function u on at least the intersection of the envelope and our relevant cobordism. It is this problem, which goes by the name of the singular Cauchy problem, that we shall sketchedly relate to spin cobordisms and hypermanifolds.

We can without any greater restraints illustrate the discussion in any chiral dimension by the quaternionic case and assume the real problem which we can always retain by decomposition. As our spectral methods now fail, it seems we must use analytic methods instead to be able to solve the problem at hand. We begin with the process of calculating the various hyperimaginary parts of a holomorphism from a given harmonic function u which we set to be the real part. To do this we can use version of a trick in Riemann surface theory as we have hypercomplex structures I, J, K , hence

$$u_i = \int I\,du = \int I(\partial_n u\,dn + \partial_i u\,dx^i),$$
$$u_j = \int J\,du = \int J(\partial_n u\,dn + \partial_i u\,dx^i),$$
$$\dots$$

x^i coordinates on Σ, which gives us our different parts modulo a constant. We can further concentrate on the case when Σ is a hyperplane $\mathbb{R}^3$ viewed in the Euclidean context as we are merely opting to illustrate how this difficult problem might be turned simple by introducing hypercomplex calculus for this particular problem. On $\mathbb{C}$ we have, as previously stated,

$$u = cont.(\frac{\Phi|_\Sigma + \bar{\Phi}|_\sigma}{2}) + cont.(\int_{\bar\zeta}^{\zeta} \partial_n \Phi\,d\zeta)$$

Σ the real axis. On $\mathbb{H}$ we take the problem given above, which reduces to $[\slashed{D}_+, \slashed{D}_-]_+\Phi = 0$. Then, by the usual identites of Dirac algebra, one can show that

$$\Phi = cont.(\frac{\Phi_+|_\Sigma + \Phi_-|_\sigma}{2}) + Im\ cont.(\int \partial_n \Phi\,d\zeta)$$

$\Sigma = \mathbb{R}^3$, *cont.* again the continuation functor. In the above Φ is a matrix valued spinor. We remark a very important feauture; what we are really doing analysis on is, after compatification of the spatial directions $S^3 \times U$, U an interval, this is important for it is linked to the special behaviour of this problem since it is defined on a blow-up resolution of the usual point-like charecteristic. E.g, in physical Minkowski space we are thus really dealing with a partial differential equation transversal to a lightcone, something that endows the differential equation on conformal equivalences of this domain with special properties. Albeit we cannot proceed in abstracto too easily with this, we can use hypercomplex methods to effectuate the calculations needed. E.g. we have for a general holomorphism Φ_+, satisfying a holomorphicity condition $\bar\partial\Phi = 0$,

$$\Phi_+ = \sum_{n\in\mathbb{Z}}(\zeta^n a_n + \zeta^n a_i ni + \zeta^n a_{j,n} j + \zeta^n a_{k,n} k)$$

The usual formulas listed in the hypermathemtics part then yield answers in a number of the explicit cases in manner remincent to the complex case, e.g. by using the explicit formula for regular, i.e holomorphic, functions. Furthermore, explicit calculations are often made easier if there is also a trace operating on the functions involved.

94

14.2. An Example of a Spin Boundary with Relations to Analytic Number Theory, Combinatorics, Theta Function Theory and D-branes.

Let S^3 bound D^4 the model of a topologically trivial set in $D^4 \subset X_\pm^{(4)}$ space-times of various helicites. If we see $X_\pm^{(4)}$ a space-time as a hypermanifold, to be more exact a hypercomplex manifold, so the S^3 inhertits a framing which we might regard as given by vectorfields $Y_i := I, Y_j := J, Y_k := K$ satisfying

$$[I, J] = K,$$
$$[J, K] = I,$$
$$[K, I] = J$$

which are not to be confused with the hypercomplex structures. Let the Hamiltonian $\hat{\mathcal{H}} = I^2 + J^2 + K^2$ be given and defined on $S^3 \cong SU(2)$ this spin boundary. We shall, en passent, display how a partition function can have a shockingly number theoretic interpretation because of the different ways in which it can equivalently be physically written as while casually mentioning some phenomenology of space-times, i.e $D3$-branes or one-particle space-times. The thought is to invert this line of reasoning a little bit later to be able to express the partition function involved in a problem of mass degeneracy of super D3-branes as a formula of analytic number theory and theta function theory. Consider the diagonal supertrace on the $\hat{\mathcal{H}} = L^2$ spectrum, then with $\beta \in \mathbb{C}$ a parameter, $q := e^{-\beta}$, one obtains

$$Z = (STR_{SU(2) \cong S^3}[e^{-L^2 \beta}])^{-1} = (STR[q^{L^2}])^{-1} = \left(\sum_{l \in \mathbb{N}} (-1)^l (2l+1) q^{l(l+1)}\right)^{-1}$$

Looking at the mathematics, remembering e.g that the super angular momentum physical system of elementary quantum mechanics it could have represented we know by the physics of that problem that we also can write this as the —unnormalized— partition function

$$Z = \Pi_{I,m\in\mathbb{Z}_+} Z_m^{(I)} \Pi_{J,m\in\mathbb{Z}_+} Z_m^{(J)} \Pi_{K,m\in\mathbb{Z}_+} Z_m^{(K)},$$
$$Z_m^{(I)} = (e^{\frac{Im\beta}{2}} - e^{-\frac{Im\beta}{2}})^{-1} = (STR_{\mathcal{H}_{+m} \oplus \mathcal{H}_{-m}}[q^{\frac{I\beta}{2}}])^{-1}$$

Z_m being the component corresponding to the m:th eigenvalue, and $\mathcal{H}_\pm$ the eigenspaces correponding to negative and positive helicity respectively. Normalizing Z_m^I this is by $Z_m^I = (1 - q^m)^{-1}$ for the m:th eigenvalue

$$Z_m^{(I)} = (1 - q^m)^{-1}$$

Hence

$$Z = \Pi_{m\in\mathbb{Z}_+} Z_m^{(I)^3} = \Pi_{m\in\mathbb{Z}_+} (1 - q^m)^{-3}$$

The m:th eigenvalue in the I,J,K corresponds to a state that winds $n/2$ times the I, J, K:th direction around S^3, i.e at infinity in the

I,J,K:th direction if we regard $\mathbb{R}^3$ compactified, while the m:th power of q corresponds to the number of times a state winds around the temporal direction, then counting the time in q-space on a circle around the origin in the complex plane. By the above we thus have

$$Z = (STR[q^L])^{-1} = (\sum_{l \in \mathbb{N}} (-1)^l (2l+1) q^{l(l+1)})^{-1} = \Pi_{m \in \mathbb{Z}_+} (1 - q^m)^{-3}$$

We can interpret

$$Z_I = Z^{1/3} = \Pi_{m \in \mathbb{Z}_+} (1 - q^m)^{-1}$$

statistically by expanding it (This is a classical function of combinatorics and number theory called the canonical partition function. It was e.g studied by Euler by function analytical methods.), hence writing Z_I as a trace in a new basis. The new Hamiltonian $\hat{H}$ would have to correspond to the charge on a stack pointed branes $\wedge(*_+ \cup *_-)$ corresponding to the spin eigenvalues $\pm m^{64}$, so it would be by additivity of the charge of tensors of Hilbert spaces,

$$\hat{H} = Q = \sum Q_i$$

Q_i the charge $Q_{\pm i} = \pm \frac{m}{2}$ on the pointlike branes in charge/momentum space. Let a prime denote an eigenvalue of an operator. Since the vacuum was relabelled by the renormalization, to $Q'_- = 0$, $Q'_+ = m$, this means that $\hat{H} = Q = \sum Q_i$ takes integer eigenvalues whilst $Q'_i \in \{0, m\}$ will sum for a fixed Q' eigenvalue in the number of ways one can write Q' as a sum of integers, where we only count the distinct ways since we have identical branes(particles), and hence give the degeneracy at each eigenvalue Q'. The number of such partitions of a Q' are called Euler partitions—but they were already known to the Greeks, which knew that they followed a pentagonal series(see below). So we write, with $d(Q')$ the degeneracy at eigenvalue Q',

$$Z_I = \sum_{Q' \in \mathbb{N}} d(Q') q^{Q'} = \Pi_{n \in \mathbb{Z}_+} (1 - q^n)^{-1}$$

with the knowledge that $d(Q')$ has to be the number of partitions of Q' by this number theoretical interpetation of Q. Still β in $q = e^{-\beta}$ has

---

[64]This is nice, as this gives en example of how arbitrary branes are and the simple perspective of a brane a spectrum, which is needed to ty things to noncommutative geometry in Connes sense later. E.g, here we have point-like branes in a Fourier space.

not changed interpretation, it's a temporal non-isospectral[65] deformation of the D3-brane, that is—in the present case— it is the parameter of a non-unitary transformation on a Hilbert space $\mathcal{H}$ which deformes it's pertaining operator algebra of linear vector space endomorphisms $Hom_{\mathbb{C}}(\mathcal{H}, \mathcal{H})$ (the term has been borrowed from finite dimensional integrable systems). We have then

$$Z_I = \sum_{Q' \in \mathbb{N}} d(Q') q^{Q'} = (\sum_{n \in \mathbb{Z}} (-1)^n q^{\frac{3n^2-n}{2}})^{-1}$$
$$= (1 - q^1 - q^2 + q^5 + q^7 - q^{12} - q^{15} + q^{22} + q^{26} + \cdots)^{-1}$$

where the numbers $1, 5, \cdots$ are the sum of vertices in nested pentagons, one vertex larger in side, which all have three sides in common, as the reader can explictly check, an this was known to the ancient Greeks. Let us concentrate on the elementary number theory/combinatorics emerging from the string field geometry so that we can put full weight on the branes later. What we will have to say about the former will not be original, rather canonical and intended to just give a standard example for those who are not already string theorists and thus aware of such matters. The physical intent is to touch where the mass degeneracy of strings resides in free field case and to reapply this technology for finding the mass spectrum for the special case of hypercomplex D3-branes. On the latter, however, some emphasis will be laid.

14.3. Strings, Fields, Mass and Degeneracy. The partition function

$$Z_I = \Pi_{n \in \mathbb{Z}_+} (1 - q^n)^{-1} := f(q)^{-1}$$

has been a much studied object in science (e.g string theory, number theory, statistical mechanics) and can be expressed by a Hardy-Ramunjan formula. One has, by using the Dedekind eta function, setting $q = e^{2\pi i t}$,

$$\eta(\tau) = e^{i\pi\tau} \Pi_{n=1}^{\infty} (1 - e^{2\pi i n \tau})$$

satisfying the well known modular identity

$$\eta(-\frac{1}{\tau}) = \eta(-i\tau)^{\frac{1}{2}} \eta(\tau)$$

---

[65]This peculiar technical term will be clear at the end of this section, see the D-branes section. It stems from that the action generated by the Hamitonian on our brane deforms the states over the spectrum, leaving the interpretation that it evolves in time while not leaving the spectrum invariant. We will also give examples of isospectral deformations of a brane. The non-isospectral behaviour is part of the Euclidean metric, which after appropriate temporal Wick rotation is isospectral in Minkowski space. This is the instability predicted already by heuristic methods at the end of Part I.

which implies (See GSW I in the references, which we lean heavily on presently)

$$Z_I^{-1} = f(q) = (\frac{-2\pi}{lnq})^{\frac{1}{2}} q^{\frac{-1}{24}} (\zeta)^{\frac{1}{12}} f(\zeta^2), \ \zeta = e^{\frac{2\pi}{lnq}}.$$

In particular, this gives for $q \mapsto 1$

$$f(q) \sim const. \ (1-q)^{-\frac{1}{2}} exp(-\frac{\pi^2}{6(1-q)})$$

from this we can, recognizing $Z_I^{(D-2)}$ to be the free partition function for bosonic strings with 2 off-shell dimensions, e.g, recognize the bosonic string in critical dimension $D = 26 = 2+D_++D_-$. By usual fomulae for Laurent series coefficents we then the degeneracy in the mass spectrum as, setting $G_D(q) = Z(q)$ the generating function in $D = 26$,

$$d_n = \frac{1}{2\pi i} \int_\gamma \frac{G_{D=26=2+D_++D_-}(q)}{q^{n+1}} dq = Res\{\frac{G_{D=26=2+D_++D_-}(q)}{q^{n+1}}, 0\}$$

, γ a small in loop around the origin of unit monodromy $1 \in \pi_1(S^1) \cong \mathbb{Z}$ in additive notation. This can be evaluated for $D = 26$ for the high mass limit, i.e large n, via e.g a saddle point evaluation as in GSW, to

$$d_n \sim (const.)n^{\frac{-27}{4}} exp(4\pi\sqrt{n})$$

or

$$\rho(m) = m^{-\frac{25}{2}} exp\frac{m}{m_0}, m_0 = \frac{1}{4\pi}(\alpha')^{-1/2}$$

We may be interested in the partition function with both super and non-super degrees of freedom on our brane. On our S^3 spin cobordant to D^4 that would have resulted by adding free superfields to the usual fields, hence multiplying the usual trace by a supertrace. Hence, with these new degrees of freedom included, we get

$$Z_I := \frac{Z_{usual}}{Z_{super}} = \Pi_{n\in\mathbb{Z}_+} \frac{(1+q^n)}{(1-q^n)}$$

We have the theta function identity

$$Z^{-1} = \theta_4(0|q) = (-\frac{ln(q)}{\pi})^{-\frac{1}{2}} \theta_2(0|e^{\frac{\pi}{ln(q)}})$$

which on our original $SU(2) \cong S^3$ generated D3-brane would have given

$$Z = Z^I Z^J Z^K = (\Pi_{n\in\mathbb{Z}_+} \frac{(1+q^n)}{(1-q^n)})^3 = (\theta_4(0|q))^{-3}$$

98

So this time we used the link the other way, from number theory to physics—to get the naive graded partition function on the spatial degrees of freedom on our super $D3$-brane. Let us now study the mass degeneracy of our $D3$-brane. We begin by displaying the stringy results in super string theory of GSW, as those are standard, and then proceed to the $D3$-brane, which will model a standard space-time, since the vanishing of fields at infinity of $\mathbb{R}^3$ will make it possible for us to smoothly compactify at infinity without obstructing the solution spectrum. Adding infinity in time as well to the future, and enforcing that solution fields fields take identical values at Euclidean time-like infinity gives then the problem the topology D^4. This becomes the same thing as $S^3 \times S^1$ if we instead just identify the extra time dimension as a circle, and this corresponds to periodic boundary conditions obtained by omitting a small closed ball $\bar{B}_\epsilon$ around the origin in the D^4 and gluing it's boundary with the original S^3 of radius 1, and then letting $\epsilon \to 0$ induce direct limits on Hilbert spaces, so that we effectively have the topology $S^1 \times S^3$(Remark: This is under the assumption of no extra pathology being present in this special case, something that is off hand hard to check without explicit calculations in both cases.)

14.4. Superstring Partition Function and Mass Spectrum in $D = 10$.
We have for $D = 10$ superstrings with effective on-shell dimension $D_+ + D_- = 8$,

$$Z_{D=10} = 16\Pi_{n=1}^{\infty}(\frac{1+q^n}{1-q^n})^8 = 16\theta_4(0|q)^{-8} \sim exp(\frac{2\pi^2}{1-q})$$

with a for the present discussion irrelevant factor 16 stemming from additional degeneracy in string theory. And so, again isolating the Laurent series coefficents in the meromorphic germ around the origin which give the degeneracy d_n, one obtains for large n

$$d_n \sim n^{-\frac{11}{4}} exp(\pi\sqrt{8n})$$

or, equivalently, for large m

$$\rho(m) = m^{-\frac{9}{2}} exp(\frac{m}{m_0}), \ m_0 = (\pi\sqrt{8\alpha'})^{-1}$$

in a continuum description of the discrete spectrum near positive infinity in mass space.

14.5. Mass Spectrum and Degeneracy of $D3$-branes.
Finally we can return to our scenario with the $D3$-branes. Now our, and above all Hitchins (See below $D3$-branes/One-particle Space-times in General), pondering about the hypercomplex structures pays off; Because of the

Jacobi identities we noticed, linking the partition function on the $SU(2)$ to the case above by, e.g,

$$Z_{S^3 \cong SU(2)} = \frac{\sum_{l \in \mathbb{N}} (2l+1) q^{l(l+1)}}{\sum_{l \in \mathbb{N}} (-1)^l (2l+1) q^{l(l+1)}} = \Pi_{n \in \mathbb{Z}_+} \left(\frac{(1+q^n)}{(1-q^n)} \right)^3$$

and the free state nature of definite mass states with no interactions (this becomes a bound state from the five dimensional perspective in the sense of an eigenvalue of the d'Alembertian), which permits us to identify the time-like dimension to have the oscillatory topology S^1, noticing the that belonging partition function must be

$$Z_{S^1} = \Pi_{n \in \mathbb{Z}_+} \frac{(1+q^n)}{(1-q^n)}$$

that we can directly write down the partition function we seek for on the appropriate $S^1 \times S^3$ topology, as

$$Z = Z_{S^1} Z_{S^3} = tr q^{p^2}, \quad p^2 = -\Box = -\nabla_a \nabla^a = -\left(\frac{\partial}{\partial s}^2 + Y_1^2 + Y_2^2 + Y_3^2 \right) = -(\partial_0^2 + \partial_1^2 + \partial_2^2 + \partial_3^2)$$

where we refer the reader to the theorems of Hitchin and our conjectures in thesection below for the notation, and where x^i are the integrated coordinates induced by the hypercomplex structures. Hence we obtain, including degeneracy from spin and helicity,

$$Z = 4 \Pi_{n \in \mathbb{Z}_+} \left(\frac{(1+q^n)}{(1-q^n)} \right)^4$$

which is the square root of the string partition function in $D_+ + D_- = 8$ on-shell dimensions or $D = 10$ with the two off-shell dimensions included, and (exactly !, even with the in fron of the product included) what it should be from intuition and theorems on in part II and III. This is then treated in the same manner as the various other string field theory partition functions the exact result for the mass degeneracy from our hypothesis, coinciding with the dimensionally reduced stringy result, is

$$d_n = \frac{1}{2\pi i} \int_\gamma \frac{G_{D=D_\pm=4}(q)}{q^{n+1}} dq = Res \left\{ \frac{G_{D_\pm=4}(q)}{q^{n+1}}, 0 \right\}$$

, $G_{pm} = Z_S^1 Z_S^3$. For large n, this also gives the stringy results

$$d_n \sim n^{-\frac{5}{4}} exp(\pi \sqrt{n})$$

or, equivalently, for large m

100

$$\rho(m) = m^{-\frac{3}{2}} exp(\frac{m}{m_0}), \ m_0 = const.$$

the constant cannot be foretold directly from our hypothesis, but must be multiple of a numerical factor 4 or 1, $g_G = \sqrt{4\pi G}$, and the other natural constants.

That we, above— in the most important case—obtained an explicit check that

$$Z_M = Z_{Strings} = Z_{X_-} Z_{X_+}$$

in the form of

$$Z_M = 16\Pi_{n=1}^{\infty}(\frac{1+q^n}{1-q^n})^8 = (4\Pi_{n=1}^{\infty}(\frac{1+q^n}{1-q^n})^4)^2 = Z_{X_-}^2$$

and thus vanishing homomorphism anomaly of the apporiate kind (first) to that order, can perhaps be considered a small breakthrough in our little program. Although we intutioned this before, and proved this on elements of the string pertubation theory series, those proofs cannot be said to have the clarity offered by the simplicity of this calculus oriented proof. But, then again, a particular case proves nothing in the general.

14.6. $D3$-branes/One-particle Space-times in General.

With good approximation at relevant backgrounds, $X_{\pm}$ are hyperkähler, that is, they are hypercomplex space-times with their Obata connection[66] preserving a metric in the conformal class. If one lets the measured helicity space-time be embedded in the stringy space-time, something that can always be done by a special case of Whitney's theorem, we can represent the remaining by a contractible set, this to have homotopy equivalent topology inducing an isomorphism of the sheaf cohomological theories of the corresponding background partial differential operators (i.e BRST operators), e.g free operators in the simplest case on $\mathbb{R}^4$, which we often choose to compactify. One can also work with two copies of the same space-time, as we chose to do in part III as opposed to part II, as long as one remembers to take this into account the physical results should not differ—or at least have not done so so far in this thesis. Several particles, say N, are then supposed to correspond to a stack of N standard copies of the appropriate space-time background, which works a little bit as a common zero level or reference point for the various space-times. Here is a simple explanation for the reader who wishes to understand this; Think about usual electrons in space-time,

---

[66]This is the name of a torsion-less connection that preserves the hypercomplex structures on a hypercomplex space-time.

they are but fields; they can have different field values, for they correspond to different particles with roughly independent physics, this by cluster decomposition of the S-matrix. One models this by having N different position configuration spaces, or space-times in more common language. This does not necessarily mean that one truly does have several space-times but that is anyway the way we model it. So we have several space-times because we want to allow for electrons with different field values. Now, we all agree that vielbeins and connections are but usual fields, and we can, for the line of reasoning, say we agree that we can interpret these fields as particles. The values of connections and vielbeins on a space-time can be—should be— *different*, because after all, they are but particle way functions. Since a space-time is minimally and uniquely determined by it's smooth topology, it's connection and it's ON frame(or metric, if one insists on being impractical), and both the vielbein and connection are but usual particles, we deduce that we must be having N different copies of the smooth topology but with different connections and vielbeins in order to model the physics. So we have a heap of space-times, because of independent geometry,— as many as the number of particles—but with the same position spectrum. String theory, i.e the addition of dimensions is to drop the last restraint. The stack of branes, then, is what is called multiparticle space-time. Let us see how a one-particle space-time looks. According to us—and as has been checked indirectly in the mathematical literature for one very simple but relevant case— we conjecture that the spatial sets in a brane can, after continuation to Eucidean metric, be identified with the level sets of the determinant homomorphism $det(1 + \mathcal{O})$, $\mathcal{O} \in \mathcal{A}$ a C^* algebra. One can easily intutively see how that comes about, for if we make the analytic continuation of the usual expressions to imaginary time β it is easy to see that the the Hamiltonian H does not generate unitary isometries of the relevant Hilbert spaces but, e.g instead generates a scaling of determinants by the possibly ill-defined determinant

$$det(e^{-2H\beta})$$

In the standard case, $\mathcal{O} = Q_{int}/Q$, Q the BRST charge operator, e.g the Klein-Gordon operator $Q = -k^2 + m^2$. In the mathematical literaure, see Hitchin[1] in the references, this figured and was proved two years ago in his paper "Hypercomplex Manifolds and the Space of Framings", who was simply considering the special case of harmonic functions as generating spatial sets and time in a hypercomplex space, something that corresponds to Weyl(massless) fermions after reduction to halfdensities in the SYM, and the case for general m seems thus to be suggested, by "translation invariance" of the physical laws, as it were, along the mass scale, if we only remember that we can write solutions of partial differential equations by using partition functions and pertaining determinants.

It would be interesting to know the behaviour of such possibly non-hyperkahler space-times, might they be the non-vacuum states one percieves/believes them to be, and if so, how many of them are tachyons? Such answers could give further checks that truly everything fits in the grand hypothesis of this thesis—something that we can of course not exhaustively check ourselves as it involves all of physics.

Traditionally the "spectral varieties" $det(1+O) = c$ are called isospectral sets for finite dimensional operators in the context of integrable systems (See Hitichin[2]). In the infinite dimensional case, it can be hard to establish any meaning to such an equation, but since we are any way wanting information on a particular space-time this is remidied by observing the level sets of the determinant homomorphism acting as background partition functions usually do

$$u(x,t) = det(1+O)_{xx'}u(x',0) = (\sum + \int)_{x'\in\Sigma}det(1+O)_{xx'}u(x',0)$$

where obvious DeWitt summation, i.e integration and/or summation, over the position spectrum $\Sigma_\pm \ni x'$, on the chosen hypersurface at zero time, has been explictly included. Normally to these level sets the temporal direction evolves, with the spatial sets lying as a 3-fold inheriting a framing from the hypercomplex 4-fold. It is easily seen that an isometry of the relevant Hilberts space of Minkowskian fields in Euclidean space-time, hence with imaginary time, would deform the determinant instead of leaving it fixed, wich would have been the case in Minkowskian space-time. Following our mild generalization of Hitchin's idea in his treatment, on the basis of our observation that it could have been expressed by a determinant, and the then following conjecture in this matter, we have the timelike differential as $du = d\det(1+\mathcal{O})u|_{\partial\Omega}$;

and the other spatial differentials are given by the action of the hypercomplex structures;

$$\eta_1 = I\,du, \eta_2 = J\,du, \eta_3 = K\,du$$

and so one obtains the hypercomplex framing. One can see these as the X_0^a-embedding, which were a component in the $\mathbb{X}$ superfield, letting ω be a 1-form be dual to X_0^a we get

$$\omega = du + \omega_I I\,du + J\omega_J du + K\omega_K du$$

103

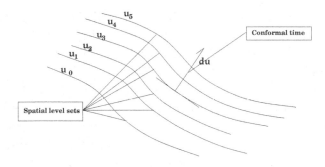

FIGURE 4. Spatial sets in a brane, as well as the decomposition in time and space over an arbitrarily chosen point. In this case, any of the sets u_i could have been chosen as $\partial\Omega$ the spatial spectrum at some given instant.

Which can be shown to be the most general decompostion of ω(Darboux's theorem.). Thus our branes arem the NC geometry related to the determinant homomorphism, the relevant charge operators being (see the quantization in part II and III), $Q = D_+D-$ in the full density string formalism, whilst usual Dirac operators apply in the halfdensity dual SYM description. The following theorems gives the afortmentioned support for the above

Theorem 14.1 (Hitchin 1998). *Let $X_\pm$ be a hypercomplex 4-fold and s a harmonic function on $X_\pm$. Let $\frac{\partial}{\partial s}$ be normal to s, and decompose $X = U \times M$, U an interval and M a 3-fold. Then (Y_1, Y_2, Y_3) dual to Ids, Jds, Kds moves according to*

$$\frac{dY_1}{ds} = [Y_2, Y_3],$$
$$\frac{dY_2}{ds} = [Y_3, Y_1],$$
$$\frac{dY_3}{ds} = [Y_1, Y_2]$$

and these are called Nahm's equations.

Theorem 14.2 (Hitchin 1998). *Let $X_\pm$ be a hypercomplex manifold generated by a framing (Y_1, Y_2, Y_3) on a 3-fold satisfying Nahm's equations. Then X is hyperkähler iff Y_i are volume preserving on the 3-folds $M_s, s \in U$.*

Lemma 14.1 (Hitchin 1998). *The triple of vector fields $\{Y_i\}$ satisfies Nahm's equations iff $d\eta_i$ are ASD w.r.t the metric $ds^2 + \eta_1^2 + \eta_2^2 + \eta_3^2$.*

Remark: As stated previously, a hypercomplex manifold is Hyperkahler iff the Obata connection is the Levi-Cevita connection of a metric in the conformal class. The metric which these canonical vacuum branes are Levi-Cevita with respect to is

$$G = e^{-2u}(ds^2 + \eta_1^2 + \eta_2^2 + \eta_3^2)$$

the volume form om the hypercomplex manifold is

$$\theta^0 \wedge \theta^1 \wedge \theta^2 \wedge \theta^3 = e^{-4u}(ds \wedge \eta_1 \wedge \eta_2 \wedge \eta_3)$$

and the natural—and preserved—volume form on $M_s = \Sigma$ the spatial slices is

$$e^{-2u}\eta_1 \wedge \eta_2 \wedge \eta_3$$

The fact that volume in configuration space is conserved on-shell is a classical theorem called Liouville's theorem. Thew above is a natural analogue on a hyperkahler space. The statement of the last lemma is equivalent to that the curvature is ASD to first order in the connection, since we recognize ω as the canonical expression for the Obata connection

$$\omega = \begin{pmatrix} 0 & \theta^3 & -\theta^1 \\ -\theta^3 & 0 & \theta^2 \\ \theta^1 & -\theta^2 & 0 \end{pmatrix} = I\theta^1 + J\theta^2 + K\theta^3$$

I, J, K automorphisms of the tangent fiber isomorphic to the quaternionic algebra imaginaries and (1,1) tensors, or, in other language, identifiable with the hypercomplex structures I, J, K on $M_s \subset X_\pm$.

Let us return and sum up. We draw a picture in string space-time to make it clear;

We sum up our thoughts on the determinant homomorphism by

Conjecture 14.1 (NC-Geometry/ D-brane correspondence, "Determinant Homomorphism Conjecture"). *Let $det : \mathcal{A} \mapsto \mathbb{C}$ be the determinant homomorphism from an appropriate C^* algebra to an appropriate field, here taken as $\mathbb{C} = \mathbb{R}^\mathbb{C}$. Then, after continuation to the Euclidean region so that the unitary action of the Hamiltonian is a scaling of the determinant instead of preserving it, we have our spatial sets induced by varieties of the form $det(O) = c$, $c \in \mathbb{C}$, where it is understood that we act by the determinant on appropriate initial value data on a spatial slice at some a priori given instant on a D-brane.*

and this gives a natural interconnection to Morse theory that will, up to the following mindmaps and flowcharts, end our thesis.

Remark: For the massless *classical* case of a scalar with only a Laplacian as charge operator— which is actually precisely what our gravitons and dilatons correspond to in this thesis—this has already been proven

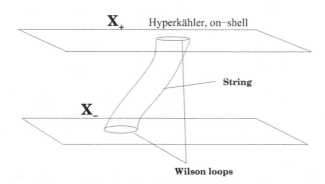

FIGURE 5. $X_{\pm}$ is a level surface of mass(i.e a foliation by the Ricci instead of the dilaton) in Kaluza-Klein space-time, $X_{\pm} = U_{\pm} \times \Sigma_{\pm}$, $\Sigma_{\pm}$ a level surface of time, and $det(1 + \mathcal{O})$ gives the branes at various instances. The above corresponds to how a Wilson loop of open strings would look like. A usual free propagating string in gravity would like the opposite, with the end points in the various space-times joining, being able to propagate arbitrarily in either space-time. Hence the closed dimension and open dimension would reverse in the above picture.

(by Hitchin in the references). Otherwise the question is open. Again we emphasize that the above includes illdefined objects (the determinants) whose cure may well prove to have some interesting effects on the evolutions of branes. As an example of the above scaling behaviour, theta functions are quasi-periodic, and they are partition functions on a cobordism in the Euclidean plane belonging to a parabolic differential equation.

106

Two Theories of Gravity and the Relation to a Third: A Generalized Maldacena Conjecture(proof: Chiral splitting theorem of string theory(D'Hoker, Phong)+ work that still has to be done)

<u>I: Dirac-YM on $Y_\pm(X_\pm), D(Y_\pm) = 4$</u>

<u>Fundamental Fields:</u> Φ super Dirac field, ω a spin connection with gauge degrees of freedom, so $\omega := \omega + \mathcal{A}_0 + \mathcal{A}_1 + \cdots$, $\mathcal{A}_i \in \{U(N_C), SU(N_C), \cdots\}$-connections. A 1-admissible theory, that is, a theory about worldlines.

$$\mathcal{L}_{Dirac} = \bar{\Phi}_\mp D_\mp \Phi_\pm, \quad Dirac - YM$$

$$\text{Form } \mathbb{X}^a \text{ from } \Phi^A$$

<u>II: H.E. on $X_\pm, D(X_\pm) = 4$</u> $\Updownarrow$

<u>Fundamental Fields:</u> $\mathbb{X}^a_\pm$ a supervector in the ON-frame, $\mathbb{X}^a_{\pm,1} = e^a_{\pm,\mu} dx^\mu_\pm = e^a_{\pm,\mu} \theta^\mu_\pm$ a vielbein, $\mathbb{X}^a_{\pm,0} = X^a_\pm$ a vector which gives an embedding via exponentiation. $\omega_\pm$ a connection with extra gauge degrees of freedom. This is a 2-admissible theory, i.e a theory about worldsheets, with fields in $X_\pm$, $D(X_\pm) = 4$,

$$\mathcal{L}_{H.E.} = *(\mathbb{X}^*_\pm D_\mp D_\pm \mathbb{X}_\mp), \quad Hilbert - Einstein$$

Stack X_+ onto X_-, include a worldsheet to be able to deform mass states. $X = X^{(5)}_+ \times X^{(5)}_- = \Sigma \times X_+ \times X_-$, $X^{(5)}_\pm$ now each include one degree of freedom more than usual, as well as gauge degrees of freedom. The extra degree of freedom is by effectively killed by world sheet on-shell(BRST) conditions.

<u>III: Strings on $X, D(X) = 10$</u> $\Updownarrow$

<u>Fundamental Fields:</u> $\mathbb{X}^a$ a NC supervector with gauge degrees of freedom. This is also a 2-admissible theory, so it is a theory about worldsheets.

$$\mathcal{L} = *(\mathbb{X}^* D^2 \mathbb{X}), \quad STRING/M - THEORY$$

107

A Picture Series Of The Above;

I: SDirac-YM on $Y_\pm(X_\pm)$

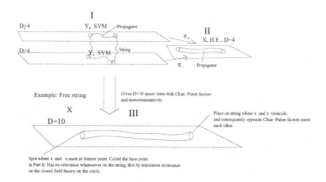

FIGURE 6. The points I, II and III above are the various steps and pertaining theories on the previous page. Both the 4 dimensional space-time and the 10 dimensional space-times above are noncommutative, and are in particular hyperkähler for a variety of more or less sensible backgrounds.

REFERENCES

[1] Hugett, Mason, Tod, Tsou and Woodhouse, The Geometric Universe, Oxford.

In particular

[2] "Hypercomplex Manifolds and the Space of Framings" by Nigel Hitchin in the above book.

[3] "Noncommutative Differential Geometry and the Structure of Space-Time" by Alain Connes in the above book.

[4] Daniel S. Freed, Five Lectures on Supersymmetry, American Mathematical Society, 1999.

[5] Quantum Fields and Strings; A Course for Mathematicians, volumes I and II, Editors Pierre Deligne, Pavel Etingof, Lisa C. Jeffery, Daniel S. Freed, David Khazdan, David K. Morrison, John W. morgan, Edward Witten.

In particular

[6] Eric D'Hoker, String theory lecture notes from the 1996-1997 year; Institute for Advanced Study, e.g. in vol. II, Quantum Fields and Strings; A Course for Mathematicians, Editors Pierre Deligne, Pavel Etingof, Lisa C. Jeffery, Daniel S. Freed, David Khazdan, David K. Morrison, John W. morgan, Edward Witten.

[7] Green, Schwarz, Witten, Superstring Theory, vol. I and II, Cambridge; 1987.

In particular volume I, where the material which we refer and lean on in the mass spectrum calculation we have done is located. Comparative material in string theory which we used for proving vanishing homomorphism anomaly of the first kind to tree level is also there.

Finally

[8] Hitchin, Segal, Ward, Integrable Systems, Oxford Science Pubications; 1999.
which we was our inspiration to regard our infinite dimensional determinants from a partly integrable systems perspective (beside the C^*-algebra perspective), in particular using the same nomenclature to describe some features of infinite dimensional systems such as our D-branes.

E. B. Torbrand

ABSTRACT. In this paper we give general principles of string field theory that emanate in a proof of the Maldecena Conjecture in a more general setting.

1. INTRODUCTION

In this paper we rederive or maybe reformulate Schwingers closed time formalism or, equivalently, prove the Maldacena conjecture. This also gives the proof and mechanism of dimensional reduction in string theory through holography.

2. MAIN RESULTS I

Theorem 2.1 (String field theory, Proof of Equivalence of Correlator Functions I, Second Quantized Part of Theorem). *Assume the outgoing Hilbert space is complete. Then We have*

$$< T[\mathcal{U}_\gamma \Pi_i \mathbb{X}_i] >= \sum_{out} | < \Pi \bar{\phi}_j^{out} \mathcal{U}_{\gamma-} \Pi \phi_i^{in} > |^2 =< \Pi \bar{\phi}_i^{in} \mathcal{U}_{\gamma+} \mathcal{U}_{\gamma-} \Pi_i \phi_i^{in} >$$

where we have defined $\mathbb{X} = \bar{\phi}\phi$, with superindices supressed.$+$ denotes the part of the closed path γ that is in positive time direction and $-$ the opposite. We must remark that ordered product above also orders outgoing states to the left and ingoing to the right. Here $\mathcal{U}_\gamma = \int_\gamma \mathcal{A}$.

Proof. In view of completeness of the outgoing Hilbert spaces and standard properties of ordered products this is so. □

Theorem 2.2 (Maldacena Theorem, Equivalence of Correlators II, First Quantized And Last Part of Theorem). *We have in a first quantized formalism that*

$$\ln(\mathcal{U}_\gamma) = \int_{\gamma=\gamma_++\gamma_-} \mathcal{A} = \int_\Sigma (\mathcal{G} + \mathcal{B})_{\mu\nu} \partial X^\mu \bar{\partial} X^\nu dz \wedge d\bar{z}$$

Thus stated equivalently a theory of wilson loops boils down to a theory of closed strings since they have the same lagrangean ex second quantization.

Proof. Using Stokes theorem for tensors(which may or may not be antisymmetric) we have the theorem by pulling the two-tensor down to a world sheet. □

Theorem 2.3 (Mechanism of Dimensional Reduction, String Field Theory). *A amplitude is given in $D = 8$ by taking the $D_+ = 4$ and $D_- = 4$ half-density amplitudes and multiplying them. Setting $X_+ = X_-$, that is the two different space-times in the holographic factorization to be equivalent this gives an amplitude for a full density in $D = 4$. Off the mass shell we should set $D_- = D_+ = 5$ and $D = 10$ or $D = 5$.*[67]

Proof. Since the product of the two amplitudes when they are each others conjugates become the square norm or full density we seek we are done. $\square$

3. MAIN RESULTS II: BRST-COHOMOLOGICAL PROOF OF EQUIVALENCE OF MODULI SPACE OF SOLUTIONS

We shall take a given setting and the reader must then understand that the results apply in more general situation. The important thing is the general thought or principle and not the proof in a specific case.

Theorem 3.1. *The solution space of $\not{D}\, \psi = 0, \nabla \cdot R = J$ is equivalent to the solution space of $(\square + \not{R}\,)\theta = 0, \nabla \cdot R = J$. Stated equivalently $N = 1$ SYM is equivalent to SUGRA after the correct identification is made. Here $R = F$ is the curvature of the connection $\omega = \mathcal{A}$.*

Proof. The proof is simple. Let us look at gravity first. By using a standard Bochner identity of Dirac operators we have that $\square + \not{R} = \not{D}^2$. Inverting the dirac operator we get: $\not{D}\, \theta = 0$. What we then do is use $\theta_\mu^a = \bar{\psi}\gamma^a T_\mu \psi$. Actually γ_{+AB}^a gives the components from $\gamma_+^a T_\mu$. This is used when doing rotations or boosts of spinors in special relativity. The standard representation of Dirac operators on spinors gives then $\not{D}_+ \psi_- = 0$. That we can go the other way around, proving the converse, is obvious. As stated under the title of our thesis we use the representation $\gamma^a = e^a + e^{*a}$ on space time vielbeins while using the standard Pauli matrix representation on spinors for the Dirac operator acting on spinors. $\square$

REFERENCES

[1] Hugett, Mason,Tod, Tsou and Woodhouse, The Geometric Universe, Oxford.
[2] Green, Schwarz, Witten; Superstring Theory Vol I and II.
[3] DeWitt, Bryce; The Global Approach to Quantum Field Theory, Oxford.
[4] Weinberg, S.; The Quantum Theory of Fields, Vol I, II and II.

---

[67]We might call this the diagonal projection for obvious reasons. Basically the pauli interpretation of wave functions in quantum mechanics is at work here.

PART II: QUANTUM GRAVITY AND STRING FIELD THEORY; ORIGINS

E.B. TORBRAND DHRIF

"We have a habit in writing articles published in scientific journals to make the work as finished as possible, to cover up all the tracks, to not worry about the blind alleys or describe how you had the wrong idea first, and so on. So there isn"t any place to publish, in a dignified manner, what you actually did in order to get to do the work."

Nobel Lecture, 1966

"[We] ...are not pleased when we are forced to accept a mathematical truth by virtue of a complicated chain of formal conclusions and computations, which we traverse blindly, link by link, feeling our way by touch. We want first an overview of the aim and of the road; we want to understand the *idea* of the proof, the deeper context."

H. Weyl, Unterrichtsblätter für Mathematik und Naturwissenschaften, 38, 178-188(1932), translation by Abe Shenitzer in the Amercian Mathematical Monthly.

"It's conceivable, although I admit not entirely likely, that something like modern string theory arises from a quantum field theory. And that would be the final irony."

S.Weinberg, March 1996

4. Introduction

The second part of this thesis concerns itself with the problem of gravity.

Roughly, from the point of view of held in this thesis, the problem can be subdivided into two problems; Problems associated on the one hand to general [non-linear] quantum theory and on the other hand gravitation theory itself. For this reason we have chosen to include some material that concerns the mathematical structure of general quantum theory, and keep it in some sense separated from gravity, since it is more general. This material then naturally interconnects the former part of this thesis with the second, and so brings us the idea of the hypercomplex in finite and infinite dimensions in mathematics, collectively termed hypermathematics- taking identical statements in terms of Fock algebra, supermathematics, etc.

This part of this thesis is not a work of mathematics, although it often deals with mathematical statements, and should not be regarded as such. The reader who expects German "Thirring" proofs is looking in the wrong place for such. Rather we are doing a blend of suggestions, heuristics and proofs valid within physical standards. Thus we are making suggestions when relevant rather than omitting, despite the fallible nature of such mathematically [without conclusive mathematical proof] unacceptable suggestions, in order to move progress forward in whatever extent it can be done. This is the tradition of physics, which is what this part of this thesis deals with.

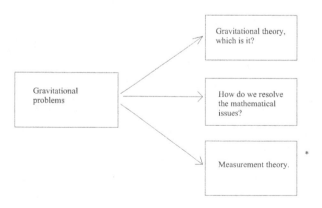

FIGURE 7. ∗:Not mentioned in this thesis.

5. Defining the Problem

In order to be able to "solve" the problem of gravity, if it is solvable, we have to subdivide it into manageable parts and define them. We must be sure that the parts are not too big and that we achieve some certain set of beforehand defined goals or criteria, otherwise we might loose ourselves in never ending academical arguments. Also, we must be prepared to disregard the theory if we find it incorrect, rather than try to save it by changing it time and time again, because then we will likely in the end run into selfcontradictory arguments, and truly not know precisely what the theory, or rather hypothesis, is.

We will call a theory for gravity "acceptable" if the following overlapping, and admittedly quite stringent, features are present in dimension four after possible compactification etc;

- It reproduces, at least at a macroscopic level, the symmetries we observe, in particular mass attracts mass for particles of positive energy and a 'metric" graviton should have spin 2.
- Hilbert-Einstein gravity is it's classical limit.
- It yields sensible and finite answers to definite problems of a reasonable nature, then disregarding problems that are associated to general perturbative theory such as convergence of the S-matrix etc. In particular the theory in question should satisfy calculationability "in principle", as in the usual status of particle physics.

It should be understood directly that string theory roughly satisfies all of the above except perhaps the last line of the last criterion.

We could consider the following as four fundamental and good ways to check such a theory[which of course satisfies the above requirements] and at least start a debate from

- It reproduces the Schwarzild metric out of quantum theory, with perturbations.
- Hawking radiation and entropy.
- Coulomb interaction and black hole scattering, together with the crucial attractive feature.
- Good strong curvature dynamics, such as sensible pair creation rates.

Thanks to work by Juan Maldacena in string theory, we at least know that string theory produces Hawking entropy and radiation.

However string theory battles with some problems too, for example extra dimensions which gives ambiguous compactification. Many persons have for some time had the impression that it is a little bit 'too

117

general", which makes it seem to be a little hard to wield in the service of physical prediction in practical situations. So it might be right but need some reformulation in various specific instances of M-theory. Also, the topological situation set by the plethora of various Dp-branes and p-branes is hard to cope with, making theory remicent to a trip to the zoo. Finally another problem, at least if it is going to be commonly accepted by most scientists, is that some people find the idea of strings "wrong' for some metaphysical reason.

What is done in this thesis is try to remedy some of these last problems partially. However, rather than taking the string "legacy" for granted and turning my back on the sceptics I derive it, out of field theory in two different ways. Once that this has been done I develop a field theoretic version of string theory in which worlds and space-times are created and annihilated, a *third* quantized formalism, that builds on the links that have in recent years been seen between string theory and Yang-Mills, among other things in connection to D-branes. This string field theory, however , makes no a priori use of string theory at any point and in particular in a systematic way gives the amplitude to any process in way exactly equivalent to the field theory S-matrix picture with operator fields, but with a "square root" of the fluctuating space-time diagrams corresponding to a spinorial diagram of field theoretic origin. So in effect a singular string diagram becomes a field theory diagram and conversely. As to the supposed matter of string theory being too general, we remedy this partly by the perspective which we obtain after having gone the long road to derive it through field theory, which gives us perspective, hence, paradoxically, by putting the "too general" in a context and generalizing further we achieve simplification since what one understands is often subjectively percieved as simpler, although it may objectively be more complicated.

Presently, although it is clear what the amplitude for transitions for space-times[each space-time will correspond to a Poincare dual of something called string based at a point, which is a point where two Chan-Paton factors coincide, turning out to be equivalent to charge dual Dirac point particles.] for different space-times of different topology which are not generated by each other I do not know how to look at the general situation of all backgrounds in the same matrix. Also, I do not know how many space-time toplogies which are generated by the S-matrix on a single background, since it also does topological operations to the background when creating copies of space-times and gluing them together in different ways.

What seems to be clear to some extent is that this version of string theory has the predictive power missing at some points of M-theory, and does not require extra dimensions either. Also the idea of strings as a self-interaction of spinor fields and space-time itself seems to be appealing, at least since it can be derived in field theory without prior knowledge of strings which should convince some sceptics. The idea makes contact with noncommutative spaces in a very natural way[symplectic manifolds and associated hypermathematics], which for finite factors and dimensions lead to "hypercomplex" objects and in infinite dimensions may or may not be noncommutative geometry in Connes meaning. The blend of hyper/supermathematics[noncommutativity], integrability, solitons, and strings/Dp-branes/fields , which for long have been part of the "string" world, together with the ability to produce some predictions might make this version interesting however.

I should warn the reader that this thesis starts whith what might seem as total asides, for example realizing a property of lagrangeans that will be crucial to us in the following, and a superficial plunge into connections between Yang-Mills theory and Einstein gravity , especially in the conformal Yang-Mills theory of the gauge group spin(n,m). This is intentional, and is so because these ingredients will be needed in the following, and we assure the reader that we return to the point once these things are gone through. Finally I warn that we really should have written String + field theory(but omitted the + so that everybody could at least recognize the title, as it came closest to what we are doing.) in the second part of this thesis, this because we will sketchedly find a duality between string theories and certain Q.F.T's in manner that was predicted by Maldacena four years ago. This shall come from an independent attempt to quantize gravity field theoretically using coordinate noncommutativity (the attempt is sucessfull to some extent as it presently seems), and exhibit a candidate for field theoretic gravity. It will also lead to a sketchy proof of the Maldacena conjecture. Much more important than the proof is the idea of the proof; The attempt will also give us a good insight into the underlying dynamics of AdS-CFT correspondences, which seem to be much more general than one would initially think, indeed those ideas will seem be the origins of string theory.

Again warning the reader for the speculative nature of this part, which concretely means that nothing in it should necessarily, without further investigation, be considered science fact, we now commence our speculation.

6. Conventions

We work with God-given units

$$\hbar = G = c = 1$$

mostly, and often in theoretical discussions omit factors 2π, i and various signs when they do not affect a purely theoretical discussion. Specifically in this part we have at several places a non-dimensionful Ricci, and from the beginning of the section entitled "Getting Down to Business" we use, after an analysis which relates to the natural contants, $\hbar = \sqrt{4\pi G} = c = 1$ instead, so we, for most purposes, set $\mathcal{G} = \sqrt{4\pi G} = 1$ a natural constant after that section. This should be contrasted to what one obtains if one does not separate the behaviour of vielbeins (see below), namely $\lambda = \sqrt{8\pi G} = 1$, which is mostly connected to pathological theories.

μ, ν, κ and other indices late in the Greek alphabet are space-time coordinate indices, i, j, k space indices, while a,b,c and other small Latin letters are ON-frame/ vielbein indices. α, β, γ, etc are supposed to be gauge Lie algebra indices. We sometimes have "Einstein" or "De-Witt" sums, i.e integration/summation over index sets with repeated indices, in various places-it should be clear from the context when such apply, and thus we sometimes even skip putting out the indices when it is obvious.

The ON space-time metric in D Lorentzian dimensions is

$$\eta = [\eta_{\mu\nu}] = dx^0 \otimes dx^0 - dx^i \otimes dx^i$$

The reader is assumed to be familiar with such facts of "Twistor" geometry such as that complexified space-time can be identified with the grassmannian of 2-planes in $\mathbb{T} \cong \mathbb{C}^4$ in the physical dimension.

A manifold is denoted M or X, sometimes with subscripts to denote homology class and/ or metric, and it should be understood that this only defines an equivalence class of manifolds. Manifolds are assumed to be complex if not otherwise stated, and we define certain direct limits associated to these as pseudo-riemannian real analytic manifolds. It should be understood that the objects we study are principally singular and smoothness etc. only holds in "general position". In general, neither of these concepts are sufficient, and instead one needs concepts such analytic spaces etc that are outside the scope of this thesis.

A anti de Sitter space is defined to be an arbitray space of constant negative curvature, in contrast with the more common situation where only the "classical" AdS space of trivial topology is meant. In this regard, it should be noted that by a mathematical theorem (Yamabe's

120

theorem), any proper Riemannian manifold is conformally equivalent to a space of constant curvature.

A p-brane is defined to be a p-dimensional object in space-time, in contrast with almost any other literature, which usually defines it to be a p dimensional object at a fixed time, this to have Lorentz invariant formulations which treat time on the same footing as space and to have better systematics. From this point of view a string is 2-d object in space-time with it's history already specified, a 2-brane. A subset, in particular a [homological] chain of dimension $\leq D-1$ is denoted Σ. Dp-branes, on the other hand, are treated as usual with regard to the meaning of p.

We use big Latin letters like X,Y,Z to denote tangent vectors and sections, and small letters to denote coordinates on a chart. An exception to this are basis sections of bundles, e.g. cotangent bundles, which have basis sections e^a or θ^a.

The following is not a convention but should be stated right away to avoid confusion:

Since we will be dealing with [holomorphic] symplectic manifolds, although not always explicitly, the above conventions sometimes produce confusion by expressions like dX for coordinate differentials. It is then undertood that we are taking the tangent as a chart, and in pulling back to x-coordinates we are going to another coordinate system which is also a tangent. These coordinate systems are then related to the manifold by the exponential map, the difference between the former and the latter is that the latter are background congruences, while the former are congruences corresponding to an interacting physical system as classical/quantum fields over the background congruences. The image under the exponential map might be called the "integrated" picture, while the former the tangent picture. We have almost no use at all for the integrated picture, but rather always assume that we work with the tangent picture, since we can always patch together a manifold by background congruences over different points as charts. Hence we never mention that we are in this picture since we always assume it. Of course this picture is only to give physical intuition, but as it would have been a likely cause of confusion we have included it here.

7. The Heart of This Thesis

Roughly we are here to describe the main point of this thesis, over which the subsequent developments depend.

Consider the classical action

$$\mathcal{L} = \frac{-1}{16\pi G} \int *R = \frac{-1}{16\pi G} \int d^4x \sqrt{h} R$$

it is well known that we would like to give it some "quantum" interpretation of it.

Consider an apparently unrelated problem; Why do the lagrangeans of quantum physics seem to omit so many terms? For Yang-Mills-Dirac theory we might ask why

$$\mathcal{L}_{GEN.} = \sum A_{\alpha\beta}(\bar{\psi}(i\not{D})^{\alpha}\psi)^{\beta} + \sum B_{\alpha'\beta'} < \mathcal{A}, (D^*D)^{\alpha'} \mathcal{A} >^{\beta'}$$

does not appear more often, the coefficients being c-numbers, as it is the most general lagrangean compatible with the symmetries of the theory.

The answers to both questions turns out to be quite complex and related, and much more interesting than only "they are suppressed by large terms of high mass dimension"; one cannot put an arbitrary term in a lagrangean and expect it to describe the evolution of a point particle if one exponentiates the lagrangean. That is, albeit one may obtain a differential equation by variation that describes solutions which are superpositions of point particle solutions one does not obtain these solutions by exponentiating the lagrangean, so Dirac's principle

$$\phi(x, \gamma) \sim e^{-i \int_{\gamma} \mathcal{L}} \phi(x, 0)$$

for small enough perturbations is not accurate for a set of lagrangeans if we assume the solutions to be point particles. Never the less one does obtain something, which however need not be a point particle solution.

Hence there are additional requirements that must be put on a lagrangean in order for it to describe a point particle. Let us try to "derive" this answer in elementary way. To find it, we must first understand what "lagrangean" is and perhaps differ between different lagrangeans. Suppose we a had a smooth analytic function f in some disc of convergence around the the real X_0 of radius R, then we could expand

$$f(X) = \sum_{n=0}^{\infty} a_n(X - X_0)^n, |X - X_0| < R \in \mathbb{R}_+$$

122

But then we see, upon identifying terms, setting $X - X_0 = \Delta X^\lambda$, that this can be written as

$$f(X) = \sum_{n=0}^{\infty} a_n (X - X_0)^n = \sum_{n=0}^{\infty} (\frac{\partial_\lambda f(X_0)}{n!})(\Delta X^\lambda)^n = e^{\Delta X^\lambda \partial_\lambda} f(\lambda)|_{\lambda=0}$$

We can look at this $e^{\Delta X^\lambda \partial_\lambda}$-part independently for a moment. Suppose it is acting on an approximate momentum eigenfunction in position-space, i.e. we roughly have a free classical configuration. Then, with λ a proper coordinate and x^μ coordinates in position space, X^0 being the "time" t we have

$$(\Delta X)^\lambda \partial_\lambda = \Delta t \partial_t + \Delta X^i \partial_{x^i} = (\int dt)\partial_t + (\int dx^i)\partial_{x^i} = -i[\int E - \int dx p_x]$$

with energy E, space momenta p_x. Substituting for the Hamiltonian we get, parameterizing the trajectory with t

$$(\Delta X)^\lambda \partial_\lambda = -i \int (H - \dot{x} p_x) dt$$

i.e, recognizing one of the usual formulas for the lagrangean $L = H - \dot{x} p_x$, that gives

$$(\Delta X)^\lambda \partial_\lambda = -i \int L dt$$

The property above, that one can generate a 1-flow [a one-dimensional flow], is the most fundamental property of lagrangean in point particle quantum theory-if we have a theory where we start with L and decide we are doing a point particle theory we must be very careful and choose only lagrangeans that correspond to such flows, i.e such that there is an equality of the above type. Let us call such lagrangeans admissible or 1-admissible, then understanding that we have an additional requirement rather than just symmetry in such a theory. At this point it might be well worth while introducing a lagrangean density $\mathcal{L}$. By imagining the relevant interactions to be "smeared" out codimensionally to the time coordinate, we get, remembering that Hodge duality gives us the codimensions for a path γ in a space-time X^1;

$$X = -\int_\gamma (\int_{*\gamma} \mathcal{L} * dt) dt = -i \int_{\gamma \times *\gamma = X} \mathcal{L} d^d x$$

---

[1]Locally this statement makes sense, globally either assume the existence of a foliation/stratification of the tangent bundle or embedd in a topologically trivial set and Kaluza-Klein continue to obtain such a subdivision. This should not be interpreted as intrinsic non-foliation having no physical consequences however, or that the K.K-continuation avoids them, rather these are instead generated as obstructions[singularities, homotopy types of spaces, solitons, etc] off the physical space-time.

The lagrangean density still describes a point particle flow, but this time "smeared out"-similarly we call this lagrangean density a 1-admissible (admissible) lagrangean density. In the following we often abbreviate largangean density to lagrangean.

What does such an L look like?

$$L = H - p_x \dot{x}$$

tells us, recognizing terms, that $\dot{x}^i$ has to correspond to some current J^i and p to this generator. So for example

$$L = H - p^i x_i = H - p^i J_i = \int \mathcal{H} |\psi|^2 - p^i \bar{\psi} \gamma_i \psi d^3 x.$$

or, in an explicitly Lorentz invariant way for an eigenstate

$$\int L = \int \int \mathcal{H} |\psi|^2 - p^i \bar{\psi} \gamma_i \psi d^3 x dx = \int \bar{\psi} i \slashed{\partial} \psi d^4 x.$$

So under assumption that the current is given by the above expressions we get a Dirac action. The important thing about the action above is that it is only linear in position space derivatives; this we can see because $L = H - p_x \dot{x}$ is only linear in p^i, H, and the rest falls by covariance.

Despite this we might wish to consider other types of functions under the integral sign, which do not generate flows, at least not in the classical 1-flow sense. So these these new lagrangeans do not give solutions to differential equations, at least not along the "time" coordinate when they are exponentiated and they do not obey Dirac's principle in the usual sense. Never the less, we could always vary the lagrangean, and so $\delta S = 0$ would give a differential equation that we might solve appropriately as functions of the "time" coordinate , so we will call these new lagrangeans variationally correct, since they could satisfy the symmetries of the theory and give sensible equations of motion. How do the variationally correct lagrangeans relate to the admissible ones? Well, lets pick an example that might help us;

$$\mathcal{L}_{NEW} = \phi \partial_t^2 \phi - \phi (\nabla)^2 \phi$$

gives the equation of motion $\Box \phi = 0$. We see directly that the solutions ϕ are superpositons $\phi_k = A_k e^{ikx} + B_k e^{-ikx}$, A_k, B_k c-numbers, where it is understood that the time coordinate in the last term may be reflected. Thus we understand, recognizing e^{-ikx} to be just the spectral representation of e^X that ϕ will be of the form

$$\phi = A e^{-i \int_{t_0}^t (\int \mathcal{L}_{LINEAR} d^3 x) dt} + B e^{i \int_{t_0}^t (\int \mathcal{L}_{LINEAR} d^3 x) dt}$$

124

where $\mathcal{L}_{LINEAR}$ is directly proportional to a derivative. And there is simply no way, at least not in a clearly visible sense, that this new lagrangean would have produced it's own solutions by exponentiation. It is important to recognize that when we do so, for example in the operator picture, this is a remnant from elementary quantum mechanics, where the differential equation is parabolic-simply because it is relativistically invalid. So they are solutions of something that looks like

$$(-\Delta + \cdots)\phi = i\partial_t \phi$$

which is very far from any discussion in this thesis.

Hence one must be careful in applying Dirac's principle, which is wrong when incorrectly applied. Let's put these ideas in a definition and a new principle;

Definition 7.1. *We call a lagrangean p-admissible if it is a pure p-derivative term lagrangean.*

A p-admissible lagrangean solution is quite often for the cases we are considering, not always, generated by flows of fundamental or 1-admissible lagrangeans, i.e

$$\phi = A e^{-i\int \mathcal{L}_1} + B e^{-i\int \mathcal{L}_2} + \cdots + D e^{-i\int \mathcal{L}_n}$$

with $\mathcal{L}_i$ 1-admissible. For example the "Pauli" lagrangean $\bar{\psi} \mathcal{F} \psi$, $\mathcal{F} = \frac{F_{\mu\nu}}{2} \frac{[\gamma^\mu, \gamma^\nu]}{2}$ corresponds to a part of $\bar{\psi} \slashed{D}^2 \psi$ if we vary ψ, which gives the differential equation

$$\slashed{D}^2 \psi = (\Box + \mathcal{F})\psi = 0$$

which has solutions given by

$$\psi(x, \gamma) = \frac{e^{-i\int_\gamma \bar{\psi} \slashed{A} \psi} P_+ + e^{+i\int \gamma \bar{\psi} \slashed{A} \psi} P_-}{2} \psi(x, 0)$$

for a path γ joining x_0 and x, where we could call the second term the anticausal evolution. Now, it must be understood, that despite that we in classical physics assume that for some initial value data the anticausal evolutions are inseparable from the causal, and that we must act by an evolution operator on a state that is a mix of causal and anticausal, this is not so a priori in quantum theory, although it may become so later in the S-matrix. There, instead, one separates the causal from the anticausal evolutions and instead call the solutions particle and antiparticle. So a solution is then truly given by

$$A e^{-i\int_\gamma \bar{\psi} \slashed{A} \psi} \psi(x_0) + B e^{+i\int_\gamma \bar{\tilde{\psi}} \slashed{A} \tilde{\psi}} \tilde{\psi}(x_0)$$

125

and again Dirac's principle did not apply. In the above ~ denotes charge conjugate, e.g in the sense of the operation PT for free fields. This is the reason for us not having "Pauli" terms in our lagrangeans-because a lagrangean not only need satisfy the symmetries of a specific theory, but also must "be" a lagrangean in the sense that it *generates a flow*, e.g. truly belongs in a Feynman path integral, since a PI depends on Dirac's principle being correct. This is also the reason why they give pathological theories when exponentiated in the operator picture, for exponentiating them is simply wrong.

Let us sum so far, before going on to this new situation:

(1) Not every lagrangean that satisfies symmetries is associated to worldlines.

(2) Not every lagrangean belongs in an exponential of quantum theory when integrated.

(3) This is so already in field theory.

We would like to generalize the above to the case where we are not considering 1-flows. Using the higher-dimensional equivalent of "differentiation", e.g the operator

$$\frac{\partial}{\partial \Sigma^\mu} = \frac{\partial^{|\mu|}}{\partial^{\mu_1} x^1 \cdots \partial^{\mu_n} x^n}$$

for the multi-index $\mu = (\mu_1, \cdots, \mu_n)$ in some local orthogonal coordinate system, in effect a representation of a Radon-Nikodym derivative when having unit or nill entries in the μ_i, it is not hard to show, following the previous more or less exactly, that

$$L = H^\sigma + \sum_{|\mu|=|\sigma|, 0 \leq \sigma_i \leq 1} p^\mu \dot{x}^\mu$$

In the above dot denotes differentiation w.r.t. to the multi-index σ, and for example

$$\dot{x}^\mu = \frac{dx^{\mu_1} \wedge dx^{\mu_2} \cdots dx^{\mu_n}}{dx^{\sigma_1} \wedge dx^{\sigma_2} \cdots dx^{\sigma_n}}$$

for a parameterization by the coordinates $\{x^{\sigma_1}, x^{\sigma_2}, \cdots x^{\sigma_n}\}$ is the corresponding Jacobian for a coordinate change. This gives terms of order $|\sigma|$ in the lagrangean, so this tells us to associate p-dimensional in space-time to such a lagrangean-but one must be careful; it is also possible to write down terms that do not make sense as flows, for example by squaring a Dirac lagrangean-then one obtains "Fermi"-similar terms, and indeed experience tells us they are pathological.

As an example of the the converse we might look at the classical lagrangean

$$\mathcal{L} = (h_{\mu\nu}) + B_{\mu\nu}) dX^\mu \wedge dX^\nu$$

126

a version of a Polyakov lagrangean, non-nill under small transversal complexifications and hermitean metrics, which after a change of coordinates to x^a and neglection of a surface term is

$$\mathcal{L} = h_{\mu\nu}\partial_a X^\mu \partial_b X^\nu \delta^{ab}\epsilon_\Sigma = -h_{\mu\nu}X^\mu \partial_b \partial^a X^\nu \epsilon_\Sigma$$

ϵ_Σ the world sheet volume element, δ^{ab} a flat world sheet metric. So indeed it has this second derivative term, namely the d'Alembertian.

Let us, as an aside, state the general Dirac's principle that truly applies, in an admittedly imprecise way since the cases must be attacked case by case.

Principle 7.1. *The approximate solution to the equation* $\delta\mathcal{L} = 0$ *is given by*

$$\phi \sim e^{-i\mathcal{L}_1}e^{-i\mathcal{L}_2}e^{-i\mathcal{L}_3}\cdots e^{-i\mathcal{L}_k} + e^{-i\mathcal{L}_{k+1}}e^{-i\mathcal{L}_{k+2}}e^{-i\mathcal{L}_{k+3}}\cdots e^{-i\mathcal{L}_l} + \cdots$$

where integrations are supressed.

It is understood that the above merely states that deformations of functions may be generated by Lie algebras of smooth, analytic or ditto germs of some suitable bundle and that this is effectuated by near on-shell conditions relating certain operators approximately in the dynamical direction(s) to operators on the spatial directions and as such is a very general statement. It is also understood that with general enough generators $\mathcal{L}$ we can generate any function possible, for example the string theory lagrangean

$$\frac{1}{2\pi\alpha'}G_{\mu\nu}\gamma^{ab}\partial_a X^\mu \partial_b X^\nu$$

is locally connected to the diffeomorphism [Virasoro] algebra on the unit circle and the on-shell condition $\partial\bar{\partial}X^\mu(z,\bar{z}) = 0$, generated by L_m satisfying

$$[L_m, L_n] = (m-n)L_{m+n} + \frac{c}{12}(m^3 - m)$$

This algebra is in a holomorphic representation for vanishing central charge represented by $L_m = e^{-mw}\partial_w$ in a small transversal complexification of S^1. If we want to get off-shell deformations of closed string space S^1 we have to go transversal to the diffeomorphism algebra and find other generators, which would imply breaking away from the biholomorphism group in this example, which has a flavor of the general thing. So, in the general case, Dirac's principle amounts to partitioning the deformations of a [function, field, algebra] space $\mathcal{V} = \bigoplus \bigotimes \mathcal{V}_i \oplus \cdots$, $T\mathcal{V}$, as

$$T\mathcal{V} \cong \bigoplus \bigotimes T\mathcal{V}_i \oplus ker(\phi)$$

127

, ϕ a homomorphism, and these factors correspond to all kind's of degrees of freedom. e.g. gauge, spin and brane degrees of freedom according to suitable conditions like for example on-shell conditions and their codimensions in field configuration space lifted to local tangents. The above lagrangeans in general now contain terms corresponding to p-flows, and correspondingly the partitions $\mathcal{V}_i$ are locally defined in the field configuration space.

Let us return to the Y.M-term of the action we were looking at,

$$\mathcal{L}_{YM} = Tr[\mathcal{F}^*\mathcal{F}] = <\mathcal{A}, \Box\mathcal{A}>$$

which is 2-admissible in the above heuristic way. Yet it is well known that this lagrangean gives sensible answers in a quantum theory, where it is certainly exponentiated. Just as as the cocycle we most easily associate with the Dirac action is the current 3-form J, which satisfies $DJ = 0$ we have for this case the field strength $\mathcal{F}$, which satisfies $D\mathcal{F} = 0$, a 2-form with codimension 2. Despite that one could view the above lagrangean in a $1 + 3$ way, the most interesting way is to look at it in a $2 + 2$ way, which associates it to 2-dimensional objects, i.e. flux tubes as in Misner, Thorne, Wheeler, so it could be viewed as generating a 2-flow.

We note then that the Yang-Mills-Dirac action can be interpreted as realizing dualities in the DeRham cohomology ring, or , to be more precise, in a cohomology rings taking coefficients in various spaces, i.e in a ring of DeRham currents and uses them to generate deformations of p-forms and associated p-dimensional objects. More clearly for the compact smooth case, writing ω for the connection this time

$$-iS = \int \mathcal{L} = -i[\omega_{\mu\alpha x}J^{\mu\alpha x} + F_{\mu\nu\alpha x}F^{\mu\nu\alpha x}]$$

$$= -i[<\Sigma^1, J^1> + <\Sigma^2, J^2>] = -i[\omega^*J + F^*F] = -i\Omega_x^*\mathcal{J}^x$$

with $\Omega = \Sigma_1 + \Sigma_2$ a sum of a 1/3-chain and a 2/2-chain (Poincare duality applies), $\mathcal{J} = J_1 + J_2$, i.e a p-dimensional object is deformed in $d - p$ codimensions, and it is in the Poincare dual space that the generators are, where it is understood that we only retain volume elements before the integration. Hence the Yang-Mills-Dirac action is also a smooth topological and geometrical object, namely the generator of a deformation of smooth geometry/topology, running diagonal over the cohomology ring in dimension 4. Actually, this is no bigger surprize, for already the work of Donaldson and others showed that for the case with vanishing fundamental group $\pi_1(X) \cong \{0\}$, which corresponds to retaining the Yang-Mills term only, that 2-forms, indeed even field

128

theoretic considerations— roughly corresponding to calculating instantonic solutions in dimension 4— gave totally new means for differential topological classification theory.

Similar heuristics that we made for the Pauli term, which we do later in this thesis, explains also why the Hilbert-Einstein action, which is 2-admissible by explicit dimension counting, is pathological as an action in a quantum theory exponential. That is, albeit the field equation obtained by it is sensible and certainly does describe classical gravity(modulo reformulations), one cannot exponentiate it naively, or coupled versions with naive kinematical terms, to obtain solutions to this same non-linear equation- if one wishes to associate these to 1-flows. To see this within the Hilbert-Einstein context is not difficult for the case of small gravitational fields; we have with bars denoting the trace-anti trace operation

$$\overline{T_{\mu\nu}} = T_{\mu\nu} - \frac{1}{2}g^{\mu\nu}T_{\mu\nu}, T_{\mu\nu}e^{\mu} \otimes e^{\nu} \in \Gamma(X, \mathcal{T}_2(X) \otimes L)$$

$\Gamma(X, \mathcal{T}_2(X) \otimes L)$ the sections of a smooth bundle of contravariant 2-tensors of space time with possible twisting line bundles L for small deformations $h_{\mu\nu}$

$$R = g^{\mu\nu}Ric_{\mu\nu} = g^{\mu\nu}\overline{\overline{Ric_{\mu\nu}}} =$$

$$= -\frac{1}{2}h^{\mu\nu}\Box h_{\mu\nu} + O(h^3_{\mu\nu}) + constants, \ g_{\mu\nu} = \eta_{\mu\nu} + h_{\mu\nu}$$

which will not generate it's own solutions as a lagrangean. This is remarkable, as the above linear approximation is remarkably accurate for a wide range in the low-energy/ long distance physical applications of gravity, indeed most of the applications of gravity might use even rougher approximations, which still show this defect. Thus, using the same reasoning used before for the scalar lagrangean, the conclusion that Hilbert-Einstein action does not generate it's own solutions to the equation determined by variation, at least not in the low-energy limit as 1-flows, is now unavoidable.

That is, if we instead just change perspective we might discover that already field theory is intimately connected with topology, geometry and above all p-dimensional objects. The reader who is subscribing to Yang-Mills is thus already subscribing to two dimensional objects roaming about in space-time, although perhaps unaware of it, from this perspective, which may or may not be correct.

And, following the above, we have the heart of this thesis; if already field theory can be used to describe 2-dimensional objects, indeed p-dimensional objects straight away, maybe there is a simpler way to describe the gravitational field, that perhaps would give order in the chaos of the p-brane zoo. So maybe we could turn the string paradigm around and discribe strings by point particles. But are these 2-dimensional objects truly the 2-dimensional objects of string theory?

Let's find out!

7.1. Summary.

- According to the above, lagrangeans are not arbitrary, but must generate flows. This constraints which lagrangeans that are admissible as quantum actions.
- These flows are then supposed to be associated to N-dimensional objects, and consequently their lagrangeans, which are associated to these N-dimensional objects, are called N-admissible.
- H.E-Gravity looks like a lagrangean that is not generating such a 1-flow, so it might be that the equation of motion is right although the lagrangean is not for purposes of exponentiation. Furthermore if, despite this, it is right to exponentiate it it should be associated to 2-dimensional objects.

7.2. Yang-Mills Theory.

Let $P(X, G)$ be a smooth principal bundle over space-time X with fiber being the group G corresponding to the Lie algebra $\mathfrak{g}$, and a let $\mathcal{A} = \mathcal{A}_a^\alpha T_\alpha e^a$ be a connection on the cotangent principal $T^*P(X, G)$, T_α generators belonging to some finite dimensional Lie algebra $\mathfrak{g}$ satisfying $[T_\alpha, T_\beta] = f_{\alpha\beta}^\gamma T_\gamma$. Call this connection $\mathcal{A}$ gauge potential and G a global gauge group, it being understood that one can have a right action by non-constant sections of P on P, which are then termed 'local" since they shift value over different space-time points, so that these automorphisms of P are called gauge group. Define for any charge g, $\not{g} = \frac{g}{2\pi}$ and let it be understood that we for unitary cases use $\frac{1}{2\pi i} g T_\alpha = -i \not{g} \, T_\alpha$ to generate transformations of some vector space which then become isometries leaving the amplitudes of quantum mechanics invariant. Let $E \to X$ be a vector bundle associated to this principal bundle, and form a $\mathbb{Z}_2$ graded extension by introducing grassmannian coordinates θ^i satisfying $[\theta^i, \theta^j]_+ = 0$, and letting these bundles take graded coefficients. Then a Dirac field $\psi = \psi_i \theta^i$ may be associated to sections of the form $\psi_i \theta^i$, i, j now space-time superindeces, it being under stood that G will have to include the spin group of the appropriate space-time in the form

$$G = G_1 \otimes Spin(n, m)$$

and thus $E \to X$ be a bundle module for such an action, G_1 another group. Let us relate this picture with the picture with the antisymmetrization that one usually does for classical Dirac fields to obtain the correct statistics for fermionic fields. We have for ψ^1, ψ^2 expanded in space-time grassmannians,

$$\psi^1 \psi^2 = \psi_i^1 \psi_j^2 \theta^i \theta^j = \psi_i^1 \psi_j^2 (\theta^i \otimes \theta^j - \theta^j \otimes \theta^i)$$

or, mapping by $\theta_i \otimes \theta_j$ corresponding to the 'time", [event ordered] space-times using the obvious duality

$$< \theta_i \otimes \theta_j, \psi^1 \psi^2 > = < \theta_i \otimes \theta_j, \psi_i^1 \psi_j^2 (\theta^i \otimes \theta^j - \theta^j \otimes \theta^i) > = \psi_i^1 \psi_j^2 - \psi_j^1 \psi_i^2$$

which corresponds to the usual picture when not using grassmannians.

7.3. Physical interpretation of the Yang-Mills Equation.

7.3.1. *A Local Study.*

Variation of the Yang-Mills lagrangean

$$\mathcal{L}_{YM} = \bar{\psi} i \not{D} \, \psi + Tr[\mathcal{F}^* \mathcal{F}] = < \mathcal{A}, \Box \mathcal{A} >$$

gives the equations

$$D^*D\mathcal{A} = J$$
$$\displaystyle{\not{D}}\, \psi(x) = 0$$

which are supplemented by two conserved quantities according to $DJ = 0$, $D\mathcal{F} = 0$, $\mathcal{F} = D\mathcal{A}$. We may begin with a local study for line bundles. We have then, looking at the YM part only,

$$d\mathcal{F} = 0$$
$$d * \mathcal{F} = J$$

We may call $\mathcal{F}$ the Faraday form and $\mathcal{M} = *\mathcal{F}$ the Maxwell form. The above equations have the important property that they are invariant under both the conformal group and the relevant gauge group. For Σ a 3-chain 'space", the first implies $\int_{\partial\Sigma} \mathcal{F} = 0$, which is equivalent to $\int_{\Omega_1} \mathcal{F} = \int_{\Omega_2} \mathcal{F}$ for homologous 2-chains Ω_1, Ω_2, implying conservation of the action

$$I = \int_{S^1}^{S^{1'}} \mathcal{F} = \int_{S^1}^{S^{1'}} tr\mathcal{F} = \{trK\}[\Omega_i] = \{c_1\}[\Omega_i]$$

for $S^{1'}, S^1$ bounding Ω_i.

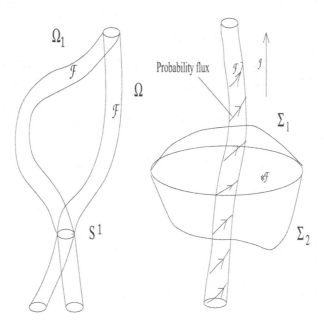

FIGURE 8. Flux sheets in Yang-Mills and associated spatial chains. The functoriality preserves the a value of the integral associated Ω_1 when it is deformed to Ω_2. Notice how the probability current evolves in a circular manner on the sheet.

132

The probability flux through a hypersurface Σ_1, $dim(\Sigma_1) = d - 1$, is

$$\Phi(\Sigma_1) = \int_{\partial\Sigma} *\mathcal{F}$$

and this is seen to be conserved, i.e

$$\Phi(\Sigma_1) = \Phi(\Sigma_2)$$

for homologous Σ_i. A "Hodge-Weyl" or "Chiral" transformation $\mathcal{F} \mapsto \mathcal{F}(\alpha), \alpha = Re(\alpha) + *Im(\alpha)$, $*^2 = -1$, is a combined rescaling and Hodge rotation, and has the property that it interchanges gradually the Faraday tensor for the Maxwell tensor. This means that current-less directions obtain current and conversely. Alternatively we can see this as a rigid rotation of the surface Σ in a direction codimensional to the current, making the current through the surface vanish at angles $\pi/2, \cdots$. The set of such global Hodge-Weyl rotations can be seen to form a Riemann surface due to complex structure of the target space identified with constant H-W rotations.

It is not uncommon that various objects can be related to the Yang-Mills because of various identities and cohomological matters. For example, if the flux sheet(tube) is on shell w.r.t. to an Einstein criterion, the Ricci form on complex space-times $\mathcal{F}$ equals the Kahler form on the sheet, i.e $\mathcal{F} = \lambda h_{\mu\nu} e^\mu \wedge e^\nu$ h a world sheet metric, λ a c-number, and if the entire space-time is on-shell ditto holds for TrR, R being the curvature form, which is associated to the case with a complexified $spin(n,m)$ Yang-Mills bundle.

7.4. **A Global Study.** Globally, properties of the target space, i.e space-time, have to be taken into account when looking at flux sheets in space-time. The flux tube Ω below would close, but it might encircle a obstruction of smooth topology of dimension p, for example a coulomb singularity. Thus $\int_\omega \mathcal{F} = c_1[\Omega]$ might be non-nill. If this is not even, this gives an obstruction to spin structure since $H_2(X, \mathbb{Z}_2)$ then has (non-trivial) generators. This can be taken into account to deform world tubes arbitrarily, by splitting the integral in a part that gives a contribution and one that does not. The same thing holds for the current J, the flux out of any closed 3-surface might not be nill, since $\mathcal{F}$ might for example not be smooth, something that is contained in violation of one of the above equations

$$\{d * \mathcal{F}\}[\Sigma] \neq \{*\mathcal{F}\}[\partial\Sigma]$$

7.4.1. *Interpreting the Chern Character and Class Physically.* We have $ch(\mathcal{F}) = Tre^{\mathcal{F}}$ in one convention. According to previous discussion, this is for the on-shell flux tubes (classical "string' sheet) the same thing as the Kahler form on the sheet modulo a constant. It might seem as we are confusing an antisymmetric field with a symmetric one, but note that we might very well abandon the convention of having symmetric metrics mathematically, with symplectic metrics as a perfect example, if we only remember the physical consequences. Interpreting the metric $h_{\mu\nu}$ as a classical background acting on first quantized fields we have then, taking ends as the "'cap" product

$$1 + \lambda h_{\mu\nu} e^{\mu} e^{\nu} + \frac{1}{2}(\lambda h_{\mu\nu} e^{\mu} e^{\nu})^2 + \cdots =$$

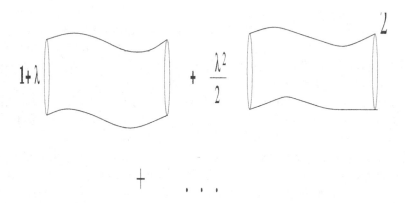

FIGURE 9. The pertubative series generated by h.

where Π^*_{1B}, B for background, should be evaluated according to the actual classical evolution, and could be interpreted as a classical correction to the evolution of a free particle.

We shall not pursue this further, due to inadequacies soon to be corrected, but note that we are essentially dealing with the S-matrix of a topological quantum field theory if we project to $\bigwedge T^*\Sigma$, Σ a compact closed even dimensional chain, and that the above also 'gives", taking an inductive step, that we can split manifolds locally as lower dimensional objects, or, physically put, that we can decompose spectra to describe fields of composite physical systems in a first quantized formalism. For example, the above says for 2-chains Σ_i generating dualities,

$$1 + \lambda < \Sigma_1, \Sigma^1 > + \frac{\lambda^2}{2} < \Sigma_1 \times \Sigma_2, \Sigma^1 \wedge \Sigma^2 > + \cdots$$

which can be put in the more familiar form

134

$$1 + \lambda < \Sigma_1, \Sigma^1 > + \frac{\lambda^2}{2} < \Sigma_1 \times \Sigma_2, \Sigma^1 \wedge \Sigma^2 > + \cdots = T e^{\int \lambda h}$$

T a time/event ordering putting $\Sigma_i \succ \Sigma_j$, $i > j$, where we define an arbitrary operator $\mathcal{O}_1$ to be latter to another $\mathcal{O}_2$ if it appears to the left of $\mathcal{O}_2$, i.e

$$\mathcal{O}_1 \cdots \mathcal{O}_2$$

gives $\mathcal{O}_1 \succ \mathcal{O}_2$. For non-closed chains the ends can be glued, and so not necessarily give something with no dynamics, although they do give a topological theory in the sense that the amplitudes are invariant under smooth deformations of the intermediate chain, since the ends can be variable. For example most of classical mechanics is governed by this, where we have conservative forces and potentials, i.e cocycles, governing the dynamics, so such a theory is not uninteresting for physical purposes since it gives a first approximate description of fields valid within classical limits. This correspondence that arises between partition functions/ S-matrices and topological functors when restricted to compact d-folds is called the topological functor correspondence, and can be used to guess the look of a partition function in some elementary cases, but more importantly it tells us that general field theory is intimately related to topology since we require similar invariances that link the two concepts.

Consider the spinorial Klein-Gordon lagrangean which belongs to such a Yang-Mills theory

$$\mathcal{L} = \bar{\psi}(i\slashed{D})^2 \psi = \bar{\psi}(\Box + \slashed{F})\psi$$

which is a "quantum" version of the classical lagrangean given by $\mathcal{L} = \lambda h_{\mu\nu} e^{\mu\nu}$, now "smeared out" codimensionally. This is realized by neglecting the kinetic term, hence obtaining

$$\mathcal{L}_{int} = \mathcal{L}_{\mathcal{F}} = \bar{\psi}\slashed{F}\,\psi$$

In the classical limit, just like

$$\int \bar{\psi}\slashed{A}\,\psi \mapsto \int \mathcal{A} = \int \phi dt \pm \int \bar{A} \cdot d\bar{l}$$

for quantum electrodynamics going over to classical electromagnetism (notice the different integration domains) this is

$$\int \bar{\psi}\slashed{F}\,\psi \mapsto \int \mathcal{F} = \int (G_{\mu\nu} + F_{\mu\nu})e^{\mu} \wedge e^{\nu}$$

135

So a "Pauli" term, a term well known to be pathological in field theory, carries heuristically over to our antisymmetric Polyakov term.

We have then a hint for interpreting the Chern character. Taking a graded version to link to the topological functor discussion we just had (We will be worrying about the "straight" Pauli version quite a lot later and so make up for this.), we have writing a unrenormalized path integral, with Ω the curvature of the total gauge group corresponding to the direct sum bundle $T^*X \oplus P(X,G)$ corresponding to factoring out the spin gauge group, G groups corresponding to K.K degrees of freedom, that for a fixed background connection ω in the standard model gauge group,

$$Z[\mathcal{F}] = TR[e^{-W[\mathcal{F}]}] = \int D\psi D\bar{\psi} e^{-\bar{\psi}(i\slashed{D})^2\psi}$$

Factoring out the G gauge degrees of freedom, which are 'dead" since we only consider a classical configuration in those fields, and introducing a deformation parameter β

$$Z[\mathcal{F}] = TR[\Gamma^5 e^{-\slashed{D}^2\beta}]$$
$$\sim TR[\Gamma^5 e^{-\slashed{D}_\omega^2\beta} e^{-\beta\mathcal{F}}]$$
$$\sim TR[\Gamma^5 e^{-\slashed{D}_\omega^2\beta}]e^{-\beta\mathcal{F}}$$
$$= SDET[\Box + \slashed{R}]e^{-\beta\mathcal{F}}$$

Where we neglected quadratic and worse Baker-Hausdorff-Campell terms in β. We can evaluate the remaining term involving the curvature of the tangent bundle; Following previous remarks and factorizing the partition function accordingly locally we can calculate it choosing the free field theory as a renormalization point for the PI. Expanding in the spectrum of the operator $\Box = D^*D = -D^\mu D_\mu$, the trick is to consider calculus on $S^1 \times S^1 \times \cdots \times S^d = T^d$ for d-folds. Let us use coordinates $x^i \in [0, L_i]$ on $T^d = \mathbb{E}/\Gamma$, Γ an appropriate lattice. Then $\phi_{n_1,\cdots,n_d} = sin(\frac{n_1\pi x_1}{L_1}) \cdots sin(\frac{n_d\pi x_d}{L_d})$ are eigenstates, with $\Box'_n = \xi'_n = \frac{\pi^2 n^2}{L_i^2}$ eigenvalues [2] in the spectrum of the free Laplacian reduced to the i:th degree of freedom. Hence, with minus for our case and plus for the usual fermionic case,

$$Z_N[\mathcal{F}]^{\pm 1} = \prod'(1 + \frac{\slashed{R}}{\Box'}) = \prod_{\mathcal{S}p(\Box)}{}'(1 + \frac{\slashed{R}}{\xi'_n}) = \prod_{1 \leq i \leq d}\prod_{n_i}{}'(1 + \frac{\lambda_i L_i^2}{\pi^2 n_i^2})$$

---

[2]Prime ' denotes eigenvalue here, but denotes a little bit later Eisenstein product.

Putting $z_i = -i\frac{\sqrt{\lambda_i}L_i}{\pi}$, this is, using the usual Euler product formula $\prod_{\mathbb{Z}_+}(1 - z^2/n^2) = \frac{sin(\pi z)}{\pi z}$,

$$Z_N[\mathcal{F}]^{\pm 1} = \prod_{1 \leq i \leq d} \prod_{n_i}{}' (1 - z_i^2/n_i^2) = \prod_{1 \leq i \leq d} \frac{sin(\pi z_i)}{\pi z_i} = \prod_{1 \leq i \leq d} \frac{sinh(L_i\sqrt{\lambda_i})}{L_i\sqrt{\lambda_i}}$$

Using a holomorphic factorization of the complexified tangent bundle TX, $TX = T^+X \oplus T^-X$, we have $\sqrt{\lambda_i} = \lambda_i^{\pm}$ corresponding eigenvalues corresponding to the holomorphic factors, which are like 1-dimensional spinors. Hence setting $x_i := 2\lambda_i^{\pm}L_i$ the corresponding dimensionless factors of the holomorphic curvature,

$$Z_N = \prod_{1 \leq i \leq d} \frac{\frac{x_i}{2}}{sinh(\frac{x_i}{2})}$$

which we recognize as the $\hat{A}$-genus on TX^+. The partition function is thus

$$Z_N = \hat{A}(T^+X)Ch(P(X,G))$$

which we know to give the index of the Dirac operator on compact orientable smooth even dimensional manifolds. We could prove that this truly gives the analytical index of the Dirac operator, but instead we move on as this is done in many places elsewhere.

Thus, as far as the Chern character concerns, we can view it as a classical contribution to the partition function with a topological interpretation. Similarly, other aspects of the Yang-Mills field have other topological uses, which we partly review in later chapters when discussing Donaldson theory. Before moving on we remark a last thing on connections between partition functions, physics, and topology. We know that

$$c = det(1 + K)$$

$K = \frac{i\mathcal{F}}{2\pi}$. This understood, it is not hard to interpret the determinant as the amplitude for something at all to happen when generated by the interaction $\frac{i\mathcal{F}}{2\pi} = \frac{\delta\mathcal{O}}{\mathcal{O}_B}$, $\delta\mathcal{O}$ a perturbation over a background operator $\mathcal{O}_B$. Since we work with closed surfaces when looking at the compact scenario we have that the above is a sum of vacuum amplitudes, i.e p-dimensional vacuum bubbles[3] in the target space d-fold. The same thing goes for the Chern character, indeed for any cocycle we equate to a cycle we get a classical vacuum bubble (this is the cycle criterion) amplitude having the correct invariance properties(this is equivalent to the cocycle criterion) under small transversal deformations. For

---

[3]Corresponding to classical p-branes. Notice the convention of p being the dimension of a p-brane bounding two spatial branes of dimension $p - 1$.

all of the functors and partition functions discussed above the field strength, which is a two-form and hence associated to 2-surfaces, is fundamental. Hence it would not be unnatural to associate the usual non-super partition function to two-dimensional objects too, although we traditionally do not do so.

7.5. Summary.

- According to the above, 2-dimensional terms and 2-admissible lagrangeans are heuristically related to Pauli terms and the spinorial Klein-Gordon equation for Yang-Mills.
- Yang-Mills theory and smooth topology are intimately related, and certain topological functors can be interpreted as vacuum amplitudes in field theories connected to a Pauli version of Dirac-Yang-Mills.

7.6. **Einstein Theory.** In this section we review briefly Einstein theory, and mention 'the" two canonical ways of deriving the equations of motion the theory briefly. After that we focus on a third way, that makes it possible to link Einstein theory to previous discussion about 2-admissibility, Pauli terms and Yang-Mills.

From Hilbert's point of view, Einstein theory is the theory governed by

(1) The action

$$S = \frac{-1}{16\pi G} \int_X *R + \int_X \mathcal{L}_{int}$$

with

(2) extremality of

$$S = \int_\gamma \left\|\frac{d}{d\lambda}\right\|^2 = \int_\gamma X^\mu X_\nu = \int_\gamma \partial_\lambda x^\mu \partial_\lambda x_\nu$$

relating the metric to the connection as an on-shell condition.

It may not be clear right now how the connection is related to the last equation, however, so let us discuss it first. Explicit variation of the latter gives $\nabla_X X = 0$ in the form

$$-\ddot{x}^\mu = \Gamma^\mu_{\sigma\rho} \dot{x}^\sigma \dot{x}^\rho$$

with Γ the connection components in the coordinate frame belonging to a Levi-Cevita connection. Despite that emphasis is usually put on the former action and less on the latter, the latter is equally important in determining the physics of gravity. Just like Yang-Mills theory is determined by two fields and two equations, with one of these equations which could be envisaged as lowering the number of fields by introducing relations among them, e.g.

$$\slashed{D}\,\psi = 0$$

which is obviously a statement of covariance, we have in Einstein theory

$$DX = 0$$

, the latter also a statement of covariance. Indeed, put in a way that most clearly displays this similarity, namely Cartan's first structure equation with $\slashed{D}$ the induced image of the usual vector space isomorphism $\bigwedge T^*X \to Cl(X)$ we have

140

$$\not{D}\,\theta = 0$$

θ the ON vielbein. Before going back to the first action, it's time to mention one thing. Mathematically, we chose an 'energy" functional above, related to the Klein-Gordon functional, not to confound the reader with the just slightly more technical 'length" functional, which would have corresponded to a Dirac functional. The extremals of the Dirac functional are among the extremals of the Klein-Gordon, although the converse does not hold, so we came out fine anyway. Notice that it would have been the length functional that we would have chosen if we would have done 1-admissible quantum theory and wanted 'quantum' lagrangeans that we can exponentiate. The two above lagrangeans are classical lagrangeans however, so they do not need generate anything, and hence escape the argument.

Let's go back to the first action. Following to the variational identities listed in the appendix variation of the lagrangean gives, with h the usual space-time metric,

$$0 = \delta(\frac{-1}{16\pi G}R\sqrt{h}+\mathcal{L}_{int}) = (\frac{\overline{Ric}^{\mu\nu}}{16\pi G}\sqrt{h}+\frac{\delta\mathcal{L}_{int}}{\delta h_{\mu\nu}})\delta h_{\mu\nu} = (\frac{\overline{Ric}^{\mu\nu}}{16\pi G}+\frac{1}{\sqrt{h}}\frac{\delta\mathcal{L}_{int}}{\delta h_{\mu\nu}})\sqrt{h}\delta h_{\mu\nu}$$

Setting $T^{\mu\nu} := \frac{2}{\sqrt{h}}\frac{\delta\mathcal{L}_{int}}{\delta h_{\mu\nu}}$ this is

$$\overline{Ric}_{\mu\nu} = 8\pi G T_{\mu\nu}$$

with $G := \overline{Ric}$ called the Einstein tensor. This is the usual Einstein field equation. Hence our two equations are, with unit gravitational constant,

$$G = 8\pi T$$
$$\nabla\theta = 0$$

the latter the usual first structure equation for vanishing torsion, where we used ∇ for a torsion-less compatible connection. This from Hilbert's point of view, an essentially lagrangean point of view.

From Einstein's point of view, we know that the stress-energy will be covariantly conserved, i.e $\nabla^* T = 0$, and that we want to equate it to some other 2-tensor of a geometrical nature. It can be shown that also G satisfies covariant conservation by contraction of the second Bianchi identity

$$\mathfrak{S}(\nabla_X R)(Y,Z) = 0$$

$\mathfrak{S}$ denoting symmetrizer in X,Y,Z. Hence Einstein equates this to T modulo a constant λ and then fixes the constant via a calculation of the physical outcome, e.g Newtonian limit, which gives him $\lambda = 8\pi G$.

141

The compatibility condition from the connection then comes from the equivalence principle, hence

$$G = 8\pi T$$
$$\nabla \theta = 0$$

We remark that Einsteins reasoning is essentially equivalent to a statement of cohomology of the operator ∇^* and the dimension of it's kernel. In a similar spirit it can be shown that, at least locally, G is the only covariantly conserved symmetric tensor that can be constructed out of the metric of second order in derivatives.

Let us derive the above equations in various reformulations to understand them better. The first way we would like to reformulate the Einstein field equation is in the moment of 4-rotation way. Consider $D\theta = \Theta = \mathcal{T}$, $\Theta = \mathcal{T}$ the torsion form. Θ has an interpretation as the twisting of a vector along parallel displacements(hence the name) so we interpret it as something reminiscent of angular momentum L in a d-dimensional version, $D\theta = L$. To obtain the 'torque" M, we take the covariant derivative, then getting

$$M = DL = D(D\theta) = (D^2)\theta = R \wedge \theta$$

$R = R^\alpha_{ab} T_\alpha e^a \wedge e^b$, T_α the appropriate Lie algebra generators of the structure group of the frame bundle on our d-fold. We know that the stress-energy is a vector valued one-form, since

$$P_a = \int_\sigma T_{ab} * e^b = \int \overline{T} \cdot d\overline{\Sigma}$$

gives the momentum flux through the hypersurface Σ, this more or less by the definition of T. To get a 1-form out of a 3-form in 4 dimensions we take the Hodge dual, thus $*R \wedge \theta$ is an appropriate vector valued one-form and we only need check that this is covariantly conserved to be able to equate them. Stating conservation in terms of ∇ instead of it's Hodge dual, we have $\nabla * T = 0$. Similarly

$$\nabla R \wedge \theta = (\nabla R) \wedge \theta + R \wedge \nabla \theta = 0$$

in view of the Bianchi identity $\nabla R = 0$ and the first structure equation $\nabla \theta = 0$ (We are assuming ∇ to be torsionless).

Hence we identify

$$R \wedge \theta = *T$$

(Again we do not bother with signs and constants). Replacing $R_{ab} e^a e^b \mapsto R_{ab} e^a e^{b*}$, i.e taking a "trace" (contraction) among two indices, this is $\not{R} \theta = (*T)_{contracted\ indices}$, where $\not{R} = \frac{R_{ab}}{2} \gamma^a \gamma^b$, $\gamma^a = e^a + e^{a*}$. But now we recognize instantly that since $\not\nabla \theta = (\nabla + \nabla^*)\theta = 0$ we have

142

$$0 = \nabla^2\theta = \nabla_a\nabla_b\left[\frac{\{\gamma^a,\gamma^b\}}{2} + \frac{[\gamma^a,\gamma^b]}{2}\right] = (\nabla^*\nabla + \not{R})\theta$$

so we have our candidate for a very "natural" stress-energy T, setting $T = \nabla^*\nabla\theta$, where we as usual are scratching constants all over the place. The important thing is that

$$(\nabla^*\nabla + \not{R})\theta = 0$$

, hence $\nabla^*\nabla\theta = \not{R}\,\theta$, is as close as can be to the spinorial Klein-Gordon equation. We can now see that we can find some interpretation of $\not{R}$ too, since $\nabla^*\nabla = \nabla^\mu\nabla_\mu$ in flat space-time. Hence $\not{R} = m^2$ up to a c-number constant. By the same token we now get a interpretation of the stress-energy, $\nabla^2\theta = R \wedge \theta = *T$ gives $*T$ to be the "moment" of rotation in the group $SO(1,3)$, as something reminiscent of a combination of force and torque, interchangeably mixed by the choice of coordinate system. So we could formulate the above as a "Newton-Einstein" equation

$$\underbrace{\nabla^2\theta}_{\sim m\ddot{x}} = \underbrace{*T}_{\sim F}$$

with $\sim$ denoting analogy.

We notice another couple of reformulations, if, following Misner, Thorne, Wheeler, we let ★ denote Hodge duality on the exterior tangent of our pseudo-riemannian acting only on the vector index, we have

$$\bigstar R \wedge \theta = e_\sigma \otimes (*R*)^{\sigma\nu\kappa}_\nu d\Sigma_\kappa$$

where $(*R*)^{\sigma\nu\kappa}_\nu = G^{\sigma\kappa}$ is well known. So the MTW construction is the same thing, except that they fancy the exterior tangent, and they also like to interpret the R acting on the θ as the moment of the gravitational field.

Let us derive the above, heuristically, from a naive first quantized 'quantum' point of view instead. We start with the Klein-Gordon equation $(\partial^\mu\partial_\mu + m^2)u = 0$, u a real scalar. Tensoring u with $SO(1,3)$ module in the defining representation to get a $SO(1,3)$ theory we get for such basis covectors e^σ that

$$(\partial^\mu\partial_\mu + m^2)u \otimes e^\sigma = 0$$

Taking a vector valued array of these equations and writing $[ue^\mu] := \theta$ we thus have

$$(\Box + m^2)\theta = 0$$

Interpreting the θ as a spinor of a $SO(1,3)$ theory we want to identify the square mass with some geometrical quantity. Letting $\mathcal{F}$ be the gauge field strength of a massless theory we get from $\nabla\theta = 0$ the

143

relation $(\Box + \mathscr{F})\theta = 0$, hence $m^2 = \mathscr{F}$ by identification. So $(\Box + \mathscr{F})\theta = 0$, i.e $-\Box\theta = \mathscr{F}\theta$. Taking the part proportional to $\bigwedge^3 T^*X$ this is

$$-\Box\theta = \mathcal{F}\theta$$

Usually, we want to split up the curvature as a perturbation over the curvature in one-particle space(mass), where we look upon the former as a variable curvature/mass induced by interaction with other particles, i.e $R = m^2 + R_B + R_{int}$, $R_B{}^4$ the multiparticle space-time background curvature, R_{int} being interaction curvature, which would then include pure gauge field strength terms in other gauge groups than $SO(1,3)$ (spin(1,3) when we totally first quantize) like various terms associated to usual Yang-Mills theories. The above split could for natural reasons be called a Zakhorov partition of the curvature, as A. Zakhorov at an early stage of physics suggested that renormalization schemes may produce space-time curvature, and a mass change effectuated by renormalization by the above precisely produces curvature, thus having some similarity to Zakhorovs idea. Later we will learn a relation to AdS spaces(spaces of constant negative curvature) for mass states at infinity, and then there will be a minus sign in front of the mass to account for a necessary double Wick rotation. In our conventions, this corresponds to going to the MTW conventions for the signature, so if one is already in those conventions one need not do the continuation. This has the interpretation that the way we handle the signature is not arbitrary, and the "Why?" of this will, hopefully, become clear later.

Hence we can sum up the content of the Einstein field equation as

$$P^2\theta = M^2\theta$$

$P = i\nabla$ a covariant momentum operator, $-M^2$ the Riemannian curvature.

A good rounding way of rounding off this section would be to rigorously prove the stated link between the above Klein-Gordon equation and the Einstein field equation. This is done in the same appendix as the variational identities used to derive the equations in the Hilbert-Palatini formalism are stated, since it breaks a little bit away from the style otherwise used in the physical part of this part of this thesis. It is anyway deeply recommended that the reader makes himself aquainted with that section straight away to be able to follow the line of reasoning used in the following.

---

[4] Still omitting numerical constants.

7.7. Summary.

- We investigated the Einstein field equation, and came to the heuristic conclusion that it is related to a Klein-Gordon equation on space-time vielbeins. We promised that this is proved in section 8.1.
- The Hilbert-Einstein action was identified with a "Pauli" term in this KG equation.

8. Quantum Theory

According to the appendix touching aspects of classical gravity we have that classical gravity is supposedly described by an action only corresponding to a variationally correct, but inadmissible, action

$$S = \int d^D x \mathcal{L}_{CLASS.} = \int d^D x < \theta, \rlap{\,/}D^2 \theta > + |\Omega|^2$$

θ being the appropriate ON-frame, Ω the curvature of the connection D. In the above and the following the fundamental fields are always the connection and the vielbein θ. A split-up of the first term reveals the Hilbert-Einstein term as follows;

$$< \theta, \rlap{\,/}D^2 \theta > = < \theta, \Box + \rlap{\,/}\Omega\; \theta > = < \theta, \Box \theta > + < \theta, \rlap{\,/}\Omega\; \theta > = < \theta, \Box \theta > + R$$

By the same remarks we could, imagining in an ad hoc manner θ to be a matter field, so that we can apply the reasoning of the section "The Heart of This Thesis", "cure" this action and make it's matter part 1-admissible by the substitution $\rlap{\,/}D^2 \mapsto \rlap{\,/}D$, i.e obtaining a gravitational version of a Yang-Mills lagrangean

$$\mathcal{L}_{CLASS.} = \int < \theta, \rlap{\,/}D\; \theta > + |\Omega|^2 = \int < \theta, \rlap{\,/}D\; \theta > - \tfrac{1}{4} \Omega^\alpha_{\mu\nu} \Omega^{\mu\nu}_\alpha$$

The classical gravity above, which realises the field equation as an identity implied by $\rlap{\,/}D\, \theta = 0$, rather than a classical equation derived by variation, has a conserved current

$$\mathcal{J}^\mu = < \theta, \Gamma^\mu \theta >$$

, to be integrated over an appropriate $(d-1)$-surface, satisfying

$$D_\mu J^\mu = 0$$

and cannot be interpreted as a probability current although it is the thing that comes closest to this in this formulation. Rather, interpret as suggested by the Dirac current a space-time vector X^a as such a probability current(also called probability vector) and a vielbein $\theta = [e^a_\mu]$ as a change of frame on such probability vectors. It is then understood that a probability to observe an event is given by the Lorentz covariant expression

$$X^a_{\mathcal{O}} X_{\mathcal{O}'a}$$

with $X_{\mathcal{O}}^{\mu}$ the probability current of the observer $\mathcal{O}$ and $X_{\mathcal{O}'}$ the probability current of the relevant event $\mathcal{O}'$. As an example, consider the lab frame case when $X_{\mathcal{O}} = (1, 0, 0, 0)$ [5] Then the above reads

$$\int d^{D-1}x \bar{\psi}(x) \Gamma^0 \psi(x) = \int \bar{\psi} \Gamma^0 \psi = \int \psi^*(x) \psi(x)$$

which is the frame canonically used in the Born interpretation of quantum mechanics and also the usual "Born" expression. For non-Dirac fields there is a similar general expression that will be reviewed later and it too has a hypercomplex nature, albeit on a bigger space.

One could now directly quantize the classical gravitational action above to obtain a (disregarding the influence of the volume term in the lagrangean) finite theory and make a continuation of space-time coordinates to attempt a SUSY version if we wished to include supersymmetry. Let us list the most basic ingredients of such a theory[6] before we move on

We have, e.g , mass off-shell field operators

$$\theta^a = e^a(x) = \int \frac{d^D k}{(2\pi)^D} e^{-ikx} a_e^*(k) \epsilon_e^a(k),$$
$$\Gamma^a = e^a + e^{*a},$$
$$\omega_a(x) = \int \frac{d^D k}{(2\pi)^D} e^{-ikx} a_\omega^*(k) \epsilon_{a,\omega}(k) + e^{+ikx} a_\omega(k) \epsilon_{a,\omega}^*(k)$$

, with ladder operators $a_e(k)$ fermionic, $a_\omega(k)$ bosonic, and the various ϵ's with various subscripts being polarization vectors. It should be understood that $\theta^a \mapsto \Gamma^a$ in the operator formalism for the θ term of the above lagrangeans, and that in the above we considered classical lagrangeans. The zero level level of the above fields must be shifted so that they vanish at infinity, so that they are covered by the Riemann-Lebesgue lemma of Fourier analysis, hence naïvely $e_\mu^a := \delta_\mu^a + e_\mu^a$, where the latter goes to zero at infinity. Mass on-shell reduction is included by inserting a delta function under the integral sign of the type $2\pi\delta(k^2 - m^2)$ to yield the usual expressions

$$\Gamma^a = \int \frac{d^{D-1}k}{2E(k)(2\pi)^{D-1}} e^{-ikx} a_e^*(k) \epsilon^a(k) + e^{+ikx} a_e(k) \epsilon^{*a}(k),$$
$$\omega_a = \int \frac{d^{D-1}k}{2E(k)(2\pi)^{D-1}} e^{-ikx} a_\omega^*(k) \epsilon_{a,\omega}(k) + e^{+ikx} a_\omega(k) \epsilon_{a,\omega}^*(k)$$

and it should be understood that such a reduction is only advisable for considerations of the S-matrix at infinity and perturbation theory of

---

[5]By convention and probability interpretation it is good to normalize X to unity.

[6]I conjecture this theory to be finite to all orders, but have not had the time to do confirming computations and checks, and rather relying on a seeming isomorphism to Yang-Mills theory in this statement. Would be very interested if someone could take upon him/her the task of doing the checks or knows anything about it.

this kind, where there are well defined mass states at infinity. Instead it is usually better to regard a continuum of such mass states (together with a possible discrete part) implicitly by simply looking the variety of fields over a specfic D-fold for S-matrices at finite time. One thing before jumping, though. When doing calculations in the above gravity the reader will notice that to get correct dynamics S/he will have to join particles with positive mass and frequency with vertices of the opposite signs. This is hard to explain in that theory from the point of view above but has to do with that the charge actually stems from the propagation of the intermediate boson corresponding to non-compact dimensions, as opposed to compact ones, with the signs multiplying the mass charges on each side of the propagator being opposite, as the boson propagation looks reversed in the noncompact dimension when looking at the other end from the point of view of the other fermionic particle. In a sense this states that the fermions only react with the antifermions, and in order to get a fermion-fermion reaction one has to reverse the momentum charge of one of the fermions, which then reflects in the vertex, mass being timelike momentum in a co-moving frame of a particle at infinity[7] and the charge in the vertex being this momentum. Later we will learn that this is a consequence of a double Wick rotation in our conventions.

Instead, we abandon the above completely—for now— to be able to have a probability interpretation for Dirac fields, which are then also supposed to be decribed by the above circle of ideas, then especially having the Einstein condition is No-condition version of gravity in mind as in appendix 8.1.3.

So here is the point where a "square root" of space-time, i.e spinors, and hypergeometry come into play and that we get the link with the previous part of this thesis, which dealt with hypermathematics. The part of such hypermathematics that we will be dealing with is well known, namely spin geometry and spin bundles, see the excellent book by Lawson and Michelson, Spin Geometry. But we have not reached the link to string fields yet, nor the creation and annihilation of worlds in such a formalism.

8.1. Summary.

---

[7]In a general enough field theory there are as many parity reflections as space-time coordinates, and in particular we can choose to reflect only non-compact dimensions or only Kaluza-Klein dimensions.

- Classical Einstein theory, which we came to the conclusion that it was decribed by a Pauli term on vielbeins in a vielbein Klein-Gordon term, was substituted for a spinorial Klein-Gordon theory.
- This spinorial Klein-Gordon theory has then solutions generated by a spinorial Dirac-Yang-Mills lagrangean.

8.2. **The Link To The Former Part of This Thesis.** To cure the final defect outlined in previous sections we must, so that we can look at amplitudes instead of probabilities, return to the $spin(n, m)$ version of the gauge theory above. Since we do not want to go to the cumbersome ordeal of working with bundles over spinor space, i.e homogenous twistor space, partly because of the difficulty of transforming back to space-time and finding the aprropriate isomorphisms of (non-exact) cohomological theories in the general case we shall simply have it understood that this is the nature of the base, i.e as a grassmannian of 2-planes, algebraically corresponding to superpositions of background states of two different spins. A deeper study would force us to leave the space-time scenario completely, with grave reprecussions for the possible physical interpretation of such a theory as it is not easy to transform back. That is, since we are in effect dealing with functions $f : Cath \mapsto Cath$ taking values in spaces looking locally as

$$Cath := \bigoplus \bigotimes \mathcal{H}_i$$

and considering their geometry, and we do not know which are the generating spaces, consider for example possible structures lying below spinorial structures such as color, we are forced to take space-time as a background and base for our bundles despite that we know better. With maps in the global situation, just to mention the twistorial case, severe constraints coming from the selfduality of the Weyl tensor to achieve integrability of such bundles(see Atiyah, Manin, Drinfelt, and Hitchin in the references) being such a criterion would inflict massive troubles, not to mention the diabolically laborious explicit formulae for the transforms in even the most simple cases. Thus we can only realistically expect to trade away the fiber and not the base at this point.

Doing this, gives us the usual Yang-Mills theory over space-time with only $spin(n, m)$ being considered for an S-matrix at infinity, and $\tilde{G}L(n+m, \mathbb{C})$ acting on the fiber for the finite case, $GL(n+m, \mathbb{C})$ being the action on space-time covectors and $\tilde{G}L(n+m, \mathbb{C})$ denoting it's lift, then understanding that the $spin(n, m)$ action induced by contravariant functors lifted from space-time diffeomorphisms will correspond to the action that preserves the notion of spin of such space-times, much like the action of the biholomorphism group in the complex case with S^1 as space (see below).

Let's return to the $spin(n, m)$ action and make the mathematical aside that links to the former part of this thesis. The diffeomorphism group has a lift to a contravariant functor on T^*X and hence on the

150

Clifford algebra $Cl(X)$ generated by $e^{a*} \pm e^a \wedge$, the sign corresponding to a choice between two different representations and a sign in the relations defining the Clifford algebra, conveniently taken care of by setting $\Gamma_+^a = e^a + e^{*a}$, $\Gamma_-^a = i(e^a - e^{a*})$, both satisfying

$$[\Gamma^a, \Gamma^b] = 2\eta^{ab}$$

Let us work with the latter, $\Gamma^a := i(e^a - e^{a*})$, for now, and have it understood that we we work with complexifications of space-time. If we work in the ON-frame we have a reduction of the structure [frame] bundle F to an $SO(n,m)$ principal bundle. We would like to define a "hyper" analogue of holomorphic transformations, akin to local conformal transformations but in higher dimensions. Let us try to spot this explicitly. If we have a diffeomorphism $f : X \to Y$ we have for x a coordinate system on X, $x^i : x^{i-1}(\mathcal{U}^i) \to \mathcal{U}^i \subset X$, $U\mathcal{U}^i \supset X$ an open cover of X, and y similarly a coordinate system on Y, that

$$(y^{-1}fx)_* = \frac{\partial f}{\partial x} = \Lambda$$

will act on the frame bundle. We may, following Kobayashi's ideas[8], define structural equivalence in terms of such compatibility, and so we define hyperholomorphism by saying that it is a diffeomorphism preserving the $SO(n,m)$ structure. We can easily see what this corresponds to in the case of the complex plane with holomorphisms f, there $SO(2) \cong U(1)$ is the reduced frame bundle, generated by

$$T_1 = \begin{pmatrix} i & 0 \\ 0 & -i \end{pmatrix}$$

and satisfying

$$\Lambda T_1 \Lambda^{-1} = \Lambda \begin{pmatrix} i & 0 \\ 0 & -i \end{pmatrix} \Lambda^{-1} = \begin{pmatrix} \frac{\partial f}{\partial z} & 0 \\ 0 & \frac{\partial f}{\partial z} \end{pmatrix} \begin{pmatrix} i & 0 \\ 0 & -i \end{pmatrix} \begin{pmatrix} \frac{\partial f}{\partial z} & 0 \\ 0 & \frac{\partial f}{\partial z} \end{pmatrix}^{-1} = T_1$$

so the $U(1)$ structure is preserved. In terms of strings, this is like the Virasoro algebra corresponding to holomorphic vector fields, preserving the action of an $SO(2)$ on a S^1 in the plane, i.e preserving spin. For the case of space-time we might take $SO(4)$ after Wick rotation, or, going over to hypercomplex manifold look at rotations, and so get that we must preserve $SU(2) \oplus \overline{SU(2)}$ or $SL(2,\mathbb{C}) \oplus \overline{SL(2,\mathbb{C})}$ after complexification when acting with the gauge algebra. Not only do we wish

---

[8]S.Kobayashi, Transformation Groups in Differential Geometry, treats the subject of automorphisms of differential geometric objects subject to various constraints.

to have automorphisms of space-time as a base to our Yang-Mills bundles to have and induce this property, but we also wish the transition functions to respect this, just like we when dealing with holomorphic transition function in holomorphic string theory for a helicity preserving reason. Hence we have the notion of a spin structure, and we must have transition functions that respect $spin(n,m) \oplus \overline{spin(n,m)}$, to similarly preserve spin. A manifold is 'patchable' in terms of spin structure when the second Stiefel-Whitney class w_2, defined to be an element of the Cech cohomology group $H^2(X, \mathbb{Z}_2)$, is vanishing. We may, for obvious reasons, want to check that such bihyperholomorphisms, let us call them spin holomorphisms, satisfy things that we associate to conformal maps. The conformal property, i.e that a metric induced by such a map is only a local conformal rescaling, is easily checked by using the equivalent criterion $\slashed{\psi}\,' = \Lambda\slashed{\psi}$, $\Lambda \in \mathbb{C} \otimes SO(n,m)$ with $v := \slashed{\psi}$, $\Lambda = c\Lambda_0$, $\Lambda_0 \in \Gamma(X_\pm, SO(n,m))$, $c \in \Gamma(X_\pm, \mathbb{C})$;

$$\overline{v'}_1 v'_2 = \overline{\Lambda v_1}\Lambda v_2 = \overline{v}_1 \Lambda^* \Lambda v_2 = \overline{v}_1 \bar{c}\Lambda_0^{-1}c\Lambda_0 v_2 = |c|^2 \overline{v}_1 v_2$$

Spin hypergeometries, or simply spin geometries, are thus the good thing to have as a base to our Yang-Mills bundles, them being generalized conformal manifolds of the correct kind. They preserve spin, give conformal manifolds, and are quite common among the 4-folds. The structure group consists of scalings times the appropriate special orthogonal group.[9] For example, for non-compact space-times of Lorentzian signature a theorem by Geroch states that a spin structure exists if and only if a global frame does. We say space-times and bundles are spin equivalent when there is a spin structure preserving diffeomorphism of the base together with an induced bundle morphism of a spin bundle. Among these hypergeometries, we have a very special kind of manifolds, the hyperkahler manifolds-these correspond to the mass-less states, and are supposed to be conformally equivalent to Ricci flat space-times, i.e flat space-times. Then we have hyperbolic manifolds, corresponding to positive mass squared, i.e by definition positive mass, and elliptic manifolds, corresponding to unstable states with negative mass squared and imaginary mass and momenta.

---

[9]Nevertheless we can very well generalize further by admitting ramified sections and associated bundles so that horizontal lifts induce non-trivial monodromy of base loops, much like the action of the group of permutations on the zeros of the polynomial

$$P(z, w_0) = z^5 + f_4(z, w_o)z^4 + \cdots + f_0(z, w_o) = 0$$

for deformations in w-space loops in a great enough neighbourhood of w_0 in $\mathbb{C}$, f_i holomorphic, would interchange the zeros $\{z_{0,i}\}$ this would interchange solution sections among local spin structures.

8.3. Summary.

- To be able to describe the quantum dynamics of space-times we resort to spin geometry and have different physical interpretations for the $spin(n, m)$ action on such a bundle module and the complement in $\widetilde{GL}(n+m, \mathbb{C})$, which is the lift of $GL(n+m, \mathbb{C})$ acting on space-time covectors. Hypermathematics, as reviewed in Part I, is thus supposed to cover this appropriate generalization of the holomorphic calculus in string theory.

8.4. **Formulation of a Supposed Theory.** Gravity, in the above approximation, which admittedly has it's limitations(but that can scarcely be circumvented within present day mathematics) is thus simply $spin(n,m) - \tilde{GL}(n+m,\mathbb{C})$[10] Yang-Mills theory, governed by the usual action

$$\mathcal{L} = \bar{\psi} i \slashed{D}\ \psi - \frac{1}{4}\Omega_{\mu\nu}\Omega^{\mu\nu}$$

A $spin(n,m) - \tilde{GL}(n+m,\mathbb{C})$ connection ω and a Dirac field ψ being the fundamental fields, then supressing obvious features such as ghosts and some constants. The "Einstein field equation" of this quantum theory then stems from $\slashed{D}\ \psi = 0$ as the spinorial Klein-Gordon equation

$$(\Box + \slashed{\Omega}\ )\psi = 0$$

which also bears the name the Einstein-Yang-Mills equation, this because of previous calculations done in background space-times. And what a divine coincidence that it should so rightfully claim this name also in this second sense, as a true "Einstein field equation" of a square root of space-time, see section 8.1 and the Einstein theory section.

Thus the only thing to a low order approximative calculation in the above theory is taking into account a sign, substituting for the correct coupling and finding the correct group theory factor, and the rest of an invariant element or S-matrix element can be 'ripped off" previous Yang-Mills calculations. Also, just to clarify, my statement is not that one can use background spin connections to include the effect of a background field, but rather that Yang-Mills theory in some extent already is a theory of gravity.

In the above we used Ω, a notation usually reserved for a general connection 2-form in this thesis, to emphasize that we can equally well have connections that run non-diagonally over space-times with different colors and electroweak quantum numbers. It is important to realize that the above gravity is probably but a low energy statement, much like we today percieve the theories governing the three other interactions, and that it does not make any pretense at being any more general than each one of these other low energy theories.

We usually have it understood that we take the SYM or NC SYM versions, this to be able to compare with supersymmetric theories such as string theories and noncommutative theories.

---

[10]Again this depends on whether one is concidering S-matrices at infinity or not, and if we are working at tree level or not, because then we can use classical theory in some extent. One simply does what one can handle in the various situations.

154

9. Getting Down To Business

In this section we shall try to probe our idea, checking if it is realistic. One of the first things we must do is to obtain various factors that we have previously neglected. The first few calculations will be very primitive, for often we do not need go into complicated matters to prove something to be wrong, and if there is something fishy about it we might as well know about this as soon as possible instead of waisting time on something incorrect.

9.1. **Problem 1: Basic Considerations.**

We have a supposedly 2-admissible lagrangean, that is supposed to be variationally correct as follows;

$$\mathcal{L} = \bar{\psi} \slashed{D}^2 \psi$$

Upon variation we have, setting $\Box = D^*D$ that

$$(\Box + \slashed{R})\psi = (\Box + \frac{R}{4})\psi = 0 \Leftrightarrow \underbrace{-(-\frac{1}{\sqrt{h}}\partial_\mu h^{\mu\nu}\sqrt{h}\partial_\nu)}_{=-p^2}\psi = \frac{R}{4}\psi$$

In the last equivalence a intermediate double Wick was performed. We recognize 1) the latter term as stemming from the spinorial Hilbert-Einstein term $\bar{\psi}\slashed{R}\,\psi$, and 2) that we can relate a $spin(n,m)$ connection to a ON-frame connection on space-time canonically, and that R above means the space-time Ricci curvature and set p to be the momentum operator. On a momentum eigenstate this is

$$m^2\psi = p^2\psi = -\frac{R}{4}\psi$$

hence

$$m^2 = -\frac{R}{4}$$

Denoting a variable mass m as M we have $M(x)^2 = -\frac{R(x)}{4}$. Now we note that the eigenvalues of $-\nabla^a\nabla_a = P^2$ correspond to constant curvature configurations on d-folds and that any metric induced by $\theta := \psi^{11}$ a is conformally equivalent to a symmetric manifold, which has $R(x) = constant$, so we have a natural choice for a gauge slice, which we call a conformal gauge slice. Hence we could write the space-time curvature at infinity as a term of negative curvature, i.e the free mass term, which would then give the asymptotic curvature of the one-particle space-time associated with a certain particle of mass m. To be able to work with the same space-time background for all particles

---

[11]This is admittedly abusive, but stems from us regarding the spinors as vielbeins for the metric on the fiber.

involved in a process one instead shifts the curvature in the one particle space-times to make them asymptotically flat by simply calling the one-particle space-time curvatures mass, i.e for i indexing the various particles

$$M_i^2 = -\frac{R_i}{4} = -\frac{R_{B,\,i} + R_{int,i} + R_{0,\,i}}{4} = -\frac{R_{B,\,i}}{4} - \frac{R_{int,\,i}}{4} - \underbrace{\frac{R_0}{4}}_{:=m_0^2} = -\frac{R_{B,\,i} + R_{int,i}}{4} + m_0^2$$

where it is understood that m_0^2 is the bare mass squared at infinity(we always take the free field case at infinity as a normalization). Having noticed that our one-particle space-times correspond to AdS spaces in the above, and that we can now use copies of a flat space at infinity to modell an entire collection of massive particles by simply having the curvature of the various one-particle space-times buffered up in the masses of the individual particles we now go on[12].

Let us study the "vielbeins" $e^a = \theta^a$, $[e^a] = \theta$. We have, for a any choice of gauge connection ω_0 the equation $D\theta = \Theta$, $\Theta = e_a \otimes \Theta^a_{bc} e^b \wedge e^b$ the torsion form. If we <u>choose</u> such a background connection that satisfies $\not{D}\,\psi = 0$ by the super/hyper correspondence induced by the vector space isomorphism $\bigwedge T^* \to Cl$, $\theta^a \wedge \cdots \theta^c \mapsto \gamma^a \wedge \cdots \gamma^c$, we get a torsionless connection for a particular θ, and thus we have made a choice of background(Again we emphasize that classically two fields,(ω, θ), are needed to state the choice of state.), i.e $D_{\omega_0}\theta_0 = 0$ or $D_{\omega_B}\theta_B = 0$. So we start with the background vielbein and this then induces the background connection, which a priori has nothing to do with the vielbeins. Picking the remaining connection as 'fluctuations" ω_1 over this background connection we obtain $\omega = \omega_B + \omega_1$. Let's see what this means, we set our standard space $\mathbb{R}^n$ first. We have $\omega_B = 0, \theta_B = 0, R_B = 0, \Omega_B = 0, \Omega = D\omega$. Then $\Omega = \Omega_1 = (d + \omega_1) \wedge \omega_1 + \omega_1 \wedge \omega_1$, hence $R_{\mu\nu\rho\sigma} h^{\sigma\rho} h^{\sigma\rho} = -4M^2$, $h = h_B + h_1$ is a "mass" squared, fluctuating from point to point directly describing the curvature of the state(one-particle space-time). Let's see if this can't be derived directly from the Einstein field equation. Remembering that all the mass is supposed to be a manifestation of curvature, which means that we are not allowed to introduce mass terms freely when equating the Einstein tensor to the stress-energy but only have the kinetic term in the stress-energy, we get

---

[12]Incidentally this this could provide, with some work, a link to the AdS-CFT correspondences that have appeared in recent years. Actually, we do so quite directly in Part III and IV, when we know more.

156

$$T^{\mu\nu} = \frac{2}{\sqrt{h}} \frac{\delta\mathcal{L}}{\delta h_{\mu\nu}}$$
$$= \{\mathcal{L} = \sqrt{h} h^{\sigma\rho} \partial_\sigma \phi \partial_\rho \phi^*\}$$
$$= \frac{2}{\sqrt{h}} \frac{1}{\delta h_{\mu\nu}} [\frac{1}{2\sqrt{h}} h h^{\sigma\rho} \delta h_{\sigma\rho} h^{\kappa\tau} \partial_\kappa \phi \partial_\tau \phi^* - \sqrt{h} h^{\sigma\kappa} h^{\rho\tau} \delta h_{\sigma\rho} \partial_\kappa \phi \partial_\tau \phi^*] = -2\overline{\partial^\mu \phi \partial^\nu \phi}$$

the bar being the usual trace-anti trace operation. Hence

$$\overline{Ric} = -16\pi G \overline{\partial\phi \otimes \partial\phi^*}$$

and setting $\phi \sim e^{-ikx}$ we get, remembering the conjugation symbol on the second ϕ term and using that $\bar{}$ is an involution

$$Ric = -16\pi G k \otimes k |\phi|^2$$

which gives

$$R = -16\pi G k^2 |\phi|^2$$

confirming previous suspicions. Incidentally the formula former to the last formula also confirms our suspicion that the Hilbert-Einstein action is only a classical action and that the Einstein term needs to be "smeared out" by a probability density. This is because we want the Ricci term $R\!\!\!/\,$ to be associated mainly with the kinematics of the field, so the above really is slightly incorrect since we are equating variations of classical lagrangeans to quantum lagrangeans, something that we really are not allowed to do. We are probably just lucky that it comes out right anyway for vielbein gravity.

According to us things go classically as

$$\bar{\psi} R\!\!\!/\, \psi \rightsquigarrow \frac{1}{16\pi G} R_{physical}$$

hence

$$\bar{\psi} \frac{R}{4} \psi \rightsquigarrow \frac{1}{16\pi G} R_{physical}$$

thus, setting $g^{-2} = \frac{1}{4\pi G}$, and setting $\frac{1}{g^2} R_{physical} = R_{mathematical}$, i.e for the Lie algebra generators

$$g T^\alpha_{mathemtical} = T^\alpha_{physical}$$

we have our constants and know where to put them. Let's check how this goes in previous calculations which were in mathematical units. Then $\frac{R}{4} = -M^2(x)$, hence by $R \mapsto \frac{R}{4\pi G}$ we get

$$\frac{R}{4} = -4\pi G M(x)^2 \leftrightarrow R = -16\pi G M(x)^2$$

precisely the result from the former formula, but with no factor $|\phi|^2$ appearing-just as it should be, since we in the above equated a classical variation to a quantum variation, and if we only had quantum versions on both sides when doing the first calculation the amplitudes would have dropped out on both sides. We directly realize that $g = \sqrt{4\pi G}$ can't apply for all cases, since that would mean that all particles have the same gravitational charge, so looking at the attractive potential generated by such a massless ω field we have with $\alpha_G = \frac{g^2}{4\pi}$ that

$$\mathcal{V} = \frac{\alpha_G}{r} = \frac{G}{r}$$

gives that it would be appropriate to associate a factor M to this expression, hence,

$$g_G(M) = \sqrt{4\pi G}M, \ \alpha_G(M) = \frac{g_G(M)^2}{4\pi}$$

would be appropriate, remembering to take into account various signs when doing calculation later.

We still worry about how we are distributing the charge among the Lie algebra generators, need it be as above or is it the momentum that is supposed to be distributed? A simple check shows, however, that such a partition of the mass would trivially have been associated with the Clifford algebra and not the Lie algebra, so we are at least right on track when it comes to that issue.

Let us return to the issue of the metric. Presently, the way we distribute the signature seems to be arbitrary, either MTW or "field theory" coventions—the latter is the way we called the set where p^2 is positive for causal momenta— so we need to investigate this further. We may agree that it needs to be motivated further than by only reasons of statistics among the D-folds. It could be understood as stating that the MTW metric is "God-given", or , better put, simply the correct one, but also that it must be achieved by a continuation—at least if our geometrical treatment of mass is to be correct. Thus the choice we made in our conventions, $\eta = -diag(1, -1, -1, -1, \cdots)$, is supposedly not accurate to have at the end of a calculation. Surely there must be as simpler reason than the one listed above—indeed there is—postivity of energy combined with vanishing of fields at infinity. In our conventions hitherto, we have written—let us for simplicty concentrate ourselves on classical field theory on a worldline—

$$\mathcal{L} = \mathcal{H} - p_k \dot{q}_k = < p, \dot{q} >,$$
$$p = (\mathcal{H}, \bar{p}), \bar{p} = [p_k],$$
$$q = (t, \bar{q}), \bar{q} = [q_k]$$

which correponds to the choice

$$p_k = -\frac{\partial \mathcal{L}}{\partial \dot{q}_k}$$

158

This can be seen from

$$\mathcal{H} = \mathcal{L} + p_k \dot{q}_k = T - U - 2T = -(T + U)$$

and we see that the Hamiltonian with our conventions is the *negative* of the usual one, which is supposed to equal the energy classicaly, so it gives negative energies. This comes from us incorrectly treating the space-time signature, which is thus directly correspondent to the sign of the Hamiltonian. Of course, we can't pick $p_k = \frac{\partial \mathcal{L}}{\partial \dot{q}_k}$ with our "field theory" conventions, which are opposite to the ones of most people using classical mechanics, because that would give

$$\mathcal{L} + p_k \dot{q}_k = T - U + 2T = 3T - U$$

which corresponds to nothing at all—except perhaps a mistake. The above is so simply because we "derived" our formuals from a Noether current, which not necessarily is associated to the Hamiltonian with a correct sign. Actually, in this particular example we should have included a minus sign before equating it to the Hamiltonian when we derived the concept of a lagrangean as something that generates deformations of functions along a worldline. Correcting for this sign, we write a new formula for the lagrangean that that is corrected to give positive energy,

$$\mathcal{L} = p_k \dot{q} - \mathcal{H} = < p, \dot{q}_k >_{MTW}$$

and that is actually also what is needed to get vanishing probability densities at infinity, since now

$$Z \sim e^{\int \mathcal{L}_{\mathbb{E}}} \sim e^{i \int \mathcal{L}_{\mathbb{M}}} \sim e^{W[\chi_1^{A_N}, \cdots, \chi_N^{A_N}]}$$

, with $\mathcal{L}_{\mathbb{E}}$ including a factor being the *negative* of the Hamiltonian, as opposed to the previous case, so that $i dt_{\mathbb{M}} = dt_{\mathbb{E}}$. One also checks the positivity of energy above by

$$i \partial_t Z \sim i \partial_t e^{i \int p_k \dot{q} dt - \mathcal{H}t} = i \partial_t e^{i \int p_k dq_k - \mathcal{H}t}$$
$$= i \delta_t (-i\mathcal{H}t) e^{i \int \mathcal{L}dt} = \mathcal{H} e^{i \int p_k dq_k - \mathcal{H}t}$$
$$\sim EZ$$

just as it should be on an eigenstate. In our previous conventions we accounted for this by a minus sign in the exponential, so we can recover the new definition of Lagrangean— and consequently metric— by moving in the sign in fron of the integral under the integral and absorbing it into the metric. Hence

$$Z \sim \exp\left[-i \int \mathcal{L}_{Old}\right] \sim \exp\left[i \int \underbrace{\mathcal{L}_{Old}}_{:=\mathcal{L}_{New}}\right]$$

159

and of course

$$\mathcal{L}_{New} = -\mathcal{L}_{Old} = -(\mathcal{H} - p_k \dot{q}_k) = p_k \dot{q}_k - \mathcal{H}$$

again corresponding to postive energy. So we see that our sign choice was made in a consitent manner, since it is equivalent if one includes opposite phases in the PI exponential and corresonding *part* of the effective action. The latter thing is very important, because it can make something attractive repulsive and back and forth.

Let us sum up: We have to use the MTW metric to get the correct physical metric, or to take into account this by having a correcting opposite phase in front of that term in the exponential in the S-matrix. We make a couple of remarks, this so that the reader remembers what the discussion in the latter part of this section relates to:

Remarks:

(1) The double Wick rotation above is canonical and supposed to be well known in the context of AdS-CFT correspondences.

(2) As AdS space here we defined an arbitrary pseudoriemannian manifold of constant curvature. Any riemannian manifold can be brought to such a form by a conformal transformation, where we assume to have such defintions so that the proper riemannians are a proper subset of the pseudoriemannians.

9.1.1. *Summary.*

- We associate, in physical units with a dimensionless Ricci and after a double Wick rotation,

$$R = -4g^2 M^2(x) = -16\pi G M^2(x).$$

- We have a split of the curvature on a one-particle space-time as

$$R = R_B + R_{int} - 16\pi G m_0^2$$

- We have gravitational charges and coupling constants

$$g_G(M) = \sqrt{4\pi G} M, \;\; \alpha_G(M) = \frac{g_G(M)^2}{4\pi}$$

9.2. **Problem 2: 2-Surfaces and Tachyons.** We would like to make a study of plausibility of the above. According to us states of positive mass correspond to states of negative curvature, and this manifests itself as the abundance of hyperbolic manifolds. We might check this in dimension 2. We restrict to the compact orientable case. From

160

previous lagrangeans we have by the interaction hamiltonian $\mathcal{H}_{int} = \mathcal{R}$, counting curvature with a minus sign for convinience, that

$$Z = \sum_{g \in \mathbb{N}} d(g) e^{\frac{R_g}{4}\beta}$$

with $d(g)$ the degeneracy at genus g and R_g the curvature at genus g. If we count manifolds with different complex structures as inequivalent, then the great uniformization theorem of Riemann surface theory gives $d(g)$ as the volume of a "Bers" ball bounded by the radius Λ_g. Hence, setting, $q = e^{-\frac{\beta}{4}}, \beta \in \mathbb{C}_+$,

$$Z = \sum_{g \in \mathbb{N}} d(g) e^{-M_g^2 \beta} = deg(0)q^{-1} + deg(1) + \sum_{g \in \mathbb{N}+2} deg(g)q, M_g^2 = -\frac{R_g}{4}$$

Large β means small q so then we see by the above that the sphere S^2 will be quite probable by cutting off the sum. On the other hand if we include all states we see that the probability for a tachyonic measurement P is

$$P = \lim_{N \to \infty} \frac{q^{-1}}{deg(0)q^{-1} + deg(1) + \sum_{2 \leq g \leq N} deg(g)q}$$

Realizing that the degeneracy for $g \geq 2$ is

$$d(g) = \frac{2\pi^{d/2}}{\Gamma(\frac{d}{2})} \frac{\Lambda_g^d}{d}, d = 3g - 3$$

which is diverging for constant Λ_g as a function of g since $\Gamma(z+1) = z!$, we get the probability P for a tachyonic measurement as $P = 0$ for all β as a naive expectation. So this check turned out ok. Actually, in higher dimension the abundance of hyperbolicity is very prominent, just as it should be physically from the above perspective. Indeed a theorem of Lohkamp states that every homeomorphism type can be endowed with a hyperbolic curvature. We also recognize by the above that if we take the entropy $S = lnZ_g$ of each indidual state we obtain $S = 4\pi GM_g^2$, which is the Hawking entropy formula, so this at least seems to be compatible with what we to some extent think that we know about entropy in gravity.

9.3. **Problem 3: The Crucial Sign.** We would like to derive the attractive feature between two particles of positive mass and energy after a double Wick rotation. First of all we notice that the process of Wick rotation is really not entirely consistent for we set $k_\mu \mapsto -ik'_\mu$, $\gamma^\mu \mapsto i\gamma^\mu$, $g_{\mu\nu} \mapsto (-i)^2 g_{\mu\nu}$, without really introducing an induced

161

metric g', which would have made all expressions invariant under Wick rotations, thus spoiling the point of the rotation. e.g m^2 is not mapped to $-m^2$, which would have been consistent. Thus we have to be careful as this is not entirely systematic. As an example, we had $(\Box + m^2)\mathcal{A} = 0$ letting $\mathcal{A}$ be the gauge boson(introducing an artifical mass m), and this gives $(-\Box + m^2)\mathcal{A} = 0$, which is not consistent. We prove the attractive feature by looking at the propagators of the gauge bosons. The relevant part of a matrix element, say,

$$-igT^\alpha \gamma^\mu \cdots \frac{-ig_{\mu\nu}}{p^2 + i\epsilon} \cdots - igT_\alpha \gamma^\nu$$

is mapped (if everything was consistent) to

$$-igT^\alpha(\gamma^\mu i) \cdots \frac{+ig_{\mu\nu}}{-p^2 - i\epsilon} \cdots - igT_\alpha(\gamma^\nu i)$$

We recognize the ϵ part to be a small compact deformation of differential operator that we could identify as a mass shift at infinity, which we know by the above that we are not consistent about. Setting $i\epsilon := 0$ and then reclaiming it with a positive sign (or simply ignoring to have the epsilon from the very start) that gives

$$-igT^\alpha \gamma^\mu \cdots \frac{+ig_{\mu\nu}}{p^2 + i\epsilon} \cdots - igT_\alpha \gamma^\nu$$

so that the double Wick in effect shifts the phase on the boson propagator. That is the same thing as turning attractive into repulsive, for one can now move this sign to either vertex from the propagator.

We can see the consequence of the double Wick rotation in the lagrangean directly, since a free propagator is the inverse of of the term between to the two fields in the quadratic term. So

$$\mathcal{A}\Box\mathcal{A} \mapsto -\mathcal{A}\Box\mathcal{A}$$

reverses the sign of the kinematic term relative to an interaction term and thus makes a repulsive interaction attractive. The last reasoning amounts to only doing a classical reasoning about the lagrangean.

As far as practical concerns of field theoretic calculations this is best remembered by associating a priori a minus sign to the end of a propagator so that the charges on the two ends of a propagator are always a priori opposite.

9.4. **Problem 4: The Schwarzild Metric.** Let us see if we can't derive the Schwarzild metric out of field theory with the above. We know that $\mathcal{J}$ is a conserved current, i.e $D_\mu \mathcal{J}^\mu = 0$, reminding of $D_X X = 0$ or $D_a e^a = 0$ from classical gravity, where we can associate the latter with a current on space-time. So let us equate $\mathcal{J}^\mu = X^\mu$ for a vector field X that is covariantly conserved, where $\mathcal{J}^\mu$ is understood to be evaluated w.r .t space Σ. We want to get the metric $h_{\mu\nu}$. Now, for the case we are considering, space-time is more or less void of currents, except for this chunk at the origin which is the star, and is evolving in time-like direction. Hence we write

$$i \not{D} \, \psi = 0, D^* D\omega = \mathcal{J} = \rho \delta^{d-1}(x) e^0$$

with ρ a temporal density set to unity in a unit mass system, and where we imagine ourselves to take averages over Lie algebra indices.

If we expand around the background that is given by $D\theta = 0$, this is

$$D^* D\omega_0 = \delta^{d-1}(x)$$

i.e for a fixed time $t := t_0$, remembering that ω does not depend on time,

$$\omega_0(\bar{x}) = \frac{-1}{(d-3)} \frac{1}{||x||^{d-3}} \frac{1}{\mu(S^{d-2})}$$

$\mu(S^d)$ the Euclidean Lebsgue volume of S^d. In dimension 4 this is

$$\omega_0(\bar{x}) = -\frac{1}{r} \frac{1}{4\pi}$$

We need to include the gravitational coupling constant. We have $g = \sqrt{4\pi G}$, $g(M) = \sqrt{4\pi G M}$, i.e

$$\omega_0(\bar{x}) = -\frac{g(M)^2}{r} \frac{1}{4\pi}$$
$$= -\frac{GM^2}{r} = -\frac{GM}{\frac{r}{M}}$$

Where we define a "new" radial variable $\frac{r}{M}$ with $[\frac{r}{M}] = [LENGTH]$ dimensions instead of $[r] = [1]$ and so get

$$\omega_0 = -\frac{GM}{r}$$

$\delta h_{00} = 2\mathcal{V}$, $\mathcal{V}$ the potential, can now be computed, hence the temporal component of the metric is

$$h_{00} = 1 - \frac{2GM}{r}$$

Remembering that a vacuum Einstein manifold is maximal in volume, and hence the volume element constant under topology preserving vacuum deformations, we get

$$h = (1 - \frac{2GM}{r})dx^0 \otimes dx^0 - (1 - \frac{2GM}{r})^{-1} dr \otimes dr - r^2 d\Omega^2$$

, our Schwarzild metric.

163

But this calculation has shortcomings, because it seems a little bit ad hoc. It is really just an application of a post-Newtonian approximation to gravity, or at least so it seems, so it is really not strange that we come out right. Indeed it would be strange if we did *not* come out right that way. So we have to put it under severe critisism. To remedy this we may proceed in different ways. Let us first notice that a perturbation $\delta e^a_\mu; e^a_\mu = e^a_{\mu,B} + \delta e^a_\mu$ gives

$$h_{\mu\nu} = h_{\mu\nu,B} + e^a_{\{\mu,B}\delta e_{a,\nu\}} + \mathcal{O}((\delta e)^2 = h_{\mu\nu,B} + e^a_{\{\mu}\delta e_{a,\nu\}} + \mathcal{O}((\delta e)^2$$

where $\{\}$ is supposed to denote symmetrizer of fields, and B is for background, as usual. Hence we identify as follows;

$$h_{\mu\nu} = h_{\mu\nu,B} + \underbrace{e^a_{\{\mu}\delta e_{a,\nu\}}}_{:=2\mathcal{V}_{\mu\nu}} + \mathcal{O}((\delta e)^2$$

Hence, if we define $\mathcal{V} = tr[\mathcal{V}_{\mu\nu}] = h^{\mu\nu}\mathcal{V}_{\mu\nu} = h^{\mu\nu}_B \mathcal{V}_{\mu\nu} + \mathcal{O}(V^2_{\mu\nu})^{13}$, we get

$$\mathcal{V} = tr[\mathcal{V}_{\mu\nu}] = tr\frac{e^a_{\{\mu}\delta e_{a,\nu\}}}{2}$$

Thus if we only retain the temporal behaviour for this average perturbation we obtain the post-Newtonian approximation, so we seem to come out fine, despite our original disbelief. Before going on, we note that the formula for $\mathcal{V}_{\mu\nu}$ seems to be a formula for a variational current, indeed this is already pointed out in the Hilbert-Einstein appendix of this part, where we did not directly see this link to the "geometrical" potentials $\mathcal{V}, \mathcal{V}_{\mu\nu}$. It should be pointed out that the above formula, which relates one kind of field to another which are truly functionally independent, is really only valid on-shell because, e.g., the post-Newtonian approximation is only valid on-shell, just like the usual formula for probablity currents in quantum mechanics is a consequence of the Schrodinger equation, another on-shell criterion. As should be clear from the discussion in Part III and IV, one can even deduce that it is the same on-shell criterion.

9.5. **Problem 5: Hawking Entropy.** Roughly, entropy is a measure of the effective action of a configuration that is prescribed as "vacuum" on a $D(d-1)$-brane. We have, classically, with a 2-admissible lagrangean,

$$S_{[ACTION]} = g^2 \int \bar{\psi} \slashed{R} \, \psi = g^2 M^2 \int \bar{\psi}\psi = g^2 M^2 \times 1 = g^2 M^2 = 4\pi M^2$$

---

[13]This latter identity is easily obtained by the usual formula for infinite geometical series by using it to obtain the inverse of a slighty perturbated operator.

To obtain the same thing more rigorously one can exhaust $\mathbb{R}^4$ from the interior and use plane waves in a normalized integral. So we have

$$lnZ = -4\pi M^2$$

but then $< U >= -\partial_\beta lnZ = 0$, hence with S being entropy and W free energy

$$S = k_B\beta W + k_B lnZ = 4\pi k_B M^2 = \frac{1}{4}k_B A$$

A the area of the hole. It has to be observed that the above holds classically but only for a 2-admissible case, which is according to us not what to do field theory on, at least not in this case, although it is what do string theory on.

9.6. Problem 6: Coulomb/ Black Hole Scattering and Assorted Scatterings.

Let us tentatively try to write the 0/4 spacetime diagram for n exterior $X^\mu(x) = \bar{\psi}\gamma^\mu\psi$(which we may, for reasons soon to be clear, call string at a base point x.). It is then understood that we can interpret X^μ as a propagator of a Dirac particle "biting" itself in the tail, so that that this would be looking at the propagator as a dynamical object bounded by two Dirac point particles with opposite charge.

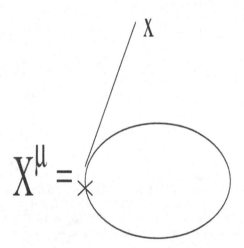

FIGURE 10. A Dirac point particle biting itself in the tail can be seen as a closed string, if we let the Chan-Paton factors be usual Dirac point particles. We call the spot where the Dirac particles, or Chan-Paton factors, coincide the base point. In this section we generalize this to hold for all particles in string field theory.

with the scalar probability density expression $\phi(x)\phi(x)^*$, now interpreted as the amplitude for a scalar selfinteraction being the predecessor in elementary quantum mechanics.

Thus we would look on a string going off to infinity as a diagrammatic expression for the probability current going off to infinity associated to a particle. The above can obviously not be made to work for gauge bosons, but we will shortly see that there is something similar. We can write the diagram for space-time probability currents, which we simply refer to as a space-time diagram, and sometimes add the suffix 0/4, then emphasizing that a "point" particle can be interpreted as a 4 dimensional soliton in dimension 4, with the 3-dimensional spatial behaviour of our our world as the behaviour of a cartesian product of wave front sets.

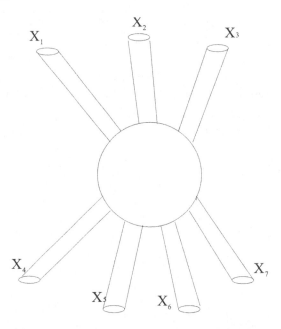

FIGURE 11. A diagram over the outgoing probability currents, i.e X fields for the restricted case above, looks like a "string" diagram. That is quite odd, as it came out of (NC) YM/SYM. The blob in the middle is supposed to be the total S-matrix working on the in states. In the above diagram at least one of the X_i must correspond to a gauge boson.

The probability for such a diagram, where we now use ψ, ϕ to denote in and out state respectively, is given by

$$X_\phi^* X_\psi = <\psi|\phi><\phi|\psi> = |<\psi|\phi>|^2$$

with $X_\phi := |\phi> \otimes <\phi| = |\phi><\phi|$ and similarly for ψ, and we call X probability vector. We notice before going on that X can be interpreted as an endomorphism of a Hilbert space, and thus the 'non-commutative" nature of X, in particular for the case of spin and $SU(2)$

$$X^*X' = X^j X'_j = |<\psi|\phi>|^2$$

which is related to a quaternionic perspective. It matters little which representation of a noncommutative theory we use, or to to focus on one specific noncommutative or hypercomplex theory, rather it is noncommutativity in general that is prominent and important-as long as we force the fields to be space-time fields(probability currents), because then we get stuck with the above expressions. The duality above can be realized by hermitean conjugation and a trace, i.e $tr[X^\dagger X']$. Notice that the above expression applies to any in state, including gauge bosons etc-it is the generalization from the spinor to (non-commutative)space-time case to the general case of any Hilbert space involved and it's (non-commutative) endomorphisms.

The diagram above would then have probability proportional to the square of a usual sum of Feynman diagrams;

$$X^*_\psi X_{S\phi} = |<\psi|S|\phi>|^2$$

and everybody nows how to calculate the latter. In view of this we can now restate the link between 2-admissibility and 1 admissibility, i.e between NC SYM and M(atrix)-theory when one looks at the supersymmetric version. Let us generalize this a little bit further, we put in factors M_{AB} that take care of symmetrizations etc that may be necessary;

$$X^C_\phi := M^C_{AB}|\phi^A> \otimes <\phi^B| = M^C_{AB}|\phi^A><\phi^B|$$

By the same token the following theorem is the resolution of the AdS-CFT correspondence, at least from the perspective depicted above. Let us define some of the ingredients;

Definition 9.1. *We set*

$$\mathbb{X} = \mathbb{X}_0 + \mathbb{X}_{1i}\theta^i + \mathbb{X}_{2ij}\theta^i\theta^j + \cdots$$
$$\Phi = \Phi_0 + \Phi_{1i}\theta^i + \Phi_{2ij}\theta^i\theta^j + \cdots$$
$$\mathcal{A} = \mathcal{A}_0 + \mathcal{A}_{1i}\theta^i + \mathcal{A}_{2ij}\theta^i\theta^j + \cdots$$
$$\mathfrak{D} = \theta(\partial\!\!\!/ + A\!\!\!/) + \partial_\theta$$
$$D = \Gamma^\mu \partial_\mu = \Gamma^{a,-}(\partial_{a,-} + A_{a,-}) + \Gamma^{a,+}(\partial_{a,+} + A_{a,+})$$
$$D_\pm = P_\pm \Gamma^a \partial_{a,\pm} = \sigma^a_\pm \partial_{a,\pm}$$

, D_- being the anti-holomorphic spin-gauge Dirac operator(we regard the 10/8-fold as being a holomorphic 5/4 fold which is a real infitisimally transversally complexified 4/5-fold. In the above we use a complex split of the dimensions only as a convenient notation, and could

167

equally well have done a real split or a foliation/stratification. In the following D denotes the Dirac-Kahler covariant superderivative given by $\gamma^a = e^a + e^{a*}$, as opposed to $\mathcal{D}$ above which is w.r.t to $\gamma^a = i(e^{a*} - e^a)$. ~ denotes charge conjugation in the below.

Theorem 9.1 (A fundamental theorem). *We have the equivalence*

$$Z_M = X^* S_M X = Z\tilde{Z} = \bar{\Phi} S_{SYM} \Phi \bar{\tilde{\Phi}} S_{SYM} \tilde{\Phi}.$$

with Z_M the partition function corresponding to a 2-admissible

$$\mathcal{L}_M = \mathbb{X}^* D^2 \mathbb{X}$$

w.r.t to the X fields in the corresponding classical/quantum action, and Z corresponding to a 1-admissible

$$\mathcal{L}_{Gauge-SYM} = \bar{\Phi} D \Phi$$

on, say, $T^+ X^{\mathbb{C}}$ the antiholomorphic part of a transversal complexification with + denoting antiholomorphic part. The important thing about the above is not the holomorphic/antiholomorphic factorization, which is there only to make things cute, but rather the split of a two-admissible theory on a even dimensional D-fold into two 1-admissible theories on half the dimension. Notice that since we know that the states at infinity(i.e the equilibrium states) are AdS this means that we expand the SYM around AdS configurations to get approximate evaluations in field theory according to previous remarks and canonical practice. See after the proof for further remarks. This theorem was conjectured by Maldacena four years ago.

First sketch of idea, for rigorous versions see Part III, theorem 1.1 and 6.1. We have for the lefthand side

$$Z_M = \int DX DX^* D\mathcal{A} e^{-iS_M}$$

S_M corresponding on a worldsheet to

$$S_{M,CLASS.} = \int \sqrt{\gamma} dz d\bar{z} d\theta d\bar{\theta} (\mathbb{X}^* \not{D}^2 \mathbb{X} + \mathcal{A} \not{D}^2 \mathcal{A})$$

, which is how string theorists prefer to do things presently. In a third quantized $D3$-brane perspective, corresponding to a cured version of SUGRA, manner this is, including a Yang-Mills term,

$$S_{M,Q.} = \int \sqrt{h} dz^D d\bar{z}^D d\theta^D d\bar{\theta}^D (\mathbb{X}^* \not{D}^2 \mathbb{X} + \mathcal{A} \not{D}^2 \mathcal{A})$$

168

Hence we have, splitting the covariant superderivative $\mathcal{D}$ as $\mathcal{D} = \mathcal{D}_+ + \mathcal{D}_-$,

$$Z_M = \int DX DX^* D\mathcal{A} e^{-iS_M}$$
$$= \int D\mathcal{A}_+ D\mathcal{A}_- e^{-i\int \mathcal{A}_+ \not{D}_+^2 \mathcal{A}_+ + \mathcal{A}_- \not{D}_-^2 \mathcal{A}_-} SDET[\not{D}^2]$$
$$= (\int D\mathcal{A}_+ e^{-i\int \mathcal{A}_+ \not{D}_+^2 \mathcal{A}_+} SDET[\not{D}_+])(\int D\mathcal{A}_- e^{-i\int \mathcal{A}_- \not{D}_-^2 \mathcal{A}_-} SDET[\not{D}_-])$$

but then we recognize the factors as

$$Z_{SYM}\pm = \int D\bar{\Phi} D\Phi D A_\pm e^{-i\int \sqrt{h_+} d^D z d^D \theta (\bar{\Phi}\not{D}\Phi + \mathcal{A}_+ \not{D}_+^2 \mathcal{A}_+)}$$
$$= \int D\mathcal{A}_\pm e^{-i\int \mathcal{A}_\pm \not{D}_\pm^2 \mathcal{A}_\pm} SDET[\not{D}_\pm]$$

If we have it understood that we have already integrated away the transversal degrees of freedom in the YM term. Thus we are, within the limitations of the ill defined path integrals above, formally done. $\square$

Now some important remarks about the above theorem. Later, in the next problem, we will be dealing with the invariances of the involved lagrangeans. What we have done above is truly to implement relative cochains via integration transversal to the SYM degrees of freedom. The SYM is on(the fiber of..), say, $T^+X^{\mathbb{C}}$, $TX^{\mathbb{C}} = T^-X^{\mathbb{C}} \oplus T^+X^{\mathbb{C}}$ in this language, but it is better to have it understood that we one is working with direct limits of relative chains (in the naive space-time picture.). Another benefit of the above theorem is that it seems to offer a resolution to the issue of extra dimensions in string theory and ambigous compactification. The world is the SYM part(this also comes independently from previous basic reasoning), so it is e.g 5 dimensional for type II comparisions. There are 4 dimensions, plus a last dimension corresponding to deformations of space-time, i.e precisely the K.K dimension of the Klein-Gordon equation, and when doing string theory we are actually doubling artifically. With hindsight, this can be viewed already either when one is considering e.g Kallen-Lehmann spectral representations of propagators, the Klein-Gordon equation or the configuration space perspective of classical multiparticle space-time and the transversal deformations that move around the solitons we perceive as particles. The last M-theory dimension is then supposed to correspond to deformations of the various string theories into each other. Perhaps we canuse this later to make the corret kind of dimensional reduction.

With this in mind we can start to calculate various crossections. We calculate first Coulomb scattering of a particle of mass m bouncing on a field created by a mass M. In the above the masses denote masses

of Dirac fields and the ω is to denote connection and associated particle. We begin with the space-time diagram corresponding to Coulomb
scattering;

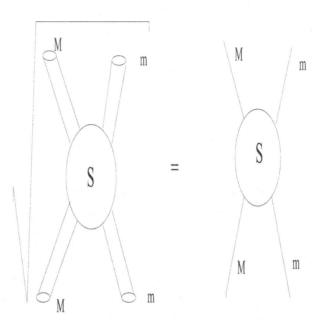

FIGURE 12. Scattering of particles of mass M and m. A "string"
space-time 0/4-diagram in non-commutative space-time corresponds to
the square of a Feynman diagram of spinorial origin. This is because X fields, which correspond to space-time vectors, correspond to
$Hom(\mathcal{H}, \mathcal{H})$, the Hilbert space $\mathcal{H}$ being an appropriate spinorial module
S, and so are 2-tensors of a spinor module.

In the above it should be understood that we really cut away half the
diagram and replace it by a fixed Coulomb source, since we assume the
M-particle to be virtually unaffected by reactions, so that it becomes
like usual Coulomb scattering with only a mass bouncing on an external
field. If we only calculate the Feynman diagram with only coupling to
the background field and no radiative corrections(corresponds to genus
nill case for space-time diagrams) we obtain a differential crossection,
including a group theory factor of 2^{14}

---

[14]R.Corrado in his Ph.D. thesis skips the center of mass motion part of the algebra. Not sure about whether that is correct, since the related transversal complexification in the physical dimension makes sense if one goes to complex Lie groups,
by taking this into account and using a complex version of a Killing form. After
some consideration, though, I think he is correct and I'm wrong, because we always
only retain unitary degrees of freedom in the S-matrix.

$$\frac{d\sigma}{d\Omega} = 2 \times \frac{\alpha^2}{4|\bar{p}|^2\beta^2 sin(\frac{\theta}{2})^4}(1 - \beta^2 sin(\frac{\theta}{2})^2)$$
$$= \frac{G^2M^2m^2}{|\bar{p}|^2\beta^2 sin(\frac{\theta}{2})^4}(1 - \beta^2 sin(\frac{\theta}{2})^2)$$

with $\bar{p}$ being 3-momentum, and β being the velocity of the incoming m particle. The more usual $M + m \to M + m$ (or $M + M \to m + m$ by crossing symmetry), being the general case described by the above diagram, can be evaluated to

$$\frac{d\sigma}{d\Omega} = 4 \times \frac{\alpha^2}{2k^2(E+|\bar{k}|)^2(1-cos(\theta))^2}((E + |\bar{k}|)^2 + (E + |\bar{k}|cos(\theta))^2 - M^2(1 - cos(\theta))^2)$$

in the CM frame, with $|\bar{k}| = \sqrt{E^2 - M^2}$. In particular the above gives for $M := m$ the amplitude for Bhaba scattering.

9.7. Problem 7: The Force Between Two Wilson Loops/Particles.
This is a computation that we make just to get a little bit of intutive feeling for what is going on, afterwhich we in Problem 8 return to the more serious issue off the symmetries of the lagrangean. We would like to compute the force between two Wilson loops, since we now know by the above that we should identify strings with such loops. Thus what we are essentially doing is calculating the force between two particles.

A Dirac quantization condition for a loop yields for the mean field strength, which we denote by B, that

$$B = -\frac{2\pi}{g}\frac{n}{A}$$

$n \in \mathbb{N}$, A the area inclosed. The minus sign is an artifact of the present convention of having a minus sign in the exponential in the path integral, and the only reason for us including is it to fix a sign that we would have been forced to fix later anyway. This gives the force F by the Lorentz force law as

$$F = \frac{g}{2\pi}(-\frac{2\pi}{g}\frac{n}{A}) = \{A = 4\pi r^2, r := \frac{r}{G(m)}\} = -\frac{Gm^2n}{r^2}$$

$n \in \mathbb{Z}$ as in $n \in \pi_2(S^2)$ and inherited by $\pi_1(S^1)$ living on the equator which is the place where the original loops, bordant to D_+ and D_- the original hemispheres, live and coincide. What could be non-trivial about the above is that the area should be the area of a sphere, hence we derive it. In the action we have

$$-i(S_+ - S_-) = -i(\int_{\partial D_+} \bar{\Phi}_+ \omega_- \Phi_+ - \int_{\partial D_-} \bar{\Phi}_- \omega_+ \Phi_-)$$

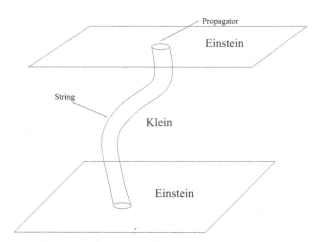

FIGURE 13. A Wilson loop propagating in a fifth dimension. Looking at the propagation of the loop rather than the propagation of the base points will double the dimension in space-time. The above is from the 10 dimensional point of view, looking at Einstein's 4-dimensional space and Kleins fifth as subspaces. Kleins fifth dimension is e.g. necessary in order to interpret mass as curvature, and implies a simple homotopy type for $X_\pm$ if we were to interpret it in the original sense of Klein, which we need not do, see Part III.

hence, either using $\partial D_+ = -\partial D_-$ from $\partial S^2 = 0$, or setting $\omega_+ = -\omega_-$ (We cannot turn both the orientation of a loop at the same time as we turn the winding direction around the loop, that would be the same thing as doing nothing at all.) that gives

$$-i \int_{S^2} d\omega$$

$\omega = <\omega_+>$ the usual Lie algebra averaged connection, i.e. in effect a $U(1)$ connection.

Please notice that the above corresponds to computing the classical part of the VEV

$$< W(C_+)W(C_-) >$$

$C_\pm = \partial D_\pm$ and differentiating in appropriate manner the natural logarithm to get the force. In the above $W(C_\pm)$ are the various Wilson loops. This computation, however, has to be questioned. We shall let it be for now, since it gave us intuition and that was what we were after rather than accurate or correct treatment, but notice that Corrado in

172

his Ph.D. thesis did some computations that seemed similar in nature, so we may have another link to AdS-CFT correspondences here.

Artificially defining C=PT, we notice that since we have to turn time on the latter sheet bounded by, say, C_+ when patching we can set $P = \tilde{}$, thus in reducing the space-time dimension we pick a definite helicity for particles involved (this is one reason for why the holomorphic calculus by way of analogy was used in the reasonings above), since that amounts to picking a particle with or without a $\tilde{}$. Doubling the dimension and having stringy coordinates, i.e having 10 dimensions, corresponds in some sense to taking into account both helicities, while only having usual space-time should correspond to having a definite helicity. And this then ties (in some sense) to the idea of a complex holomorphic space-time in a most natural way.

9.7.1. *Summary and Important Points.*

- Gravity was found to be field theoretically described by a Pauli theory associated to 2-surfaces, a theory which we could take a square root of, leading to a Yang-Mills theory, and thus reduce the configuration space dimension to half the dimension. Gravity was thus described by SYM on AdS spaces (notice that any Riemannian manifold is conformally equivalent to a space of constant curvature), with Wilson loops corresponding to usual strings from the 10-dimensional perspective. These spaces, which are Poincare duals to the string base point, are then supposed to correspond to massive particle states at infinity. Simple checks were made to try to check plausibility so far.
- Spaces/particles were subdivided into tachyonic, massless, and usual states according to if they are, after having been reduced to constant curvature and with a double Wick rotation included, of positive curvature, zero curvature or negative curvature respectively. Spaces are particles according to the above perspective, as we identify with a certain space a certain geometry; the particles are solitonic configurations on the various one-particle space-times.
- The smooth topology on a space decides it's curvature by deciding the connections and metrics which are valid, hence the topology induces the mass spectrum.
- In order to be able to describe a geometry of probability densities (space-time) by use of half-densities(spinors, quantum mechanics) we work with probability vectors(e.g matrix spinors). These are naturally giving string space-time a noncommutative

173

coordinate structure since string space-time corresponds to the 2-admissible lagrangean picture.

- A picture dual to SYM in a certain dimension would according to above be string theory in double the dimension in a certain manner described above. A sketchy proof of this was presented.

9.8. Problem 8: The Symmetries of The Actions Involved.

9.8.1. *Problem 8, Part A: Introduction and Formalism.* This section is organized as follows; First we define the actions and component fields from the 2- and 1- admissible perspectives. Subsequently we prove that various symmetries are satisfied, concentrating on the lagrangeans that are not so well known and putting little energy on the Y.M. term, to which we assume the reader familiar.

We proceed in general manner. Let

$$\mathbb{X} = \sum \mathbb{X}_{\mu_1 \cdots \mu_k, \alpha_1 \cdots \alpha_l} dx^{\mu_1} \wedge \cdots dx^{\mu_k} dy^{\alpha_1} \wedge \cdots \wedge dy^{\alpha_l}$$

where μ_i is a space-time index, α_i brane indices, x space-time coordinates, y brane coordinates. The α indices can be made to transform with an arbitrary weight relative to space-time, and conversely for the μ indices relative to the brane. The field $\mathbb{X}(x,y)$ depends on both x and y coordinates, which in the tangent picture corresponds to the base points of the exponential maps on the target space-time and the possibly singularly embedded brane respectively. For example for Y-branes, the α indices transform with half the weight of space-time indices, while for X-branes, i.e. usual D-branes, they transform with full weight. It will be important for us to have the generality of all admissibility types, for we will want to model all families of theories for D-branes of arbitray(but will restrict to integer in this thesis) dimension. That is, after all, one of the things which will make our hypothesis less arbitrary. Before expounding how these, say, rational weights are effectuated, we must clarify a couple of things;

(1) We associate dx^μ to θ^μ supercoordinates of full weight while associating dy^α to θ^α with arbitray, say half for now, weight. "Usual" formalism is obtained by taking either space-time or the brane Y , where we shall use Y for a brane that transforms with arbitray weight w.r.t space-time, and replacing the integration over it by integration over a point $*$, i.e. e.g. setting

$$\int \epsilon_X = Id \otimes P_{\bigwedge^0} \quad *_{\bigwedge T^*X},$$
$$\int \epsilon_Y = Id \otimes P_{\bigwedge^0} \quad *_{\bigwedge T^*Y}$$

$P_{\bigwedge^0}$ the projection onto the exterior cotangent space of zeroth degree. Hence we retain usual Berezin integration times the identity, so that valuation over a point of the zeroth degree of the Hodge dual corresponds to superintegration and restriction. One proves easily that these interpretations mean that while a diffeomorphism will induce a contravariant functor on forms it

175

will be covariant on space-time-coordinates, just as it should be in a superformalism.

(2) For the most general situation one must interpret $\wedge$ as symmetrizer of arbirary type, fixed in the given situation, which may e.g. be associated to a specific vacuum labelled by various more or less exotic total charges. Since each term in the lagrangean, i.e. each tensor valued current, has a well defined trace functional to the appropriate field over which the tensor algebra is defined, this is a quite natural generalization. Another generalization, actually quite necessary for our purposes, is to introduce a-indices, as to obtain an arbitrary a-tensor. Dualities for such indices are then implemented as usual, in particular in between terms that arise multiplicatively such as in the super-Polyakov term. Thus we are in effect dealing with manipulations of the tensor ring over an infinite dimensional space when regarding our generalized field theories, where the tangent space is associated to the one particle Hilbert spaces, and doing these manipulations in such a way that they have a closed action within different factors of different irreducible splits of this tensor algebra, which all belong to different spectral representations in various charges of this tensor algebra (See also Part III, Geometric Quantization of D-branes).

Let us define as follows;

Definition 9.2. *A supersymmetry is an isometry of a $\mathbb{Z}_2$-graded metrized module acting non-diagonally on the grading.*

We will be interested in structures similar or covered by this defintion such as quadratic vector spaces, pseudohermiteans, etc. In the following we shall often resort to "simplex" integration over a point, this to make solid contact with the canonical situation in supersymmetric physics, a context which we mainly intend to stay within. In such circumstances we can replace $dx^\mu = \theta^\mu$ and use Berezin integration. We can coin the above "generalized" supersymmetry, GEN SUSY or suchlike , in particular for the case where we let elements of non-pure degree in the homology ring of a space induce a functional in the obvious way on a element in the exterior cotangent,

$$< \Sigma, \Omega > = < \Sigma^p, \omega_p > = \sum_p < \Sigma^p, \omega_p >, \omega_p \in \Gamma(X \times Y, \bigwedge^p T^* X \times Y)$$

so that we have states with parts belonging to branes of different integer dimension.

176

Next we want to have the aformentioned tranfromation properties, remincent of generalized equivariance of some sort, in this context. Namely; a linear transformation on the cotangent space of the brane Y should lift to an endomorphism of , say, double the weight on space-time X and conversely a transfromation the space-time tangent TX lifts to something with half that weight on TY. Here is another point where hypercomplex manifolds enter the picture(this will be quite usual latter); The $\cdots, SO(3), SO(4), SO(1,3), SO(5), SO(6), \cdots$, of space-times of diverse dimension lift to $\cdots, SU(2), SU(2) \oplus SU(2)$, $SL(2, \mathbb{C}), USp(4), SU(4), \cdots$ acting on the relevant unit spheres bounding a unit ball in such a manifold and corresponding spinorial modules, which are themselves assumed to be tangents of the relevant Y-branes. Thus if $\Lambda_X : T_pX \mapsto T_pX$ is a linear endomorphism on X space-time we set

$$[\Lambda_X]^{\frac{1}{2}} : T_pY \mapsto T_pY$$

and oppositely. We may use a similar nomenclature and call these hypermanifolds, with the hope that we may be a little bit more specific in the indivdual cases. For some matters in the analysis, geometry and toplogy of such entities, see Part I; Hypermathematics. As concerning $\mathbb{X}(x, y)|_{x=*}$ it should be understood that while X_0^a (the subscript denotes the exterior order in $\bigwedge T^*X, \bigwedge T^*Y, 0 = (0,0)$, beginning with order in dx^μ and then degree in dy^α) this can be understood as a embedding coordinate in the "integrated" picture— provided we use the exponential map centered at the point x. We still percieve x as the usual space-time point, having nothing to do with X_0^a, which is a vector field over $x = *$. Thus x is part of a choice of background zero points for the sigma model map, one for each space-time degree of freedom, the remaining choice of these zero points being the constant terms in X_0, while the rest of X_0 is a choice of perturbation, and they have nothing to do with each other.

9.8.2. *Problem 8, Part B: Symmetries of the Actions.* Let us define the action more closely for the 2-admissible and 1-admissible cases. We have

$$\mathcal{L}_M = \mathbb{X}^* D\!\!\!/\,^2 \mathbb{X}$$

as our relevant part, with

$$\mathbb{X} = [\ \underbrace{\mathbb{X}_0^a}_{=\text{"embedding"}}\ +\ \underbrace{\mathbb{X}_{(1,0)\ \mu}^a}_{=e_\mu^a=\text{"vielbein"}}\ dx^\mu + X_{\mu\nu}^a dx^\mu \wedge dx^\nu + \cdots]$$

177

The reason for the identification $\theta^\mu = dx^\mu|_*, \theta^\alpha = dy^\alpha|_{*'}$ will soon be apparent. For now we can set D to be the usual extended gauge covariant derivative. It would be nice to show that this coincides with the usual supercomponent derivation when restricted to a point in one of the factors X, Y. We set

$$D_\theta = dx^\mu \frac{\partial}{\partial x^\mu} + dy^\mu \frac{\partial}{\partial y^\mu} + i_{X \oplus Y} = D_{\theta,X} \oplus D_{\theta,Y}$$

where e.g. on X

$$D_\theta = dx^\mu \frac{\partial}{\partial x^\mu} + i_X = \theta^\mu \partial_\mu + i_X,$$
$$i_X = x^\mu (dx^\mu)^*$$

The latter acts, as is well known, as a graded differentiation on $\bigwedge T^* X$ and so we can set $\rho(dx^\mu) = \frac{\partial}{\partial \theta^\mu}$ a convenient isomorphism. From the usual Cartan formula for Lie derivatives we have

$$[d, i_X]_+ = \mathcal{L}_X$$

and thus by the exactness conditions $d^2 = 0$, $i_X^2 = 0$ we have

$$D_\theta^2 = \mathcal{L}_X$$

in particular, setting $X^\mu = 1 \ \forall \ \mu$, one has on usual functions, denoting the usual covariant superderivative along a coordinate (x^μ, θ^μ)

$$D_{\theta\mu} = \theta^\mu \partial_\mu + X^\mu i_\mu$$

the latter with no Einstein sum, the relation

$$D_{\theta,\mu}^2 = X^\mu \partial_\mu = \partial_\mu$$

If we restrict to integration over a point $* \in Y$ for one of the factors we get

$$D_{\theta\mu} = dx^\mu \frac{\partial}{\partial x^\mu} + dy^\mu \frac{\partial}{\partial y^\mu} + i_X + i_Y$$
$$= \underbrace{dx^\mu}_{\theta^\mu} \frac{\partial}{\partial x^\mu} + \underbrace{dy^\mu \frac{\partial}{\partial y^\mu}}_{:=0} + i_X + \underbrace{i_Y}_{:=0}$$
$$= \theta^\mu \partial_\mu + \frac{\partial}{\partial \theta^\mu}$$

We stress that the $dx^D d\theta^D$ of usual Berezin integration is, from the point of view of integration of forms, truly an overcounting, since the form integration has both aspects included automatically. The invariance which this product of integration measures enjoys is thus the same as the invariance of the classical intergration measure of pseudo-Riemannian geometry $\theta^1 \wedge \theta^2 \cdots \wedge \theta^D = \sqrt{sgn(G)|G(X)|} d^D x$, θ^a ON-frame vielbeins.

<u>Lorentz invariance:</u>
We have

$$\mathcal{L} = \epsilon_X \epsilon_Y * \mathbb{X}^* \slashed{\partial}^2 \mathbb{X}$$
$$\mathbb{X}^* = *\mathbb{X}^\dagger$$

$*$ Hodge star, $\dagger$ Hermitean conjugation/transpose.(It is really important to remember in these contexts that since we are dealing with compact manifolds the Hodge star is associated to an (anti)isomorphism of Hilbert spaces as implemented by Poincare duality, since otherwise it would be hard to understand how these lagrangeans directly induce the string lagrangeans we observe in Part III.) We see directly by $[\delta \Lambda, \slashed{\partial}\,] = 0$, $\delta \Lambda$ a small enough space-time Lorentz transformation that we have Lorentz invariance. From $\slashed{D}^2 = \Box + \slashed{R}$ we get by

$$\delta * \mathbb{X}^* \slashed{D}^2 \mathbb{X} = 0 + \mathbb{X}^*[\delta \Lambda, \slashed{R}\,]\mathbb{X}$$

a conserved current.
<u>Spin invariance:</u>

This an additional symmetry symmetry that we can have by adding Y indices to $\mathbb{X}$. This works out in the same manner as the previous Lorentz invariance. Also, the Lorentz invariance lifts down to a spin invariance by the usual lift and conversely, so if we have both space-time and brane indices both both Lorentz invariance and spin invariance must be assured simultaniously. By another token we must always have this invarianc if we want to be able to rewrite the space-time vector indices in spinorial form and conversely when we can, since

$$\mathbb{X}^a = \bar{\Phi}^{A'} \Phi^A \Gamma^a_{AA'}$$

where A, B are used for vector indices on the brane Y. The (S)YM lagrangean has Lorentz invariance induced by spin invariance and this can be readily checked. One identifies the constant part of $\mathbb{X}^a_\mu$ as $\psi^a_{0,\mu}$ in string theory, since

$$[\Gamma^\mu_\pm, \Gamma^\nu_\pm] = \pm 2 G^{\mu\nu}(X),$$
$$\Gamma_\pm = \theta \pm \theta^*$$

in particular for background fields denoted by B or 0 as a subscript, the latter a convient coincidence in string theory as compared to our formalism since background fields are constants in the present formalism, we have

$$[\Gamma^\mu_{+,B}, \Gamma^\nu_{+,B}] = [\psi^\mu_0, \psi^\nu_0] = 2 G^{\mu\nu}_B(X)$$

In the above Φ^A_1 are the usual Dirac fields in spinorial notation.

Supersymmetry:

Space-time Supersymmetry:

By an easy calculation one shows that we have the metric $[\ Id\ ]$ on the space-time exterior algebra running diagonally w.r.t degrees. Hence, for DeRham currents

$$\mathbb{X}^a = \mathbb{X}_0^a + \mathbb{X}_\mu^a \theta^\mu + \cdots$$

we have

$$|\mathbb{X}|^2 = *\mathbb{X}^*\mathbb{X} = <\mathbb{X}, \mathbb{X}> = *[(*\mathbb{X}^t)\mathbb{X}] = \mathbb{X}_0^t\mathbb{X}_0 + \mathbb{X}_{1\mu}^t\mathbb{X}_1^\mu + \mathbb{X}_{2\mu\nu}^t\mathbb{X}_2^{\mu\nu} + \cdots$$

Thus, in the usual way of elementary quantum mechanics, we expect that the isometries of the Hilbert space—this time running non-diagonally on the form degree— will give us a symmetry of the lagrangean and conversely. Thus let

$$\delta\mathbb{X} = \delta\Lambda\mathbb{X}$$

be such a transformation, antisymmetric over the form degree in $\bigwedge T^*X$,

$$
\begin{aligned}
\delta\mathbb{X} &= \delta\Lambda\mathbb{X} \\
&= \begin{pmatrix} 0 & \epsilon\Lambda_1^0 & \epsilon\Lambda_2^0 & \cdots \\ -\epsilon\Lambda_1^0 & 0 & \epsilon\Lambda_2^1 & \cdots \\ -\epsilon\Lambda_2^0 & -\epsilon\Lambda_2^1 & 0 & \cdots \\ \cdots & \cdots & \cdots & \cdots \end{pmatrix} \begin{pmatrix} \mathbb{X}_0 \\ \mathbb{X}_1 \\ \mathbb{X}_2 \\ \cdots \end{pmatrix} \\
&= \begin{pmatrix} \epsilon_{-1}\Lambda_1^0\mathbb{X}_1 + \epsilon_{-2}\Lambda_2^0\mathbb{X}_2 + \cdots \\ -\epsilon_1\Lambda_1^0\mathbb{X}_0 + \epsilon_{-1}\Lambda_2^1\mathbb{X}_2 + \cdots \\ -\epsilon_2\Lambda_2^0\mathbb{X}_0 - \epsilon_1\Lambda_2^1\mathbb{X}_1 + \cdots \\ \cdots \end{pmatrix}
\end{aligned}
$$

where the subscript on ϵ_i an infinitesimal grassmannian indicates it's degree, with negative degree being dual degree in the space-time metric, i.e. for example

$$\epsilon_{-1} = \epsilon\ \theta^*$$

ϵ a small real. Hence, defining in an ad hoc manner $\delta\!\!\!/\,R = 0$, we have

$$\delta\mathcal{L} = (\delta\mathbb{X})^*\!\!\!/\,R\ \mathbb{X} + (\mathbb{X})^*\!\!\!/\,R\ \delta\mathbb{X} = -\mathbb{X}^*\delta\Lambda\!\!\!/\,R\ \mathbb{X} + \mathbb{X}^*\!\!\!/\,R\ \delta\Lambda\mathbb{X} = 0$$

we have a second symmetry, with

$$dim O(2^D) = \frac{2^D(2^D - 1)}{2}$$

generators, the latter by $2^D = dim(\bigwedge T^*X)$. We now turn to symmetries on the world volume.

180

Worldvolume Supersymmetry:

Further exanding we now get

$$\mathbb{X}^a = \mathbb{X}_0^a + \mathbb{X}_\alpha^a \theta^\alpha + \cdots$$

It is easy to see the generalization to world volume of arbirary dimension once one sees how this would look on a string. For a string we settle for two space-time basis covectors and two string sheet covectors. Thus

$$S = \int X^* \not{D}^2 \mathbb{X},$$
$$\mathbb{X} = [\mathbb{X}_0^a + \mathbb{X}_{(0,1)}^a dz + \mathbb{X}_{(0,\bar{1})}^a d\bar{z} + \mathbb{X}_{(0,2)}^a dz \wedge d\bar{z} + \mathbb{X}_{(1,0)\mu}^a dx^\mu + \cdots]$$

$* = *_{\wedge T*X_{//\Sigma}} \otimes *_{\wedge T*\Sigma}$, and thus by duplicating previous reasoning we have yet another symmetry. We thus have integrals over $X \times Y$, $Y = embedded\ brane$, $X = target\ space$, with the latter generalizing Berezin integrals, and behaving like

$$\theta^{\mu'} = |\frac{\partial z'}{\partial z}|^2 \theta^\mu, \quad \theta^\mu // T^*\Sigma$$

for a diffeomorphism $z \mapsto z'$ of Σ. From the ten dimensional perspective this is

$$S = \int \epsilon_X \epsilon_Y * \mathbb{X}^* \not{D}^2 \mathbb{X}$$

and an endomorphism $\delta\Lambda$ on T_pY, $p \in Y$, is lifted to $(\delta\Lambda)^2$ on T_pX. A standard case is when we evaluate over only a point in, say, the brane, and hold on to the space-time dependence is what we get retaining only the constant term, thus considering

$$T_{X \times p}^*(X \times Y)$$

which is the usual space-time configuration space (momenta times coordinates) plus additional supercoordinates of some arbitrary number. Finally, by setting

$$S = \int \epsilon_{X_+} \epsilon_{X_-} \epsilon_{Y_+} \epsilon_{Y_-} * \mathbb{X}^* \not{D}^2 \mathbb{X}$$

we can split this theory into two parts, hence obtaining

$$S_\pm = \int \epsilon_{X_\pm} \epsilon_{Y_\pm} * \Phi^* \not{D}_\pm \Phi$$

to worry about in terms of halfdensities on the factors of various helicities. Linearization over the brane, i.e evaluating over an average point

181

$y = *$, gives

$$S_\pm = \int d^D x^\pm d^D \theta^\pm \Phi^* {D\!\!\!/}_{\,\pm} \Phi$$

Having wrestled with the 2-admissisble $\mathbb{X}^* D^2 \mathbb{X}$-term and it's 2-admissible kins for quite a while we can put a little time on the Yang-Mills term. We have

$$\mathcal{L}_{YM} = \mathcal{A}^* {D\!\!\!/}^{\,2} \mathcal{A}$$

in Dirac-Kähler mode. Setting, paralleling the treatment of the $\mathbb{X}$-field,

$$\mathcal{A} = \mathcal{A}_0 + \mathcal{A}_{(1,0)\mu} dx^\mu + \mathcal{A}_{(0,1)\alpha} dy^\alpha + \cdots$$

it is easy to verify supersymmetries. What, however, is not necessarily true is that the PI easily reduces in the manner sketched above. Hence define

$$\int \mathcal{A}_+^* {D\!\!\!/}_{\,+}^{\,2} \mathcal{A}_+ = \int \epsilon_{X_+} \epsilon_{Y_+} \mathcal{A}_+^* {D\!\!\!/}_{\,+}^{\,2} \mathcal{A}_+$$

to be the SYM contribution on X_+, Y_+. $\pm$ is a subscript to the co-variant derivative here, and not to the gamma matrices, and means restriction of the derivative to the appropriate space-time. This can also be achieved by setting

$$ {D\!\!\!/}^{\,2} = \begin{pmatrix} {D\!\!\!/}_{\,+}^{\,2} & 0 \\ 0 & {D\!\!\!/}_{\,-}^{\,2} \end{pmatrix} $$

on

$$ \mathcal{A} = \begin{pmatrix} \mathcal{A}_+ \\ \mathcal{A}_- \end{pmatrix} $$

and then letting the integration on the inactive brane, space-time, etc result in the identity opertion, i.e.

$$\int d^D x^\pm d^D \theta^\pm \mathcal{L}_- = \int d^D x^\pm * \mathcal{L}_-$$

9.8.3. *Summary.*

- The actions involved have gauge, Lorentz, space-time supersymmetry and world volume supersymmetry as defined in the text.

182

9.9. Problem 9: Conformal and Weyl Symmetries.

One easily checks Y-world volume Weyl invariance

$$[*(\mathbb{X}^* \slashed{D}^2 \mathbb{X})]\epsilon_X \epsilon_Y$$

for the case of space-time rescalings in $D(Y) = 4 = D(X)$ inducing transformations as follows;

$$\mathbb{X} \mapsto e^{-\sigma(y)}\mathbb{X},$$
$$h \mapsto e^{2\sigma(y)}h,$$
$$\slashed{D}^2 \mapsto e^{-2\sigma(y)}\slashed{D}^2|_X$$

then having it implicit that we are taking Weyl transformations which are over a space-time point but never the less are also varying over the Y-brane world volume. Then

$$\mathcal{L}' = \epsilon_X \epsilon_Y e^{(D-4)\sigma(y)}[*(\mathbb{X}^* \slashed{D}^2|_X \mathbb{X})]$$

and the same holds for world volume depending rescalings on space-time X. General Weyl invariance only holds for the interaction term , but this, of course, also depends on the type of $\mathbb{X}$-field considered, which must be such that it tranforms as above for these purposes in $D(X) = D(Y) = 4$, since the free term contains obstructing differentiations. The Hilbert-Einstein lagrangean is covered by the interaction term in the above. For the Y.M. lagrangean one has in identical manner

$$\mathcal{L}' = \epsilon_X' \epsilon_Y' |\Omega'|^2 = e^{(D-4)\sigma(y)}\epsilon_X \epsilon_Y |\Omega|^2$$

hence again the correct invariances. Incidentally, this implies upon putting together $X = X_+ \oplus X_-$, that our lagrangean

$$G_{\mu\nu} DX^\mu DX^\nu = G_{-,\mu\nu}D_+ X_-^\mu D_- X_+^\nu + G_{+,\mu\nu}D_- X_+^\mu D_+ X_-^\nu$$

retains Weyl invariance in $D = 10 = 2 + 2 \cdot 4 = 2 + D_+ + D_-$. The additive 2 is from world sheet propagation, whilst the multiplicative 2 corresponds to the two factors. We can check conformal invariance by assuming

$$J^\dagger(G_{X_+} \oplus G_{X_-})J = (e^{2\sigma(x_+)}G_{X_+} \oplus e^{2\sigma(x_-)}G_{X_-})$$

for

$$J = [\frac{\partial f}{\partial(x_+, x_-)}] = \begin{pmatrix} \frac{\partial f_+}{\partial x_+} & 0 \\ 0 & \frac{\partial f_-}{\partial x_-} \end{pmatrix}$$

where $J_\pm$ are Jacobians of conformal tranformations on the space-time helicity factors. We are thus, in effect, having a quaternionic calculus, with the soul difference that we do not necessarily identify points in the two spaces as their conjugates, in remincence of how fermion field conjugates $\bar{\Psi}$, which only excpetionally satisfy $\bar{\Psi} = \Psi^* \Gamma^0$.

183

Part III shows this to be even a necessary consequence, this by consistency among field defintions defined in that part and arbitrariness of background. We shall see same thing happen when we construct string world sheets in Part III by world lines in the off-shell space-times. Thus, since we know that we have both Lorentz invariance and Weyl invariance on the factors, we also have conformal invariance. One procceds similarly for combined transformations of both the background target X and the brane Y $f : X \times Y \mapsto X \times Y$, $X \times Y = (X_+ \times X_-) \times Y_+ \times Y_-$ as above, where the string space-time corresponds to $X = X_+^{(5)} \times X_-^{(5)}$.

9.9.1. *Summary.*

- We have both conformal and Weyl invariance on the space-time/brane factors of dimension $D(Y_\pm) = D(X_\pm) = 4$.

9.10. **Problem 10: Superconformal Symmetry.** Here comes one of the rewards of having identified a convinient way of writing the superconformal mathematics. Since we have identified the supertransformations as various induced transformations of the usual calculus of forms on the space-time and brane exterior algebra, superconformal transformations of both branes and backgrounds must thus automatically be symmetries of the lagrangean in view of the above.

9.11. **Problem 11: Hawking Radiation.** In the $\mathbb{X}, \mathbb{X}^*$ path integral we have for integration measure

$$[D\mathbb{X}][D\mathbb{X}^*]$$

if we can show that $D\mathbb{X}_1 D\mathbb{X}_1^* = [D(G_{\mu\nu})]$ then Hawking radiation would follow, since we know by the Hilbert-Einstein appendix that

$$\int \mathcal{L}_{int} = \int \mathbb{X}_1^* \not{R} \, \mathbb{X}_1 = \int \frac{R}{16\pi G} \sqrt{sgn(G(X))|G(X)|} d^D x$$

and thus expanding around x the background expansion point for the sigma model map, and only retaining the corresponding term,

$$S = \int \mathcal{L}_{int} = \int \mathbb{X}_1^* \not{R} \, \mathbb{X}_1 = \int \frac{R}{16\pi G} \sqrt{sgn(G(x))|G(x)|} d^D x$$

which is the usual field theoretic Hilbert-Einstein lagrangean, which we historically know to imply Hawking radiation by field theoretic considerations. Let us work out the functional determinant for a transformation of variables. By $\mathbb{X}_1^a = e_\mu^a dx^\mu$ we have, since

$$\delta G_{\mu\nu} = (\delta e_{\{\mu}^a) e_{\nu\}a}$$

that

$$\left[\frac{\delta G_{\mu\nu}}{\delta e_\sigma^a}\right] = \left[\frac{\delta(\delta e_{\{\mu}^a) e_{\nu\}a}}{\delta e_\sigma^a}\right]$$
$$= [\delta_\mu^\sigma e_{\nu b} + \delta_\nu^\sigma e_{\mu b}]$$

which seems hopeless, as this should have equaled identity if the coordinate change was sucessful. However, remembering that $\mathbb{X}^*, \mathbb{X}$ might be unrelated, and must at last be treated in unrelated manner, we get on one of the helicities

$$\left[\frac{\delta G_{\mu\bar{\nu}}}{\delta e_\sigma^{\bar{a}} \otimes e_{\bar{\rho}a}}\right] = \left[\frac{\delta e_\sigma^{\bar{a}} \otimes e_{\bar{\rho}a}}{\delta e_\kappa^{\bar{b}} \otimes e_{\mp b}}\right]$$
$$= [Id \otimes Id] = [\bar{Id}]$$

where we have used the bar notation to distinguish the indices of the two space-times, the convention being that bars belong to the positive

185

helicity space-time when contravariant. Hence

$$\mathcal{D}\mathbb{X}_{1\mu}\mathcal{D}\mathbb{X}_{1\bar{\nu}}^* = [\mathcal{D}G_{\mu\bar{\nu}}]$$

on for example the space-time X_-, where $\mathcal{D}$ is the infinite dimensional Feynman measure in the corresponding Feynman integral. And so, on a constrained surface

$$\bar{\mathbb{X}}^{\bar{\mu}} = \mathbb{X}^{\mu}$$

in the total product field configuration space we have

$$\mathcal{D}\mathbb{X}_{1\mu}\mathcal{D}\mathbb{X}_{1\bar{\nu}}^*$$

and so Hawking radiation must follow—provided that the correct kinematic term exists—since the partition function resticted to $\mathbb{X}_1$ fields is

$$Z_{\pm} = \int [DX_{\pm}][DX_{\mp}^*]e^{-S[\mathbb{X}_{\pm},\mathbb{X}_{\mp}^*]}$$
$$= \int [D(G_{\mp,\mu\nu})]e^{-\int \mathcal{L}_{H.-E.}+\mathcal{L}_{Kin.}}$$

We now need to realize this by adding the correct kinematic lagrangean. Foe example, we know that we are supposed to have a kinematic term

$$\mathbb{X}^*\square\mathbb{X}$$

and artificial mass terms $\mathbb{X}^*m^2\mathbb{X}$, corresponding to background curvatures on one-particle space-times, which in the low energy limit should be approximated by e.g. a massive scalar lagrangean, hence

$$\mathcal{L}_{Lowenergy} = \mathcal{L}_{H.-E.} + \mathcal{L}_{massive\ scalar}$$

We can actually retain the explicit form of this lagrangean by letting this scalar behaviour be the norm of the vielbeins, (or rather, ON frame), thus for small perturbations of the norm, which we may denote by Φ, defined by

$$e^{\Phi}\mathbb{X}_B = (1 + \Phi + \cdots)\mathbb{X}_B$$

remembering the factor 4 that normalizes the Hilbert-Einstein term in mathematical units and taking a mass-less case,

$$\mathcal{L} = (\frac{R}{4} + \partial_{\mu}\Phi\partial^{\mu}\Phi)\sqrt{sgn(G(x))G(x)}d^Dx$$

Since the background choice is arbitrary, we have

$$\mathcal{L} = (\frac{R}{4} + \partial_{\mu}\Phi\partial^{\mu}\Phi)\sqrt{sgn(G(X))G(X)}d^Dx$$

or, redefining the lagrangean of the low energy theory with a factor of 4,

$$\mathcal{L} = (R^G + 4\partial_\mu \Phi \partial^\mu \Phi)\sqrt{sgn(G(X))G(X)}d^D x$$

which will be handy for comparision with low energy string lagrangeans in Part III. In the above we added the superscript G to the Ricci scalar to emphasize that the low energy classical Ricci is subject to an on-shell criterion equivalent to the Levi-Cevita condition, so that the Riemann is functionally dependent on the metric via the usual formula for the Levi-Cevita connection of Riemannian geometry.

9.12. Problem 12: A Further Symmetry of The Partition Function.

We would like to show symmetry under (gravitational) charge conjugation, so we want massive fields that interact with other massive fields to interact in the same manner as massive antifields interacting with massive antifields in the "gravity" above. This is, in a naive manner, most easily done. We can denote it in the same manner that we would denote world sheet parity $^-$. Then we have

$$Z_M = \overline{(Z\bar{Z})} = \bar{\bar{Z}}\bar{Z} = Z_M$$

where we used that $^-$ is a $*$-operation[15] in the second equality and an involution, satisfying $\bar{\bar{A}} = A$, in the third. Hence this naively falls. Because of lack of space, time, and that we have other thinghs to attend to, we cannot do further checks, but point out that in the crucial sign calculation the same thing is visible by simply continuing the mass charges attached to vertices by a minus sign, which then cancel out in the diagram, since an intermediate boson must start and end on a fermion line vertex.

9.13. Problem 13: Pair Production Problem.

Since we know that we have a Hilbert-Einstein action, and that e.g. the dilatons are candidates for low energy scalars, we have all the players needed for pair production, for example in usual conventions and before a Wick rotation, a low energy lagrangean of the type

$$\mathcal{L} = (\frac{R}{16\pi G} + \partial_\mu \Phi \partial^\mu \Phi + m^2\Phi^2)\sqrt{sgn(G(X))G(X)}d^D x$$

depending on the conventions of the reader, and to this one must add suitable interaction terms, such as a phi-fourth term. Hence pair production must fall, since such matters are usually calculated on the basis of lagrangeans of preciseley this type.

As an aside, in the following it will be "nice" to have some systematic way of fixing which lagrangean we have in ou conventions. We do this by staring in the "usual" conventions, listed in the beginning of this part under "Conventions", and then Wick rotating, so that this would give us a term proportional to $(\pm i)^k$ in front of a term proportional to $k:th$ order derivatives.

---

[15]That is, $*$ is such that $(AB)^* = B^*A^*$.

188

9.14. **Problem 14: Mass, Momenta and Different Kinds of Additivity.** According to us, we would naively have, in view of the Hilbert-Einstein appendix,

$$p^2 = \sum p_i^2$$

which is not quite correct. We interpret p_i^2 as square momenta associated, as suggested by the appendix and Problem 1, to the direct sum $\bigoplus M_i^{(D)}$ of D-folds. Direct sums are used as we model a space-time by it's tangent space over each space-time, and thus mean a direct sum of vector spaces. We wish to correct this, taking into account the momentum interference terms, which are bilinear in momenta. That is to say, we wish to correct this formula to a formula that is based on additivty of momentum. One sets for a total momentum operator, in canonical fashion for charge operators,

$$P = P_1 \otimes 1 \otimes 1 \otimes \cdots + 1 \otimes P_2 \otimes \cdots = \sum P_i$$

where we used abbreviated notation in the last equality. Correspondingly one has

$$|\mathbb{X}(x_1)\mathbb{X}(x_2)\mathbb{X}(x_3)\cdots\mathbb{X}(x_N) >= \mathbb{X}(x_1) \wedge \mathbb{X}(x_2) \wedge \mathbb{X}(x_3) \wedge \cdots \mathbb{X}(x_N)$$

which is the usual formula for forming multiparticle states for fields of the above type. Associating these operations on the fields and belonging operators with the mass addition thus correesponds to adding the masses of the momenta of "ripples"[16] (solitons) on the same standard copy of a space-time that we issue as a background to the different space-times in multiparticle space-time. On the other hand

$$p^2 = \sum p_i^2$$

corresponds to adding totally unrelated and independent space-times. Later we shall see how this last formula applies anyway in some cases associated to cluster decompostion of branes by cutting D-brane diagrams, as an approximate total formula, modulo a constant zero point momentum, for the codimensional momentum of a collection of branes cobordant to the same manifold or diagram. For example usual energy will be an example of this, as momentum transversal to a collection of 3-folds bounding a four-dimensional diagram. This is compatible with the other formula in the low energy limit, just like kinetic energy is additive in an external Galilean observers frame when viewing a multiparticle system, despite that this is by addition of pure quadrics in

---

[16]We do not want to have too "big" bumps travelling on the background, because then perturbation theory will not work.

momenta— modulo the aformentioned zero-point. See the last section of the N-admissibility and Feynman diagram section for this development.

9.14.1. *Summary.*

- There are several ways to add mass, but this is usually done by adding the momentum first and then squaring the momentum, and this applies even to geometrical multi-particle masses by consistency.
- Cluster decompostion of branes yields another mass additvity formula in the low energy limit, having the property that it adds pure quadrics of momenta, something that is compatible with the former in the low energy limit.

9.15. **Problem 15: Properties of Entropy.** It would be good to show that our formulas for entropy satisfy what we would expect in terms of properties of entropy. For example we want them to show positive semidefinite entropy and that compostions of closed systems leads to an increase of entropy. These properties of entropy follow readily from the formulas for entropy and additivities of mass. From general relativity we know that the area of the horizon of combined holes is always at least as large as the sum of areas of holes. If one uses the standard formula for additvity of masses of holes, this means that they actually equal. We shall see that we get even increased mass for our holes by our formula for the mass of composite objects, which is the one of elementary particle physics rather than the usual pythagorean formula. Explictely, for two resting holes of mass M_1 and M_2, we get by our formulas above

$$S_{TOT} = \frac{k_B A}{4} = k_B \pi (2GM)^2 = \underbrace{(4\pi G)}_{:=g_G^2} M^2 = g_G^2 (M_1^2 + M_1^2 + 2M_1 M_2)$$

$$= \frac{A_1 + A_2 + 2\sqrt{A_1 A_2}}{4}$$

where we percieve the holes more or less as usual particles. In contrast, for holes on independent space-times, which originate from cutting a five-dimensional diagram without identifying outgoing space-times and dropping a constant zero point codimensional momentum, this is

$$S_{TOT} = 4\pi G(M_1^2 + M_2^2) = \frac{A_1 + A_2}{4}$$

which is the usual formula which falls from pythagorean additivity of mass. Hence we have established increase of entropy when holes join. By positive semidefiniteness of M^2 also positive semidefinteness of the entropy falls. Of course holes might bifurcate too—although this is improbable macroscopically—and this corresponds, for example, to giving away a quantum. Since we have for the free energy

$$F = U - TS$$

this means an increase of energy, something that makes such an excitation of free energy unprobable as compared to a dexcitation as viewed by a observer which is not part of the background, since such events have probability given by the usual Boltzmann law as

$$e^{-F\beta} = e^{-U\beta} e^S, T^{-1} = \beta$$

which clearly shows that an increase of entropy is what is probable. The quotient of the probability of detecting a new state as compared

191

to an old one is, thus,

$$\frac{P_{New}}{P_{Old}} = e^{(S_{New}-S_{Old})} = e^{\Delta S}$$

which clearly is greater than zero only when $\Delta S \geq 0$. Thus it is not so that entropy never decreases, especially not when looking at only a part of a system in a multicomponent system, for then lightbulbs would never shine (Looking at entropy only is the same thing as only looking at the component of the system which is called background.). Rather it is really just a matter of probability and probable flows and not quite prohibition.

9.15.1. *Summary.*

- We have the usual properties of entropy in our supposed gravity.

9.16. Problem 16: Properties of Vacua for Relevant D-branes.

This is a brief list of some properties relevant to our D-branes, especially near the vacua we usually use, such as Einstein configurations.

- As mentioned before, a transformation Λ on T_pX ; $\Lambda : T_pX \mapsto T_pX$ lifts to $\Lambda^{\frac{1}{2}} : T_qY \mapsto T_qY$ on the Y-brane. This makes use of a section $s \in \Gamma(X_\pm, \mathbb{Z}_2)$ which defines the square root, called a spin structure. When we go to relevant vacua, the Levi-Cevita criterion becomes a new criterion holding on these vacua. Thus classically, when restricted to be on these vacua, we have in effect one less fundamental field by taking away the connection, and can regard the gravitational connection as dependent of the metric instead of being a free variable.

- Hyperkähler property; When a hypercomplex manifold has a Levi-Cevita connection in the conformal class of the metric it is called Hyperkähler. Intutively, this has the interpretation of gravtational connection being reminicent to a pure gauge, since this means for example for complex manifolds that the metric is closed at a trivial connection when projected onto the antisymmetric ground state tensors, i.e reminicent to a vacuum equivalent state. This coincides well with a well known property of (Weyl) positive selfdual space-times, which are Hyperkähler. The criterion for existance of a so called canonical twistor structure is Weyl selfduality of a space-time.

- It is well known that the many features of the toplogy of 3-folds (space) are generated by cobordisms to 4 (space-time) and conversely. Much accordingly, such folitations and subdivisions of space-time in into space and time go canonically via the hypercomplex nature of situation (See Part IV). Since the time like direction becomes what we call paths, as in path integrals, of space-time this means that a lot of the topology of 3/4-folds is governed by stitching along paths. Conversely, we can regard a Feynman diagram as giving a situation of 4-dimensional topology.

9.17. Problem 17: N-admissibility.

9.17.1. *2-admissibility.* Previously we heuristically related different lagrangeans and actions—when a theory seemed to be generating some kind of flow—to various branes of different dimensions. We made the dimension of these "lines of flow" a prefix to this admissibility and so called a lagrangean/action belonging to an N-brane or D(N-1)-brane N-admissible. In the following we shall try to discuss some of the aspects of the wide variety of facets of this concept, but shall never the less delimit us not to be dealing with the explicit sewing and exhibition of such D-brane diagrams. Straight away we must recognize a certain ill-definedness, or rather multiple definedness, in this concept; A lagrangean may be p-admissible for several p, e.g for both $D_\pm = 4$ and $D_\pm = 1 = dim(\gamma_\pm = *\Sigma_\pm)$, $*$ a space-time Hodge star, an example related to the duality of the path formulation and space-time formulations of quantum field theory. We have our main hypothesis;

- Not every lagrangean generates a flow, and not every lagrangean that generates a flow is associated to all admissibility types. Indeed, from some lagrangeans on can read off a certain definite admissibility type, which is subject to the above dualities.

Of course, we shall in general be unable to to put together anything resembeling a deep investigation of the correctness of this hypothesis. Never the less, we concentrate on our favourite example—gravity—and illustrate it, and give part of the generalization as we latter on attack general admissibility.

Let us review the basic features of the 2-admissible case. We had, in the above, the perspective that this could be viewed as the splitting of a partition function

$$Z_{2-adm} = Z_{1-adm}Z_{1-adm},$$
$$Z = s\det[D^2], Z_{1-adm} = s\det[D]$$

where we associate different positions to the parts. Naively, this casn be viewed aw a consequence of the above determinant split. We associate space-time fields with the former and spinorial fields with the latter 1-admissibilie scenario, which make up th string space-time entities as as tensor products. Let us spin a little bit further on this idea, then with the understanding that one would e.g. have include various renormalization devices and cutoffs to make the above well defined, although we do not do so in the following, since we are only in a prenature stage of sketching our ideas. The objective is to gain suffcient insight into the 2-admsiible case to be able to prove an unusual, and extended, verison of Maldacenas conjecture.

194

9.17.2. *N-admissible case.* It is clear how to generalize the above. Say we set an operator $P_\omega(\mathcal{O})$, $P_\omega(\zeta)$ a polynomial over $\mathbb{C}$, $\omega \in \mathbb{C}^n$ holomorphic coordinates in holomorphic cefficents of this polynomial, spectrally defined by $\rho(O) = \zeta$ on $Spec(O) \cap eigenvalues(O)$. Then we have a factorization

$$P_\omega(O) = \Pi(\zeta - \zeta_i(\omega))^{n_i}$$

up to at most an irrelvant constant, n_i the multipicity at ζ_i. Hence locally on the relevant part of the O spectrum

$$
\begin{aligned}
Z &= \det P_\omega(O) \\
&= \det \Pi(\zeta - \zeta(\omega))^{n_i} \\
&= \Pi \det(\zeta - \zeta(\omega))^{n_i} = \Pi_i Z_i^{n_i}
\end{aligned}
$$

where we can associate different positions with the various factors, whilst in the above regarding the restriction to the case when they coincide. Hence, locally, modulo singular spectrum, $Spec(O)$ is an n-fold covering of a certain set, $n = \sum n_i$. For the case of 2-admisibility we had only $n = n_1 = 2$, and zeo otherwise. We directly encounter a possible nemisis, which can only be taken care of by being careful in the individual cases, namely we can possibly confuse this with the existance of extra degrees of freedom(before identification), such as e.g. color degrees of freedom.

So let us take a second glance on how this works out for the case of strings related to field theory. We can set a space-time X_+ with vectors of the form $X^{AA'}$, $'$ meaning positive helicity, and similarly a space-time X_- of the form $X^{A'A}$. We have for our favourite terms

$$\mathbb{X}_\pm^a = \underbrace{X_\pm^a}_{\text{embedding } X_\pm} + \underbrace{X_{\mu,\pm}^a \theta^\mu}_{\text{vielbein } X_\pm} + \cdots,$$

$$\Phi_\pm = \underbrace{\Phi_{0,\pm}}_{\text{embedding } Y_\pm} + \underbrace{\Phi_{1,\alpha,\pm} \theta^\alpha}_{\text{vielbein } Y_\pm} + \cdots$$

We set actions pointiwisely given by

$$
\begin{aligned}
*\mathcal{L}_{Strings,SUGRA} &= \mathbb{X}^* \slashed{D}^2 \mathbb{X} \\
&= \mathbb{X}_+^* D_- D_+ \mathbb{X}_- + \mathbb{X}_-^* D_+ D_- \mathbb{X}_+, \\
*\mathcal{L}_{SYM} &= \Phi^* D\Phi = \Phi_+^* D_+ \Phi_- + \Phi_-^* D_- \Phi_+
\end{aligned}
$$

and so obtain the determinant split as above. So our systematic key to obtaining the admissibility type is to use the determinants in the above way. Modelling the total string space by Stein manifolds our problems will even get the correct cohomology type for tyhe problems considered, which must be equivalent to the problems on the space-time factors.

9.18. Problem 18: Statement of a Definitive Hypothesis. The hypothesis we would like to set forth is, naively, given by actions

$$S = < \Sigma, \mathcal{L} >$$

, $\Sigma \in H(X_\pm, \mathbb{C}) = \oplus H_i(X, \mathbb{C})$, of general SUSY in the sense that it remains invariant under non-diagonal transformations in the exterior algebra degrees. It includes $\mathbb{X}_G$ the most general X-field possible, which is a direct sum with elements in all relevant tensor modules, and Φ_G, ω the superdirac field and connection treated similarly. If we assume the dimensions of the involved branes is bounded from above by the dimension D, it is obtained by taking $D \mapsto \infty$. Hence we are supposed to obtain all theories on all branes, with diagrams of total arbitrariness, having components of varying D the dimension of the involved D-branes. The idea is then to percieve the various theories as flowing to each other, or to have some theories as more probable (or excited) than other, e.g. those with a high degree of symmetry fall under this category. The theories of much symmetry become weighted in an average of theories obatined by transformations on an arbitrary set of theories including those theories symmetric under these transformations. Never the less, we want to specialize to be able to calculate anything at all; We do this by looking for theories which have symmetries, since we know these to be wighted and hence probable.

The theory which has a correct invariance is then given by leading terms

$$*\mathcal{L}_{Strings,SUGRA} = \mathbb{X}^* \slashed{D}\,^2 \mathbb{X}, D = 10 = 2 + 2 \times 4$$
$$*\mathcal{L}_{H.E.} = \mathbb{X}^*_\mp D_\pm D_\mp \mathbb{X}_\pm, D_\pm = 4$$
$$*\mathcal{L}_{\pm,SYM} = \Phi^*_\pm D_\mp \Phi_\pm, D_\pm = 4$$

and in accordance with the usual technical terminology we instead refer to the elements in this triplet triality as theories, then understanding that they are really part of the same theory. We have written $10 = 2 + 2 \times 4$ in the above to emphasize that two of the dimensions in the ten dimensional theory are really off-shell, and that we have accounted for how to obtain the dimensions of the $D = 10$ theory. Antisymmetry in the exterior algebra of the background field configuration space corresponds to minimal spin polarization, and this too can be generalized away, and leads to study of quantum theory on general vacua, which is outside of our ability and time limits.

Thus, in effect, our hypothesis is the most general quantum mechanical system possible, including all theories at all dimensions as it is, reduced to various effective theories as very probable configrations which the theories fluctuate about. As a limited example we can take the renormalization group for deformation spaces of various theories. Since dimension itself would then be but another parameter in the space of theories we look upon we sometimes emphasize that it is only a most probable dimension by saying effective dimension for D at the most probable theories in the dimension spectrum, which arises at $D = 4$ and $D = 10$ respectively when having the correct admissibilities.

Now we turn to the problem of gravity instead of the most general theory possible.

9.19. **Problem 19: Quantization of Theories, Computational Checks.** See Part III in the appropriate sections. For a check of one of the determinant splits see Part IV and the mass spectrum calculation.

9.20. **Problem 20: A Sketch of a Proof of Maldacenas Theorem.** Finally we have all the parts needed for a proof, or perhaps rather derivation, of Maldacenas theorem. We shall state(and prove) a warm-up theorem first;

Theorem 9.2. *Let Z_M be the stringy partition function in $D = 10 = 2 + D_+ + D_-$, pertaining to*

$$\mathcal{L} = \mathbb{X}_G^* D^2 \mathbb{X}_G$$

and Z_{SYM} to

$$\mathcal{L}_{SYM} = \Phi_G^* D \Phi_G$$

Let us pick as our total off-shell string space-time $X^{(}10) = AdS^4 \times S^6 = AdS^4 \times (\mathbb{R}^4 \hat{\times} \mathbb{R}^2)$, ˆ denoting compactification, $\mathbb{R}^2$ corresponding to a string world sheet with two helicities, and $\mathbb{R}^4$ to a fibration of trivial monodromy on the AdS^4. Then, at tree level,

$$Z_{M,tree} = |Z_{SYM,tree}|^2$$

Proof. Let us recollect the parts of our previous reasoning necessary to put this together. Firstly we must perform an exercise of partition functions, the goal is to separate the—say—negative helicity space-time stacked onto the postive space-time. In the above Z_M corresponds to an on-shell object, so if it contains essentially only information from 2×4 degrees of freedom, i.e. the two factors $X_\pm$ living on codimensional directions to the world sheet. We know for

$$\mathbb{X}_{string}^a = \mathbb{X}_+^{\bar{a}} \oplus \mathbb{X}_-^a$$

That the $\mathbb{X}_\pm^a = \mathbb{X}^{AA'} \sigma_{AA'}^a$. Let us use this noncommutativity in the proof by using the tensor product structure; We have firstly

$$\mathcal{L}_M = \mathbb{X}^* D^2 \mathbb{X} = \mathbb{X}_+^* D_- D_+ \mathbb{X}_- + \mathbb{X}_-^* D_+ D_- \mathbb{X}_+$$

Hence

$$Z_{M,tot} = \det[D^2] == \det[D_+ D-] \det[D_- D_+] = Z_{M,+} Z_{M,-}$$

where the parts live on $D_\pm = 4$ space-time. This is the part which intutively amounts to the probability interperetation. The next step is to separate the spinorial behaviour.

$$Z_{M,\pm} = \int [d\mathbb{X}_{\mp}^*][d\mathbb{X}_{\pm}]e^{-\int_{\mathbb{X}_{\pm}^{(D_\pm)}} \mathbb{X}_{\pm}^* D_{\mp} D_{\pm} \mathbb{X}_{\mp}}$$
$$= s\det[D_- D_+]$$
$$= s\det[D_+]s\det[D_+]$$
$$= Z_{SYM,\pm} Z_{SYM,\mp}$$
$$= \int [d\Phi_{\mp}^*][d\Phi_{\pm}]e^{-\int \mathcal{L}_{SYM,\pm}} \times permutation - \mapsto +$$
$$= |Z_{SYM}|^2$$
$$= \int [d\Phi^*][d\Phi]e^{-\int \mathcal{L}_{SYM}}$$

which is ok. In the last line the integral is over a standard space-time of dimension 4. Thus, so far everything is acceptable, and we have shown that, under the assumptions above, we have coinciding partiation functions. The next step is beyond elementary algebra; We must truly show that we must have the same vacuum solution spectra on the space-times, which is a apriori non-trivial in view of the different dimensions. We can do this using the theorem that a Stein manifold has the homotopy type of a CW complex of half the real dimension, and set $X_\pm$ corresponding to this, which can do by aid of the trivial mondromy above. But then, since the extra dimensions are contractible, the cohomology of an arbitrary differential operator must be equivalent. Thus we must have equivalent vacua. $\qquad\square$

We have already checked that some invariances coincide, for example we have checked Weyl invariance of

$$\int \mathbb{X}^*(\not{R} + \not{F}\,|_{A:=0})\mathbb{X} = \int \sqrt{h}|d^4x|R_{abcd}h^{bc}h^{ad}$$

albeit in a differnt form of this same lagrangean. For example for the pure YM term we have, again, this invariance by

$$\mathcal{L}_{pureYM} = \sqrt{h}|d^4x|\Omega_{ab}\Omega_{cd}h^{ab}h^{cd}$$

transforming by $h^{ab} \mapsto e^{2\sigma}h^{ab}$ to

$$\mathcal{L}'_{YM} = \sqrt{h}e^{-4\sigma(x)}|d^4x|\Omega_{ab}\Omega_{cd}e^{4\sigma(x)}h^{ab}h^{cd} = \mathcal{L}_{YM}$$

We must also check the usual super Dirac term, then, again looking at the interaction term just as in the Hilbert-Einstein case we have by

$$\Phi^A = e^{-\frac{3\sigma(x)}{2}}\Phi^A$$
$$\omega = e^{-\sigma(x)}\omega$$

that

$$\mathcal{L}'_{SYM,int} = \sqrt{h}|d^4x|\Phi^*\omega\Phi$$
$$= \sqrt{h}e^{4\sigma(x)}|d^4x|e^{-4\sigma(x)}\Phi^*\omega\Phi$$
$$= \mathcal{L}_{SYM,int}$$

199

and again we have Weyl invariance.

We have now gone through some elementary checks of our hypothesis. I think that, at this point, it would be good to shift point of view, and I hope that it does not confuse the reader. Having propagated for the idea that gravity is a "masked" SYM theory stemming from a Pauli theory and my ideas on how things are supposed to correspond between string theory and SYM, thus being "pro" for these ideas, it would be nice if I could be permitted to adopt a very "con" point of view. What we should now doubt, in my opinion, is not that the above substitutions give parts off the unity in the the $X^* S_M X$ S-matrix, but rather that these things truly describe gravity. So we are not questioning whether one can rip off QED crossections and correct them with group theory factors to get Yang-Mills crossections in some cases, but we are simply very much critisizing the idea that gravity is decribed by a (NC) (S)YM theory on a space asymptotically looking like an AdS space. And furthermore the stated link between M-theory and SYM is not clear, what one should do is to compare amplitudes etc between the two theories systematically in some set of problems. We should also compare β functions, effective actions and asymptotics, especially when going towards the classical regime, so that we truly know that we come out in equivalences between the mentioned theories. Fortunately, because of the Maldacena conjecture four years ago, there has been a great deal of comparisons of precisely that type. In this vein we are doing checks in part III and IV.

9.21. Comparisons of M-Theory and SYM. Since I am not an accomplished M-theorist, I cannot undertake a systematic and deep comparison of SYM and M-theory. Hence I would like to refer the reader to better sources. Instead of naming many I shall only mention some, this not to confuse the reader with too many names. R.Corrado; Some Aspects of The Connection Between Field Theories And Gravity, Ph.D thesis, UT Austin. Corrado, who seems to have gone the M-theory way-and presents a tour de force as a Ph.D. thesis-seems to see through M-theory, and ties to large extent together the picture above, although some features are missing, probably because it is a little bit too an unorthodox insight(from the current string perspective). But strong suspicion if not belief seems to be there, for he opens up with quoting Weinberg, March 1996,

It's conceivable, although I admit not entirely likely that something like modern string theory arises from a quantum field theory. And that would be the final irony.

C.G Callan, C. Lovelace, C.R. Nappi, S.A. Yost,"String Loop Corrections To Beta Functions", Nucl. Phys. B288 (1987) 525 touch the

201

issue of beta functions and Weinberg reviews the beta functions of the supersymmetric standard model in the quantum theory of fields III. Maldacena, who the derived the above ideas in a totally different manner (although also heuristic) together with Russo touches both the large N limit/AdS-CFT and non-commutativity in J.M. Maldacena, J.G. Russo, "Large N Limit of Non-commutative Gauge Theories", hep-th/9908134. A recent paper by Witten and Seiberg states that some stringy corrections can be obtained via non-commutativity (a la Connes) in SYM, furthering support for the above hypothesis. So according to us, strings and SYM are really just dual descriptions of the above thing, probably valid within different ranges, and that would seem to be the string paradigm. We shall, anyway, be concerned with comparing our theories and making checks in Part III and IV, but it has then to be understood that those comparisions are made by someone with very limited exposion to M-theory. On the other hand, there may be differences to usual M-theory, and there is propably no way to better check that this truly implies M-theory than by simply unknowingly deriving the theory that is implied, and then compare the two sets of rules and ideas, eventhough one does so with limited skill assets.

I derived the above story straight out of field theory and elementary considerations as an undergraduate, unaware of that it would lead to strings[17], but it is facinating to see how one can take a totally different way via M-theory, which is found in references concerning AdS-CFT and noncommutative geometry(not in the sense of Connes, but in the sense of noncommutative coordinates). It is also fascinating how easy it is to give a proof of the Maldacena conjecture once one sees what it truly means and how we finally get a down to earth physical interpretation of strings that we can all understand, even those of us who are sceptic about strings. Equally facinating is how integral space-time coordinate non-commutativity is to this "holographicity" property from the above point of view, which lately has been in broad discussion, but also how each involved concept tends to be overexagerated or given a too important place by recent authors,[from the present point of view, that is], who seem to have ideas which are reminicent to parts of the picture presented, which may or may not be correct.

---

[17]Hence the names hypermathematics and hypergeometry.

10. IN RETROSPECT

In retrospect, for this part, we would like to mention that there is a vast number of issues that these and recent developments seem to shed light on, as well as that there has been an enormous number of physicists, both historically and presently, who have had keen intuition about gravity that turns out to be "true" (in whatever sense our scribblings can ever be), and is a subset of the present realm of ideas presented so far. For example, taking totally arbitrary examples, the notion of a discrete time as percieved by t'Hooft, is really resolved by the issue of probability vectors(see above) as an interpretation of the points of the target space, something that is already implicit in quantum mechanics. This idea is, at least from the above perspective, also part of the fundamental story that makes quantum theoretic space-time gain it's so called noncommutativity. Zakhorovs idea of quantum field theory creating curvature as a consequence of self-energy can be seen in the split of mass and curvature done above, and Kleins fifth dimension is used to deform the mass states at infinity in AdS space and generate transversal transformations of space-time. We also have Polchinski's D-branes, and the interactions perceived between string theory and gravity in recent years, as well as the notion of identifying the Chan-Paton factors of a string with point particles, thus admitting a natural step between field theory and String theory. The "Maldacena" correspondence, the theorem sketched above, making the awkward extra dimensions harassing string theory disappear naturally has turned string theory and field theory to dual descriptions of the same thing in a certain range, thus putting string theory on a firmer basis and realistically capable of doing physical prediction, at least if the hypothesisesed proof of this theorem a la our thesis is correct, is another example of such a link.

Just for the record - it might very well be that I'm wrong about something in the above and then you, the reader, should not despair but just correct it. However, since my ideas coincide to such a big extent to the string world, I think that at least they are probably partially right. But then again it may well be that the string universe is wrong, and then also my ideas would be wrong, since they imply string theory out of field theory.

I'm very glad that I did not get these ideas three years ago, for then no one would have understood them, probably not even me in the end. It is not easy being an undergraduate who gets a very involved idea, even if it is right. I hope that, inasmuch as this can be done, others will take up the purely field theoretic approach that I have taken, and put the various steps taken on more rigorous ground, as they do comprise a more systematic, economic and above all extremely simple way to hang string theory, coordinate noncommutativity, AdS etc on and get a physical interpretation of that world with whatever consequences that might imply. Above all I wish we could lift things from this shaky beginning(?) of an end(?) to a rigorous end, but that is probably more destined to be the work of a generation of physicists rather than a single or few individuals—should we be correct. In some sense I begin walking this road in a limping manner in Part III and Part IV, which, however, also have shortcomings in many ways.

Let us briefly sum up the logic of this thesis again, so far;

10.0.1. *Brief Summary of The Logic in This Thesis So Far.*

(1) Gravity is 2-admissible and described by

$$\theta^\dagger \slashed{D}^2 \theta = \theta^\dagger (\Box + \slashed{R})\theta$$

where $\theta = \mathbb{X}_1$. In the above

$$\mathcal{L}_{H.-E.} = \theta^\dagger \slashed{R} \, \theta = R\sqrt{h}d^d x$$

is the Hilbert-Einstein term.

(2) Hence by

$$det(\slashed{D}^2) = det(\slashed{D})det(\slashed{D})$$

cure this theory. This corresponds to the probability interpretation in quantum mechanics, which we have to partly dismiss, since we forcibly want to deal with $\mathbb{X}$ fields in gravity, which are full densities rather than half.

(3) Make necessary physical interpretations in order to get factors etc right, exhibit the formulas necessary, and take care of the checks and proofs around this. In particular prove that this gives a noncommutative space-time with strings roaming around in the $D = 10$ picture, and that the determinant split really works.

(4) Now the link to noncommutative mathematics becomes clear, in particular we the link to hypermathematics, which was custom made for noncommutative space-time.

11. Checklist

The below is a checklist of what we demanded to be checked in the introductory section. It is a direct quotation of our introduction, with notches where we have succeded.

"We will call a theory for gravity "acceptable" if the following overlapping, and admittedly quite stringent, features are present in dimension four after possible compactification etc;

- It reproduces, at least at a macroscopic level, the symmetries we observe, in particular mass attracts mass for particles of positive energy and a 'metric' graviton should have spin 2. $\checkmark$
- Hilbert-Einstein gravity is it's classical limit. $\checkmark$
- It yields sensible and finite answers to definite problems of a reasonable nature, then disregarding problems that are associated to general perturbative theory such as convergence of the S-matrix etc. In particular the theory in question should satisfy calculationability 'in principle', as in the usual status of particle physics. $\checkmark$ (See Part III)

It should be understood directly that string theory roughly satisfies all of the above except perhaps the last line of the last criterion.

We could consider the following as four fundamental and good ways to check such a theory[which of course satisfies the above requirements] and at least start a debate from

- It reproduces the Schwarzild metric out of quantum theory, with perturbations. $\checkmark$
- Hawking radiation and entropy. $\checkmark$
- Coulomb interaction and black hole scattering, together with the crucial attractive feature. $\checkmark$
- Good strong curvature dynamics, such as sensible pair creation rates. $\checkmark$

Thanks to work by Juan Maldacena in string theory, we at least know that string theory produces Hawking entropy and radiation."

Please see the appropriate sections and problems for the appropriate solutions of these problems. We thus have what we chose to call an acceptable candidate for gravity in our opening section. Our job in Part III and IV is make further checks of this candiate, which is rather a triplet of theories.

12. APPENDICES

12.1. Appendix; Variational Identities and Some Proofs.

12.1.1. *Variational Identities.* The variational identities used in section 4.6, Einstein theory, are

$$\delta h^{\mu\nu} = -h^{\mu\sigma}h^{\rho\nu}\delta h_{\rho\sigma},$$
$$\delta|h| = |h|h^{\mu\nu}\delta h_{\mu\nu},$$
$$\delta Ric_{\mu\nu} = \nabla_\sigma \delta\Gamma^\sigma_{\mu\nu} - \nabla_\mu \delta\Gamma^\sigma_{\sigma\nu}$$

, $|h|$ denoting determinant of the metric h. These are best derived using elementary matrix identities, except for the last which is suitably derived in a normal frame and then proved to be a tensorial identity.

12.1.2. *Proof of Equivalence of Hilbert-Einstein theory and a "Pauli" theory.* In this subsection we prove, in independent manners, relations concerning Hilbert-Einstein theory as promised in section 4.6, related to classical Bochner techniques in mathematics. We prove an equivalence between Hilbert-Einstein theory and a "Pauli" theory by

(1) Proving their lagrangeans to be equivalent.
(2) Proving that equivalent equations of motion are satisfied.

Although 1) is sufficient to prove equivalence of theories, making 2) logically redundant, since 1) implies that on-shell conditions are then related by a functional coordinate change in a field configuration space, equivalence of equations of motion is proved in a manner independent of the first proof.

Theorem 12.1 (Equivalence of Lagrangeans). *With $\theta = [e^a_\mu dx^\mu]$, $\slashed{R} = \frac{R_{\mu\nu}}{2}\frac{[\gamma^\mu,\gamma^\nu]}{2}$ we have*

$$\mathcal{L}_{H.-E.} = R\sqrt{h}d^d x = \theta^\dagger \slashed{R}\, \theta$$

$\dagger = *t$, t being duality on TX, and $*$ denoting Hodge star on $\wedge T^*X$.

Proof.

Lemma 12.1.

$$\theta^\dagger \slashed{R}\, \theta = \theta^\dagger R_{\mu\nu} e^\mu e^{\nu*}\theta$$

Proof. Straightforward manipulation gives

$$[\gamma^\mu,\gamma^\nu] = [e^\mu,e^\nu] + [e^\mu,e^{\nu*}] + [e^{\mu*},e^\nu] + [e^{\mu*},e^{\nu*}]$$
$$= [e^\mu,e^\nu] + [e^{\mu*},e^{\nu*}] + e^\mu e^{\nu*} - \delta^{\mu\nu} + e^\mu e^{\nu*} + \delta^{\mu\nu} - e^\nu e^{\mu*} - e^\nu e^{\mu*}$$
$$= [e^\mu,e^\nu] + [e^{\mu*},e^{\nu*}] + 2(e^\mu e^{\nu*} - e^\nu e^{\mu*})$$

206

hence retaining volume forms only

$$\theta^\dagger \frac{R_{\mu\nu}}{2}\frac{[\gamma^\mu,\gamma^\nu]}{2}\theta$$
$$= \theta^\dagger \frac{R_{\mu\nu}}{2}\frac{1}{2}[e^\mu,e^\nu] + [e^{\mu*},e^{\nu*}] + 2(e^\mu e^{\nu*} - e^\nu e^{\mu*})\theta$$
$$= \theta^\dagger \frac{R_{\mu\nu}}{2}(e^\mu e^{\nu*} - e^\nu e^{\mu*})\theta$$
$$= \theta^\dagger R_{\mu\nu}e^\mu e^{\nu*}\theta$$

where antisymmetry of $R_{\mu\nu}$ was used in the last line. $\square$

Lemma 12.2.

$$R\sqrt{h}d^d x = \theta^\dagger R_{\mu\nu}e^\mu e^{\nu*}\theta$$

Proof.

$$\theta^\dagger R_{\mu\nu}e^\mu e^{\nu*}\theta$$
$$= [e_a^*(*e_\mu^a dx^\mu)\}[R_{bcd}^a(e_a \otimes e^b)e^c \wedge e^d][e_a e_\mu^a dx^\mu]$$
$$= *e^a R_{bcd}^a e^c e^{d*} e^b$$
$$= \delta^{ab}R_{bcd}^a * e^a \wedge e^c$$
$$= R_{bcd}^a \delta^{db}\delta^{ac}\epsilon_X$$
$$= R_{bcd}^c \delta^{bd}\epsilon_X$$
$$= R\sqrt{h}d^d x$$

$\square$

Hence the above two lemmas give the desired assertion.

$\square$

We now go on to proving equivalence of equations of motion in independent manner. By $G^{\mu\nu} = \frac{1}{\sqrt{h}}\frac{\delta\mathcal{L}}{\delta h_{\mu\nu}}$ we have

$$\delta\mathcal{L}_{H-E.} = G^{\mu\nu}\sqrt{h}\delta h_{\mu\nu} = G^{\mu\nu}\sqrt{h}\frac{\delta h_{\mu\nu}}{\delta e_\sigma^a}\delta e_\sigma^a,$$
$$\delta\mathcal{L} = \frac{\delta\mathcal{L}}{\delta e_\mu^a}\delta e_\mu^a$$

Should we, as above, be correct in equating the vielbein 'Pauli' lagrangean $\mathcal{L}$ to a Hilbert lagrangean, then falls that the variational coefficents above should equal. Conversely falls upon integration of the variation that the lagrangeans will equal relative to a fixed background. Let us prove this as a check;

Theorem 12.2. *The equations of motion generated by the two above lagrangeans are equivalent, i.e*

$$G^{\mu\nu}\sqrt{h}\frac{\delta h_{\mu\nu}}{\delta e_\sigma^a} = \frac{\delta\mathcal{L}}{\delta e_\sigma^a}$$

Proof. We have $\frac{\delta h_{\mu\nu}}{\delta e_\sigma^a} = e_\mu^b\delta_\nu^\sigma\delta_{ab} + e_\nu^b\delta_\mu^\sigma\delta_{ab}$ by $\delta h_{\mu\nu} = \delta(e_\mu^a e_\nu^b\delta_{ab}) = e_\mu^a\delta e_{\nu a} + e_\nu^a\delta e_{\mu a}$. Hence

$$G^{\mu\nu}\sqrt{h}\frac{\delta h_{\mu\nu}}{\delta e_\sigma^a} = G^{\mu\nu}\sqrt{h}(e_\mu^b\delta_\nu^\sigma\delta_{ab} + e_\nu^b\delta_\mu^\sigma\delta_{ab}) = 2\delta_{ab}e_\rho^b G^{\rho\sigma}\sqrt{h}$$

207

For the righthand side we have, adopting the convention of raising and lowering with the constant δ_{ab} only;

$$\frac{\delta}{\delta e_\mu^a}(\theta^\dagger \not{R}\, \theta) = I + II$$

with I being

$$
\begin{aligned}
I &= \frac{\delta(e_a^\mu R_{b\mu}^{a\ \ \nu} e_\nu^b)}{\delta e_\mu^a}\epsilon_X \\
&= \frac{(\delta e_a^\mu R_{b\mu}^{a\ \ \nu} e_\nu^b + e_a^\mu R_{b\mu}^{a\ \ \nu} \delta e_\nu^b)}{\delta e_\mu^a}\epsilon_X \\
&= (R_{ab}^{\mu\nu} e_\nu^b + R_{a\nu}^{b\ \ \mu} e_b^\nu)\epsilon_X \\
&= 2R_{a\nu}^{\mu\nu}\epsilon_X = 2Ric_a^\mu \epsilon_X = 2\delta_{ab}Ric^{b\mu}\epsilon_X = 2\delta_{ab}e_\nu^b Ric^{\nu\mu}\epsilon_X
\end{aligned}
$$

and II stemming from variation of the volume term;

$$
\begin{aligned}
&\frac{\delta\sqrt{h}}{\delta e_\mu^a}d^d x \\
&= \tfrac{1}{2}\sqrt{h}h^{\mu\nu}\frac{\delta h_{\mu\nu}}{\delta e_\sigma^a}d^d x \\
&= \tfrac{1}{2}\sqrt{h}h^{\mu\nu}(e_\mu^b \delta_\nu^\sigma \delta_{ab} + e_\nu^b \delta_\mu^\sigma \delta_{ab})d^d x
\end{aligned}
$$

Hence, remembering that $h := -h$ in the case of Minkowskian signature,

$$
\begin{aligned}
\frac{\delta}{\delta e_\mu^a}(\theta^\dagger \not{R}\, \theta) &= I + II = (2\delta_{ab}e_\nu^b Ric^{\nu\mu}\sqrt{h} - \tfrac{R}{2}\sqrt{h}h^{\sigma\nu}(e_\sigma^b \delta_\nu^\mu \delta_{ab} + e_\nu^b \delta_\sigma^\mu \delta_{ab}))d^d x \\
&= 2(\delta_{ab}e_\nu^b Ric^{\nu\mu}\sqrt{h} - \tfrac{1}{2}R\sqrt{h}h^{\mu\nu}e_\mu^b \delta_\nu^\sigma \delta_{ab})d^d x \\
&= 2\sqrt{h}(Ric_a^\mu - \tfrac{1}{2}Rh_a^\mu)d^d x \\
&= 2\sqrt{h}G_a^\mu d^d x
\end{aligned}
$$

equaling the lefthand side. $\qquad\square$

Hence by integration of the variation from a fixed background we have proved equivalence of lagrangeans a second time. Before departing we mention a couple of applications. The usual "Pauli" term equation, which is simpler than the Einstein field equation, corresponds to varying a 4-form instead of a 0-form and thus having the shift of the integration wheighting included for free in the variation of the 4-form. For vacuum space-times, i.e Einstein space-times, this is of course not any greater simplification, since the original situation in the Einstein formalism is already simple, but for non-vacuum space-times this is a considerable simplification in considering classical theory. Also the above implies that we can use Yang-Mills theory to attack classical gravity together with the quantum theory[18] obtained by direct quantization of the classical theory. Among other things this seems to imply, by remarks on how to find the solutions to the spinorial Klein-Gordon equation in the section "The Heart of This Thesis", that there is an explicitly finite field theory to all orders describing this direct quantization of classical gravity[19]. To remind the reader; one cures a "Pauli" theory, wich has an inadmissible quantum lagrangean, by using the Dirac lagrangean, which is the correct object to exponentiate in a quantum theory of this kind, and then forming superpositions of solutions. This is also at least the most crucial, and very simple, explanation to why previous trials at quantization of classical gravity by direct quantization of the classical H.E. action by regarding the space-time metric as a fundamental field went wrong at orders higher than 1-loop; people were in "Pauli" trouble.

We mention a couple of consequences of the above. Let us define the mass-curvature of a space-time X as the expectancy value

$$< \not R > = \int_X \theta^\dagger \not R \, \theta$$

with θ normalized, where we might call $\not R$ the Ricci operator, and define the mass as the square root of this. This formula has Weyl invariance in $D = D_\pm = 4$, as well as lorentz and gauge invariance invariance in arbitrary D. In the above mathematical units are used and a minus sign truly differs the square mass from the mean curvature after a double Wick rotation, which is the thing to do canonically to get correct statistics among manifolds. We may view a classical space-time as being given by a pair (ω, θ) with θ an orthonormal frame (vielbein) and ω a connection.

Theorem 12.3. *Let $X_{(\omega,\theta)}$ a be a Dirac on-shell space-time with connection ω and vielbein θ. Then X is Ricci-flat if and only if it is compatible, i.e*

---

[18]Which is, however, false as a true theory of quantum gravity in the sense of a theory that deals with half-densities. There is a distinct step between classical and quantum gravity, having nothing to do with the process of quantization but rather with the stucture of quantum theory as a saga of half-densities, with the latter, quantum gravity, corresponding to a space-time "square-root". Probability densities satisfy classically the heat equation, a Euclideanized Schrödinger equation, yet we would not consider them as fundamental in quantum theory, instead we look at amplitudes, which are "roots" of probability densities.

[19]Making this correction in the graded version should correspond to obtaining a primitive but correct and finite supergravity, which then would correspond to a graded version of a direct quantization of the classical theory. See however the footnote above concerning the quantization of classical gravity.

$$\nabla \theta = 0$$

Proof. We have by the Dirac on-shell condition

$$\theta^\dagger (\slashed{D})^2 \theta = |\slashed{D}\ \theta|^2 = 0$$

hence rewriting the lefthand side

$$0 = \theta^\dagger (\slashed{D})^2 \theta = \theta^\dagger D^* D\theta + \theta^\dagger \slashed{R}\ \theta = |D\theta|^2 + \theta^\dagger \slashed{R}\ \theta$$

Thus imposing Ricci flatness $\theta^\dagger \slashed{R}\ \theta = 0 \ D\theta = 0$ falls, and conversely by the above compatibility $D\theta = 0$ implies Ricci flatness. $\qquad\square$

Let us return to our elementary expose after the above comment on classical gravity and the theory obtained by direct quantization of it.

Let us call a space-time to be of semidefinite type if it's Ricci curvature is either positive semidefinite or negative semidefinite. Then, with the understanding that everything is assumed to be smooth, we have

Corollary 12.1. *For $X_{(\omega,\theta)}$ Dirac on-shell of a semidefinite type we have that X is compatible if and only it is mass-less, i.e has vanishing mass-curvature.*

12.1.3. *The physics of the classical gravity above and the associated "direct" quantization.* Some remarks should be done in order to make things comprehensible in the above and the following, e.g the no-constraint nature(in the sense that it does not give an additional constraint) of the Einstein field equation as interpreted above. The Einstein field equation of a space-time background as realized above is an identity and much less an equation, and is implied by the statement $\slashed{D}\ \theta = 0$. Demanding the same statement of a geodesic $X = \frac{d}{d\lambda}$, i.e $\slashed{D}\ X = 0$ gives the motion of a point particle on this same background, so it is the same statement that governs the space-time background as well as the motion on the space-time. The converse, on the other hand, has been realized much earlier, indeed can be found in Misner, Thorne, Wheeler and was used among other things to derive the motion of a point particle by the Einstein field equation instead following the high curvature just around the particle, i.e the same statement was used both to derive the dynamics of a background *and* the particles on it-indeed a remarkable thing. It is almost seems remarkable that no-one thought of turning things around and using the field equation of a point particle in the sense of SYM to describe a space-time. The main, and intially difficult, obstruction to this is, from the classical point of view of field theory, that gravity attracts equal charges while SYM deals with a spin 1 boson which endows them with quite unlike symmetries.

210

The above theorems are to some extent statements of a 1-particle space-time in a classical gravity. For multi-particle space times mass-curvature as described above is additive, corresponding to the direct sum classical configuration space $\bigoplus TX_i$, X_i different manifolds, in a manner similar to the additivity of squared frequencies in classical physical systems with several independent degrees of freedom

$$m_{total}^2 = \sum m_i^2$$

something derived by simply applying the rules of elementary quantum mechanics to the above[20]. Of course, usually one never thinks of space-time in this manner but rather one imagines it to be only one space-time, namely a background space time, but the configuration space perspective is an important perspective if there is any interest in quantum matters. The world is not 4 dimensional, rather $4n$ dimensional with n the number of particles, as there are n particles free to move in 4 directions, something well known in classical mechanics and this reflects in a quantum theory of gravity as well in field theory. No interactions, i.e an independent particle approximation, means the split $TX_{total} = \bigoplus TX_i$, X_i different manifolds, and a corresponding diagonal split of the connection on the total configuration space with obvious implications for the parallel transport over the total space if this is an irreducible split of a structure Lie algebra, or more generally Lie subalgebras of the automorphism Lie algebra of the fiber of the relevant vector bundles. Having small interactions corresponds to having small off-diagonal entries in the appropriate connection over such a split, which is usually quite large in the sense of many summands and products, e.g the universe, to take an example of a dynamical system with many components, just like having a small non-diagonal Hamiltonian in statistical mechanics. Particles, when percieved as such, thus become highly localized solitonic configurations associated with the vielbeins, with something reminiscent to "probability flow" being

$$dh_{\mu\nu} = e_\mu^a de_{\nu a} + e_\nu^a de_{\mu a}$$

in such a $SO(n,m) - Gl(n,m)$ theory. Parallel transport of the fiber of the relevant vector bundle in, say, $O(n,m)$ or $U(n,m)$, corresponds to preserved amplitudes along a trajectory while transformations transversal to this indicate instability. Such behaviour, by appropriate generalization of the behavior of negative square mass, might be

---

[20]This formula is deviating by factors proportional to the inner products of particle momenta $i \neq j$, and this is accounted for slightly later in the quantum theory section by correcting and specifying the naive formula for how to form multiparticle states

called tachyonic. Since we are used to looking at S-matrices at infinity we often only look at isometries of the fiber in the vector bundles. The latter formula also shows, beside the obvious physical interpretation of manifolds in the Kähler cathegory as "vacuum" configurations(upon inclusion of exterior products), shortcomings of such a theory, since we want to associate the entries of h to amplitudes, and is one of many ways to come to a motivation for investigation of a theory corresponding to a 'square root' of space-time, i.e spinors. These Kähler configurations become when one is considering what is supposed to be the "accurate" theory hyperkähler, see Part IV. Incidentally, this is also possible to perceive in the classical theory and corresponds to the vacuum in the scale of stable, i.e amplitude preserving, holonomies in the scale below

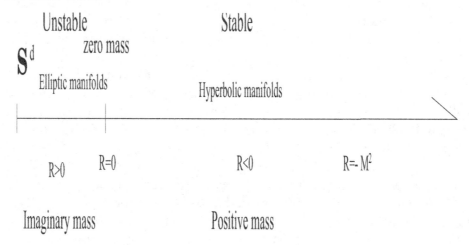

FIGURE 14. A scale containing the manifolds at dimension D. The mass scale is after a double Wick rotation made to insure correct statistics among the D-folds. Any of these could be hyperkahler, which then corresponds to them having maximal volume because the mass charge density is equally distributed on the D3- brane. ERRATA: The diagrams that are elliptic should be stable, and the ones with non-trivial topology and hyperbolic curvature are unstable if the the norm of the Ricci scalar divided by 4 is more than the squared mass of the background.

It might be worth to make some remarks on the dynamics of the scale above.[21] The following theorem, related to reasoning of the previous

---

[21]Errata; The scale above should have stable and unstable interchanged.

section, shows some of the behaviour of this scale for the case of negative square mass. Those correspond to unstable tachyonic space-times which are similar to bound configurations decaying to lower states;

Theorem 12.4 (Bochner). *Let $X_{(\omega,\theta)}$ be compact connected with positive Ricci curvature, then*

$$\pi_1(X) \cong \{0\}$$

It's content is that there are really no interesting geodesics on such states. The states of positive square mass, which thus roughly correspond to "free states" are actually in enormous abundance as compared to tachyonic(elliptic) states, and by this token, just to take an example, fixing demands on positive Ricci curvature with positive lower bounds together with the statement of vanishing fundamental group will fix things uniquely in some cases. Here is an example;

Theorem 12.5 (Sphere theorem). *Let M be compact with vanishing fundamental group satisfying*

$$0 < \frac{C}{4} < R < C$$

C some real constant. Then M is homeomorphic to the sphere S^n.

The above matters of instability also show in the behaviour of geodesics in tachyonic space-times; an extremal geodesic need not be minimal in those cases.

Smooth vacuum (Einstein) positive selfdual space-times are also scarce in dimension 4, $\mathbb{R}^4$, $K3$ and various quotients obtained by the action of some discrete group being an exhaustive list. It is well known that a Kähler property corresponds to being of extremal volume, see e.g Novikov et al.[1] for complex manifolds. Something similar holds for the hyperkähler case; The hyperkähler property implies them to correspond to manifolds of extremal volume, satisfying something similar to a 4-dimensional version of Fermat's principle, and are actually for the case of Minkowskian signature maximal in volume, much like the physics of the surface of an inflated balloon with equal rubber thickness and no extra pressure applied at any point.

Because of the abundance of hyperbolic states (the "usual" states) they form an interesting category to study, indeed there are entire books devoted to such study, e.g S.Kobayahi's Hyperbolic Complex Spaces for a complex example.

It is a theorem, Yamabe's theorem (See Milnor[1]) to be precise, that any smooth D-fold in dimension $D \geq 3$ is conformally equivalent to a

213

manifold of constant scalar curvature[22], something that is important to the classification above, indeed this implies that the mass-curvature defined above will coincide with this "canonical" constant curvature by functoriality of the construction. This is also important in some applications of quantum theory partially outside of the scope of this thesis that require the choice of a good gauge slice in some applications of gravity. It is also a theorem, Lohkamp's theorem, that any homeomorphism type in $D \geq 3$ admits a hyperbolic metric in the Ricci scalar.

---

[22]In dimension 2 this is contained in the great uniformization theorem of Riemann surface theory, in dimension 1 trivial.

The following references are mostly books, and subdivided into mathematics and physics references in book format. None of them are meant to be complete, but rather are supposed to be a list containing 1-2 books in the various areas relevant to this thesis from my book shelves. Subsequent to this follows a selection of research papers of relevance, mostly in physics. I hope that it might serve some help to other undergraduates.

References

[1] Royden, Real Analysis, Prentice Hall; 1998.
[2] Rudin, Principles of Mathematical Analysis, McGraw-Hill; 1976.
[3] Ahlfors, Complex Analysis, McGraw-Hill; 1979.
[4] Chirka, Dolbeault, Khenkin, Vitushkin, An Introduction to Complex Analysis, Springer; 1991.
[5] Remmert, Classical Topics in Complex Function Theory, Springer; 1998.
[6] Weil, Elliptic Functions according to Eisenstein and Kronecker, Springer.
[7] Cartan, H., Théorie élémentaire des fonctions analytiques, Hermann.
[8] Hormander, Complex Analysis in Several Variables, North-Holland; 1990.
[9] Krantz, Function Theory of Several Variables, Wadsworth and Brooks.
[10] Farkas and Kra, Riemann Surfaces, Springer; 1982.
[11] Nag, The Complex Analytic Theory of Teichmuller Spaces, Wiley Interscience; 1988.
[12] Bell, Brylinski, Huckelberry, Narasimhan, Okonek et al, Complex Manifolds, Springer; 1997.
[13] Kobayashi, Hyperbolic Complex Spaces, Springer; 1998.
[14] Dubrovin, Fomenko, Novikov, Modern Geometry, volumes I,II,III, Springer, 1984, 1985, 1990.
[15] Jost, Riemannian Geometry and Geometric Analysis, Springer; 1998.
[16] Lawson and Michelson, Spin Geometry, Princeton University Press; 1989.
[17] Donaldson and Kronheimer, The Geometry of 4-Manifolds, Princeton University Press; 1989.
[18] Kobayashi, Transformation Groups In Differential Geometry, Springer; 1995.
[19] Serre, Lie Algebras and Groups, Lecture Notes in Mathematics, Springer; 1965.
[20] Kac, Infinite Dimensional Lie Algebras, Cambridge Univerity Press; 1990.
[21] Novikov, Topology I, Encyclopeadia of Mathematical Sciences, Springer.
[22] Bourbaki, Topologie Generale; 1974.
[23] Bourbaki, Algèbre.
[24] Jones, Subfactors and Knots, Conference Board of The Mathematical Sciences No 80, American Mathematical Society; 1991.
[25] Neukirch, Algebraic Number Theory, Springer; 1999.
[26] Connes, Noncommutative Geometry; 1990.
[27] Choquet-Bruhat, DeWitt-Morette, Dillard-Bleick, Analysis, Manifolds and Physics, volumes I and II, North holland; 1982.
[28] Manin, Gauge Field Theory and Complex Geometry, Springer; 1997.
[29] Ward and Wells, Twistor Geometry and Field Theory, Cambridge; 1990.
[30] Atiyah, K-theory, W.A. Benjamin Inc.; 1967.
[31] Karoubi, K-theory, Springer; 1978.

[32] Milnor, Collected Papers, Publish or Perish Inc; 1994.

[33] Besson, Lohkamp, Pansu, Petersen, Riemannian Geometry, Fields Institute Monographs, American Mathematical Society.

[34] Wells, Differential Analysis on Complex Manifolds, Springer; 1980.

[35] Vershinin, Cobordisms and spectral Sequences, Translations of Mathematical Monographs, American Mathematical Society.

[36] Arnold, Ordinary Differential Equations, Springer, 1992.

[37] Hormander, The Analysis of Linear Partial Differential Operators, volumes I,II,III and IV, Springer.

[38] Hitchin, Segal, Ward, Integrable Systems, Oxford Science Pubications; 1999.

[39] Yano, Bochner, Curvature and Betti Numbers, Oxford University Press.

References

[1] Dirac, The Principles of Quantum Mechanics, Oxford Science Publications; 1930.

[2] Feynman, Leighton, Sands, The Feynman Lectures on Physics, volumes I,II and III, Addison-Wesely; 1965.

[3] Feynman, The Feynman Lectures on Gravitation, Perseus Publishing.

[4] Sakurai, Modern Quantum Mechanics, Addison-Wesely; 1994.

[5] Peskin and Schroeder, An Introduction to Quantum Field Theory, Addison Wesely; 1995.

[6] Hatfield, Brian, Quantum Field Theory of Point Particles and Strings, Addison-Wesley.

[7] Itzkyson, Zuber, Quantum Field Theory, McGraw-Hill.

[8] Adler, Quaternionic Quantum Mechanics and Quantum Fields, Oxford; 1995.

[9] Weinberg, The Quantum Theory of Fields, volumes I, II and III, Cambridge; 1996, 1996, 2000.

[10] Daniel S. Freed, Five Lectures on Supersymmetry, American Mathematical Society, 1999.

[11] Polchinski, String Theory, volumes I and II, Cambridge, 1998.

[12] Green, Schwarz, Witten, Superstring Theory, volumes I and II, Cambridge; 1987.

[13] Quantum Fields and Strings; A Course for Mathematicians, volumes I and II, Editors Pierre Deligne, Pavel Etingof, Lisa C. Jeffery, Daniel S. Freed, David Khazdan, David K. Morrison, John W. Morgan, Edward Witten.

In particular Eric D'Hoker, String theory lecture notes from the 1996-1997 year, Institute for Advanced Study, to be found in vol. II or on the web.

[14] Ketov, Conformal Field Theory, World Scientific; 1995.

[15] Schwarz, Elementary Particles and the Universe, Cambridge; 1991.

[16] Penrose, Rindler, Spinors and Space-time, volumes I and II, Cambridge; 1984, 1986.

[17] Misner, Thorne, Wheeler, Gravitation, Freeman; 1970.

[18] Gockler, Schucker, Differential Geometry, Gauge Theories and Gravity, Cambridge.

[19] Nakahara, Geometry, Topology and Physics, Graduate Student Series in Physics, Institute of physics publishing, 1990.

[20] Dirac, General Theory of Relativity, Princeton landmarks in physics.

[21] Chandrasekhar, The Mathematical Theory of Black Holes, Oxford Classic Texts

[22] Hugett, Mason,Tod, Tsou and Woodhouse, The Geometric Universe, Oxford.

Part III: Quantum Gravity and String Field Theory:
Studies at Depth

Few— but ripe.

Johann Friedrich Gauss

13. Introduction

In this part we undertake the deeper, and less heuristic, study of gravity by using the intuition we developed in part II, and the concepts that emerged such as n-admissibility, generalized supersymmetry, one particle space-times/ multiparticle space times(which we shall now identify with D3-branes and stacks of D3-branes respectively.) e.t.c.. Although we did some extensive checks in Problem 1-20 of part II these were many times but heuristic and very intuitive, so we shall find it nice to be able to rely on more traditional and straightforward methods.

As some checks of our quantizations we would like to do some calculations and see that they come out reasonable, and hopefully equivalent in both theories as well as in string theory. Now we are truly lucky, since a) various conjectures such as the emergence of string theory/AdS-CFT correspondences etc and b) we have consiously strived towards a tie up of string theory, NC geometry and twistor geometry, other authors have often made the checks in pertaining areas, e.g for the AdS-CFT conjecture, whose assumed underlying mechanism we display more fully in this part, in the beginning of the string theory era (mid 80's), or , as lately has been, between various noncommutativities and string theory. So all we will have to do in such cases is to wrap things up, ty them to what we have done, and thus put the individual parts of physics figuring in this thesis together. We shall take a number of more or less arbitrary choices; Checks of β-functions, mass spectrum, moduli of vacua, string theoretic calculations to 1/2-loop and general multiloop, string/weak coupling limits, T,S,U duals, exact and non-pertubative results. Inasfar as these matters we shall be brief and instead simply state results and interconnection, thereby referring to the various authors in the literature, and the references therein, which are responsible for those computations, and mostly only mention the relevant aspects that provide the ties.

On our own we shall be concerned with the explicit and less heuristic quantization of both theories in the physical dimension and relating them to string theory in $D = 10$, but brake off at 1-loop, as this will be sufficent to see that we are more or less right, and that our quantizations and scheme truly seem to generate string theory, thus providing the foundational basis of what we have chosen to call string field theory, i.e the combination of field theory and string theory as an interrelating whole rather in the usual sense of a field theoretic alternative to the first quantized world sheet formalism. It is in this course that we also prove Maldacena's theorem.

The systematics are as follows; We start off with more or less re-stating what we have done in PartII in a clearer and more condensed matter, and how string theory is supposed to arise. We then move on to the aformentioned checks, mostly checked and well proved in string theory, done by other authors in string theory where we on our own only have to check that we come out equivalent. Subsequently we perform the quantizations of the theories, inasfar as amplitudes concerning concentrating on graviton-graviton scattering in both theories, and check both in full generality and in the specific example of this scattering that the amplitudes of both theories coincide, and how they correspond to elements in the string pertubation theories. In particular we exhibit the field redefinitions to get the relations between the three theories in $D = 4$ and $D = 10$ respectively. Finally we mention some other things that other authors have done and do a couple of physical predictions.

13.1. Isomorphism and Defining Formulas.

13.2. A Brief Summary of Formulas.
We have in $D_\pm = 4$ the physical dimension on one of the helicities

$$\mathcal{L} = *(\ (*\mathbb{X}_+^\mu) D_- D_+ \mathbb{X}_-^\nu (G_{\mu\nu} + B_{\mu\nu})\ )$$

$*$ denoting Hodge star, wich will act as (anti)isomorphism of Hilbert spaces when we identify the appropriate exterior module of appropriate degree with it's dual. $\mathbb{X}$ has natural Chan-Paton factors given by

$$\mathbb{X}^{\mu,\alpha} = \bar{\Phi}\Gamma_\mu T^\alpha \Phi$$

which can be expressed more naturally, e.g for $M^a_{A\alpha A'\beta} = \sigma^a_{AA'} T_{\alpha\beta}$ as

$$X_- = (\bar{\Phi}_- \sigma^a T^{\alpha\beta} \Phi_+)\ \sigma_a \otimes T_{\alpha\beta}$$

or simply

$$X_-^{a\alpha\beta} = \bar{\Phi}_- \sigma^a T^{\alpha\beta} \Phi_+$$

One obtains the colored averaged space-time superfield by performing a trace over colors[23]. In the above we used the ON frame, and this will be central to the field redefinitions needed, since we will express everything firstly in that frame for our two quantization, whilst the stringy standard writes it in the coordinate frame, just as in the lagrangean above. Rewriting the lagrangean in that fram, which we may

---

[23]Implictly in this lies reducing the number of "identified" space-times. Coincident space-times is a more common word, describing the same thing, in the literature.

by invariance if we have gauge covariant superderivatives, we have the other form of our lagrangean as

$$\mathcal{L} = *(\ (*\mathbb{X}^a_+)D_-D_+\mathbb{X}^b_-\eta_{ab}\ )$$

for diagonlizable $G + B$. In the remaining we shall generally set the B-field to be vanishing, as this will not affect the present analysis to any greater extent for small such fields. We have components X_0, X_1, of exteriour degree 0 and 1 respectively being the embedding and vielbein in that order

$$f = \text{``embedding''} = exp_p(X_0)$$
$$X_1^a = e^a_\mu\theta^\mu = e^a_\mu dx^\mu$$

Where it is understood that in the above what truly is meant is a possibly singular embedding f, i.e. a higher dimensional sigma model map. exp_p denotes the exponential map w.r.t to the background over a point p, and this is in congruence with what we previously stated at the end of the conventions section of Part II.

For *real* Euclidean space-times the color averaged space-time $\mathbb{X}$ superfields take hypercomplex values by the above.

13.2.1. *Discussion of Part II.* We suspect that our hypothesis, introduced in part II, generates string theory as a 2/10 admissible theory, then including the two stacked space-times corresponding to various helicities, as well as the two off-shell dimensions corresponding to one off-shell dimension emanating from mass fluctuations added to a $D3$-brane of either helicity, i.e the world sheet. In order to show that we have an isomorphism of theories it suffices to show that all correlators are equivalent in our two hypothesis, as well as compared to string theory. We shall mainly be doing this in the same manner that one would usually prove two algebras to be be isomorphic—i.e by proving that their generators are isomorphic. In this direction we have, if we use an

Axiom 13.1. *We set for the remaining of this thesis, well known facts about quantum theory path integrals to be an axiom, in particular the fact that the worldline formulation of quantum field theory path integrals works will be condidered to be such an axiom. We also assume the axioms of quantum theory, in particular we assume additivity of similar charges in a quantum mechanical system of several components to a total charge of pertaining kind. The only exception to these axioms will be the Born interpretation of wayfunctions, as we will want to work with more general models (e.g the classical field theory of hydrodynamical*

222

*flows turned quantum by reusing the mathematics of quantum theory)
where there is not necessarily such an interpretation.*

the theorem

Theorem 13.1. *A field theoretic interpretation and treatment of our
formulae above implies*

(1) *World sheet conformal algebra on a string.*
(2) *Correlator functions and amplitudes corresponding to a string
progressing forward in stringy space-time D=10, up to the process of summation over spin structures and world sheet moduli.*

In the above the string space-time is mass off-shell and Matrix valued.

Torbrand Dhrif 2000. We start by the first statement and then proceed
to the second. The proof proceeds by using the world-line formulation
of path integrals and SYM in particular.

We have by our formulae for the correlator of an $\mathbb{X}$-field

$$< \mathbb{X}^a > = < \Phi^A(x_-)\Phi^{A'}(x_+) > \sigma^a_{A'A} = x_- \bullet - - - - - - - - - \bullet x_+$$

where we remembered the Feynman diagram interpretation of the
latter, and x_- is on the negative helicity space-time, whilst x_+ in the
positive helicity space-time, and we agreed that except for field values these space-times are identical, so that we could talk of only one
space-time instead of copies of it, which justifies the Feynman diagram
interpretation used. But then, since in the world-line formalism, we
know that a worldine can be treated as a 1-dimensional field theory
imbedded into space-time(and will give the correct SYM answers), it
must fall that this propagator can be precieved also as a path in space-
time with a field theory on it and not only a graphical device to aid
memory. But then we have a string at a fixed instant, in the sense of
having lagrangean content pulled back to a path in space-time and a
1-dimensional field theory on this interval. We must now show that
this indeed implies that we have conformal algebra on this supposed
string when it evolves. Let us identify the positive helicity space-time
with the negative helicity space-time, i.e do what people just slightly
incorrectly call letting the D3-branes being coincident, so that we set
$x_+ \mapsto x_-$, then the above propagator closes. Since we want the $\mathbb{X}$ field
to be uniquely valued this means that we can describe the string by a
an interval $[0, 2\pi]$, with periodic boundary conditions. Hence we have
NS conditions for the string corresponding to the $\mathbb{X}$ field when we make
the necessary identification to regard $\mathbb{X}$ on a single $D = 4$ space-time.
But then we can model this as an S^1 in the complex plane, or half that

223

circle when we do not identify endpoints. As for Ramond conditions, we note that we can also demand that the field be uniquely defined on the direct product of space-times to be uniquely valid, this before identification, something that must correspond to that it suffices that the spinors, which are taking values in $Y_\pm = \mathbb{T}_\pm \cong \mathbb{C}^2 \cong \mathbb{H} \cong \mathbb{R}^4$, are uniquely valued, $X_\pm = Y_\pm/\mathbb{Z}_2$, and since they must have degree half that of $\mathbb{X}$ fields they must lie , as a 1-fold covering of S^1 in $Y_\pm$, where the NS fields corresponded to a two-fold covering of the same circle. Hence they can only image half the circle in space-time, and must thus be antiperiodic when we choose the value midway as the origin. Now we must also be certain that we are not only dealing with the correct field theories and supposed string boundary values but also on the correct domains in space time, hence we must prove this. We do so by noticing the higher dimensional sigma model structure of our theories, which permits us to do reasoning about a domain by reasoning concerning a range, since the domain and range of fields are apriori equivalent when there are no interactions as we are mapping background conguences to interacting congruences. In 'usual' field theory we do not quite deal with field theory on $\mathbb{R}^4$, so no contradiction truly arises from $\mathbb{R}^4/\mathbb{Z}_2$ here. That we deal with $\mathbb{R}^4$ is certainly in some sense so, but only by abbreviation and convention, for the Green's functions are truly corresponding to Dirichlet problems on future space-time $\mathbb{R}_+ \times \mathbb{R}^3$ taking boundary value data on the slice at zero time $\{0\} \times \mathbb{R}^3$, and this is seen to coincide with the model that falls by the identifications of directions through the origin above, with some specific direction chosen as future. This is very important, and corresponds to the $i\epsilon$ in momentum space propagators. In the above, if we choose to just work with rays through the origin without any choice of future direction this is achieved by blowing up the singularity at the origin by simply taking away a small D^4 around the origin, which will then give the same toplogy. This is realised a follows; The identified rays at an equator, of $S^3 \times U$, U an open interval, will be the rays at spatial infinity at the infinte future and the given intant. We note the the condition that fields at spatial infinity at all instances must vanish, hence be equivalent. The corresponding rays are precisely the identified rays on the equator. But then it must also fall that have the same topology for the partial differntial equations.

We now turn to the issue of generating timelike translations now that the important issue of correct boundary value prescriptions on correct domains has been checked.

Let the Hamilonian generate temporal translations on the $\mathbb{X}$-field. We can do this by regarding a *usual* propagator

$$< \Phi^A(x_-)\Phi^{A'}(x_+) >= x'_- \bullet - - - - - - - - \bullet x_-$$

in, say, the negative helicity space time, as above, corresponding to the lagrangean $\mathcal{L}_{SYM,\pm} = \bar{\Phi}_\mp D_\pm \Phi_\pm$. But we *know* that this also has the interpretation of path in space-time by the worldline formulation again, say in the interval $[0,T]$. So if we let the end points of our original string sweep forward in their respective space-times we see that we have a world sheet extending in the volume given by the image of $[0,2\pi]\times[0,T]$ in the direct sum of space-times, where we now take direct sums of fields, one of each helicity. But then also follows, rewriting the intrinsic charge operators living on the field theories on respective intervals, by using the standard relation

$$Q_{total} = Q_1 \otimes 1 + 1 \otimes Q_2$$

Q_i the charge operator pertaining to both field theories on the various intervals, on the tensor product of Hilbert spaces on the two intervals. Hence also, noting that $-(D_i)^2$ would be the charge operator, this by the determinant argument in the n-admissibility section (Problem 16, Part II, see the below lemma.), which tells us directly that the space-time theory has to be proportional to a second order superderivative and the fact that we only have one coordinate, which thus determines the second order superderivative uniquely, that we have

$$Q = -D_1^2 - D_2^2$$

Hence $\mathbb{X}^a$ fields living on the product $[0,2\pi] \times [0,T]$ have vacuum defined by, e.g, $0 = Q\mathbb{X} = -D_+D_-\mathbb{X}_-$ and so are superharmonic. Since we know that we have superharmonic functions on an annulus or a halfannulus(if we do not identify the end points of the supposed string along the propagator from positive helicity space-time to negative helicity space-time we get halfannulus) in the plane, superconformal algebra also must fall, since the lagrangean on the worldsheet is implied, modulo total variational derivatives, by the equation of motion, and so must give us the string lagrangean. We can write down the result of such an integration,

$$(D_\pm\mathbb{X}^a_\mp)(D_\mp\mathbb{X}_{a,\pm})$$

listed above. We note that in the above the lefthanded fields need not be the conjugates of the righthanded, since they are constructed of spinors of different kind which are functionally independent, as the

225

reader may explictly check. Assume that we started with offshell space-times instead, so that we had e.g $D_\pm = 5$ (as for the standard Klein-Gordon case), then we obtain by analogous reasoning strings in $D = 10$. Now repeating the same argument gives, the massless superlagrangean listed a section ago, hence the usual local superconformal algebra on the world sheet must fall for that case too.

To ty together the proof we must now show that we truly must have charge operators $Q_i = -D_1^2$ on the intervals. It can be most easily seen how this becomes so if we work backwards, and this is ok since we will be working with equivalences. Extracting the superconnection from the super Yang-Mills lagrangean we have

$$\Phi_\mp^* D\Phi_\pm = D\mathbb{X}_\pm$$

on the timelike interval. Now we can either realize directly that this must be half the lagrangean, or better mathematically simply operate with D_1 a second time. The same goes for the spatial interval, which by the worldline formlism combined with the Feynman path interpretation of a propagator had to be interpreted in the same way. Since this would require an inductive step at this point if we continued that line of reasoning we cannot continue rigorously in this particular way, so for this reason we refer to the below lemma. But, under the assuption that the lemma is accurate, both dimensions have a set of charge operators which are as claimed, and so we are done, if we are allowed to use basic facts about path integrals, which is not selfevident.

As for the second part, the reasoning proceeds as follows; we note that since we have by $M_{\alpha\beta} = T_{\alpha\beta}$, $M_{AA'}^a = \sigma_{AA'}^a$ the appropiate symmetrizers of fields the accurate Chan-Paton factors, with their usual rules follow by by field theory arguments. Since the vertex operators we would now write down on the worldsheet theory by the above are the same as the ones we know from the string theory NS vertices we must conclude, after appropriate procedure of quantization and that, up to summation over moduli and spin structure, the correlators and amplitudes must be the same.

To conclude the theorem, we note that $\mathbb{X}^a$ was supposed to be a supertangent vector space, hence e.g. X_0^μ takes values in $U(N_c) \otimes \cdots$ after exponentiation in the integrated picture.

Under the axiom listed prior to the statement of this theorem and the lemma below the above now falls.

$\square$

The theorem above could be interpreted as stating that the probability interpretation, which forces us to use half-densities in terms of spinors, implies matrix strings. We will notice that this has a converse effect when we calculate string theory amplitudes—it will mean that we do not have square entites at the end of a calculation to get the correct result. We will go through this repeatedly when we calculate amplitudes later.

The above used lemma is

Lemma 13.1. *Assume the above lagrangeans to be of theories on Hilbert space $\mathcal{H}_\pm$ to be such that they induce the partition function*

$$det(iD_\pm)$$

Then it also falls, up to possible illdefinedness of infinite dimensional determinants, that the partition function it induces on the product $\mathcal{H}_+ \otimes \mathcal{H}_- \subset \bigoplus_{0=p=r+s}^\infty \bigotimes_1^r \mathcal{H}_+ \bigotimes_1^s \mathcal{H}_-$ is of the form

$$det(-D_+D_-)$$

Proof. Up to god forsaken ill definedness, cured by an arbitray and good enough renormalization choice of the reader,

$$det(iD_+)det(iD_-) = det(-D_+D_-)$$

$\square$

A small deviation from the last homomorphism property, which is of course at least not impossible, could be called homomorphism anomaly or suchlike. It is also expected to differ for different renormalization schemes, perhaps related by a field redefinition in some of them. The naive hope, which is perhaps too much of a hope, is that it vanishes. We will not be dealing with homomorphism anomalies at any greater extent in this thesis, that is breakings of relations of the form $\phi(x)\phi(y) = \phi(xy)$, ϕ a homomorphism of an appropriate algebraic structure, in this thesis, but we will show it to be explictly vanishing in a special case where the partition functions both each separate factor and the superstring function can be calculated and checked to be equivalent by methods of complex analysis. We shall also check this in the string pertubtion theory series, where it will vanish by an extension of the Chiral Splitting Theorem. One must differ between homomorphism anomalies of the first kind, pertaining to $det(D^2)\text{"="}det(D_+D_-)det(D_-D_+)$, leading from string theory to Hilbert-Einstein gravity, and those of the second kind, stemming from $det(D_+D_-)\text{"="}det(D_+)det(D_-)$, belonging to the reduction from full densites in Hilbert-Einstein gravity on each space-time to half densities in SYM.

Hence by the above falls that, after summation over moduli and spinstructues(the field theoretic "how" or rather "why" of this matter will more clear when we prove another similar thorem later) that if we include summation over spin structures and moduli that we have $D = 10$ string theory, where we shall at least discuss that issue. We shall, for the remainder of this thesis, refer to the theory on each $D3$-brane as quantum Hilbert-Einstein gravity, abberiviated to Hilbert-Einstein gravity, although this is certainly not a good name. For now we assume implicitly that string theory here is an average over moduli and spin structures at it's pertubation theory series, so that we can make comparisions to usual string theory.

14. Effective Actions

14.1. **Introduction.** As we already know that we start off with the accurate lagrangean in $D = 2$ world sheet formulation in noncommutative space-time

$$\mathcal{L}_M = \mathbb{X}^\mu D_+ D_- \mathbb{X}^\nu (G_{\mu\nu} + B_{\mu\nu}) + R$$

where we have included the Hilbert-Einstein term but not anything more of the non-free terms, pertaining to the Weitzenbrock-Bochner formula in part II, which when pulled back to the world sheet is topological and recognized as the Euler charecteristic, modulo a constant which has to do with the gravitational interaction, which we neglect in this discussion. We know, since the space-time loops must close on the space-times of various helicities when we want to talk about a single field $\mathbb{X}$ field, that we must in effect be dealing with type II strings. We want a chirally symmetric theory on obvious grounds since we have formulated the theory above in a symmetric manner.

Although it may—or may not— be that stringy calculations so far are wrong in $D = 10$ we shall not concern ourselves with remedying them in this thesis(the pathology being the dimension), as we must delimit ourselves somehow. We shall thus merely quote the results in $D = 10$, where one is supposed to be having one off-shell space-time too much, as it is there that it seems that most easily available material is, and we can of course not trust the mechanisms of dimensional reduction that string theorists have used. Accurate procedures shall be discussed in this thesis. Note that the off-shell space-times induce after compactification of the factors the topology $X_+^{(5)} \times X_-^{(5)} = S^5 \times S^5$, which, after continuation to physical space-time, is $AdS^5 \times S^5$. It is here that the most famous Maldacena conjecture lives, and we shall opt for a proof, as well as the accurate mechanism of dimensional reduction, which some readers probably already see.

We quote some results in $D = 10$ of type IIB which at least seems to be the theory which interests us, in the inaccurate $\mathbb{R}^{10}$ topology. Throughout this part we lean almost exclusively for string theory calculations and results on J. Polchinski, vol II, in particular the opening chapters and the chapter on physics in dimension 4, and Eric D'Hoker's string theory lecture notes from the year 1996-1997 at the Institute for Advanced Study throughout the following, in particular chapter 4, 5 and 9, where material on calculating amplitudes is located, something that we will be in need of later.

14.2. String Theory Effective Actions in $D = 10$. Polchinski II gives

$$S_{IIB} = S_{NS} + S_R + S_{CS},$$
$$S_{NS} = \frac{1}{2\kappa_{10}^2} \int d^{10}x \sqrt{-G} e^{-2\Phi}(R + 4\partial_\mu \Phi \partial^\mu \Phi - \frac{1}{2}|H_3|^2),$$
$$S_R = -\frac{1}{4k_{10}^2} \int d^{10}x \sqrt{-G}(|F_1|^2 + |\tilde{F}_3|^2 + |\tilde{F}_5|^2)$$
$$S_{CS} == -\frac{1}{4k_{10}^2} \int C \wedge H_3 \wedge F_3$$

Φ is the dilaton and of course not a Dirac field eventhuough our notation unfortunately coincides, H a 3-form torsion field strength, related to the antisymmetric tensor $B_{\mu\nu}$ by

$$H_{\mu\nu\kappa} = \partial_\mu B_{\nu\kappa} + \partial_\nu B_{\kappa\mu} + \partial_\kappa B_{\mu\nu}$$

F_i are the field stengths of Lie algebra valued forms denoted by $A_{(i-1)}$, as a special case of component fields previously discussed, taking pure degrees of the exterior cotangent algebra, related to the fields in the effective actions by

$$\tilde{F}_3 = F_3 - C_0 \wedge H_3, \tilde{F}_5 = F_5 - \frac{1}{2}C_2 \wedge H_3 + \frac{1}{2}B_2 \wedge F_3$$

The $NS - NS$ action are, according to Polchinski, the same as in IIA SUGRA, while the $R - R$ and CS (Chern-Simons) are supposed to be more or less similar.

For type I strings, which we really cannot yet "fit in" in our progressing "field theory $\Longrightarrow$ strings" quantization of gravity, and thus may prove irrelevant, we have

$$S_I = S_C + S_O,$$
$$S_C = \frac{1}{2\kappa_{10}^2} \int d^{10}x \sqrt{-G} e^{-2\Phi}(R + 4\partial_\mu \Phi \partial^\mu \Phi - \frac{1}{2}|\tilde{F}_3|^2),$$
$$S_O = \frac{1}{g_{10}^2} \int e^{-\Phi} tr_\nu |F_2|^2,$$
$$\tilde{F}_3 = dC_2 - \frac{k_{10}^2}{g_{10}^2}\omega_3$$

and ω_3 a CS 3-form

$$\omega_3 = tr_\nu(A_1 \wedge dA_1 + \frac{2}{3}A_1^3)$$

14.3. Naive Limit in $D = D_\pm = 4$. These above effective actions are, as previously stated, true only when considering the total space-time string geometry in both factors, so they may not be accurate. We can however, for now, perform a naive dimensional reduction by simply omitting terms that do not make sense and furthermore go limits an circumstances where we are close to the classical theory by e.g picing a vanihing torsion. We obtain for the IIB strings above

$$S_{NS} = \frac{1}{2\kappa_4^2} \int d^4x \sqrt{-G} (R^G + 4\partial_\mu \Phi \partial^\mu \Phi),$$
$$S_R = -\frac{1}{4k_{10}^2} \int d^4x \sqrt{-G} |d\phi|^2$$

ϕ an unknown scalar.

For the type I theory we have on the other hand

$$S_C = \frac{1}{2\kappa_4^2} \int d^4x \sqrt{-G} (R^G + 4\partial_\mu \Phi \partial^\mu \Phi),$$
$$S_O = \frac{1}{g_4^2} 2g_4^2 \int d^4x \sqrt{-G} tr_\nu |F|^2$$

where we explictly point out that we did *not* introduce an Einstein metric, but instead used the fact that we know that the lagrangeans are to retain their Weyl invariance in $D_\pm = 4$, as that was one of the most characteristic charecterizations of the four dimensional lagrangens considered in part II. The above gives us a possible identification of the unknown scalar above, namely the gauge potential $\mathbb{A}$, which in the low energy limit should also become a scalar, something that is consistent with the appearence of the dilaton as the low energy limit of scalings of the $\mathbb{X}$-field, whose scaling behavior we must separate, as we e.g only want to consider orthonormal vectors as vielbeins. This seems to confirm our sucpicion for the action relevant, since we there also had a pure Yang-Mills term, which we firstly guessed our way to and later proved had to be so by the most general admissible lagrangean (Problem 20, part II.)

14.4. Summary.

- We cited the effective actions in type I and type II, $D = 10$ string theory. After naive dimensional reduction for vanishing B-field, these coincided with the action cited, with the explicit factor of 4 that figured on elementary considerations in part II. In particular we saw that the dilaton interpreted as the norm of e.g. vielbeins came out as a scalar in the effective actions.

15. String Theory β-functions in $D = 10$

As the β-functions ought to depend on the way we compute effective actions, this because we may see them as variations of an effective lagrangean, and we do not have the time to remidy any possible shortcomings in previous formulae of $D = 10$ string theory, it suffices to give the results for e.g the bosonic string in arbitrary D, which corresponds to only retaing the $\mathbb{X}_0$-field in components and replacing superderivatives with usual derivatives, then understood that it may or may not be physically relevant in physical space in $D_\pm$ on-shell, $D = 2 + 2D_\pm = 2 + D_+ + D_-$;

$$\beta_{\mu\nu}^G = \tfrac{1}{2}R_{\mu\nu}^G - \tfrac{1}{8}H_{\mu\alpha\beta}H_\nu^{\alpha\beta} + D_\mu D_\nu \Phi,$$
$$\beta_{\mu\nu}^B = -\tfrac{1}{4}D^\alpha H_{\alpha\mu\nu} + \tfrac{1}{2}(D^\alpha\Phi)H_{\alpha\mu\nu}$$
$$\beta^\Phi = \frac{(D-26)}{6} + l^2(2D_\alpha\Phi D^\alpha - 2D_\alpha D^\alpha\Phi - \tfrac{1}{2}R^G + \tfrac{1}{24}H^2)$$

, with $2l^2 = \alpha'$ the Regge slope. These can be obtained as the equations of motion of

$$S(G, B, \Phi) = \frac{1}{2\kappa^2}\int d^D x \sqrt{-G}e^{-2\Phi}(R^G + 4D_\mu\Phi D^\mu\Phi - \frac{1}{2}|H|^2)$$

For the bosonic string in $D = 26$, doing a Weyl transformation $G_{\mu\nu} \mapsto G'_{\mu\nu} = e^{-\frac{4\Phi}{D-2}}G_{\mu\nu}$ this gives an explictly Weyl independent Ricci term in an action of the form

$$S(G', B, \Phi) = \frac{1}{2\kappa^2}\int d^D x \sqrt{-G'}(R^{G'} - \frac{4}{D-2}D_\mu\Phi D^\mu\Phi - \frac{1}{12}e^{-\frac{8\Phi}{D-2}}|H|^2)$$

In string theory G' is called the Einstein metric while G is called the string metric. The factor 4 in front of the dilaton term is again accurate when compared to the formulas of Part II, where we associated it with Bochner-Weitzenbröck identites.

We can from this, noting how the dimensional reduction from $D = 10$ space-time $X = \Sigma \times X_+ \times X+$, $\mathbb{X}_\pm$ on shell, try to obtain the on-shell space-time effective action by simply putting $D = D_\pm = 4$ and simply discarding terms of too high dimension. We notice at once, before listing the formula obtained, that when interrelating also the metrics, me must have

$$(G_{\mu\nu,\ X_+\times X_-}) = \begin{pmatrix} 0 & (G_{\mu\nu,\ -}) \\ (G_{\mu\nu,\ +}) & 0 \end{pmatrix}$$

, this by helicity preserving reasons. We now list the effective action naively obtained. It is

232

$$S(G, B, \Phi) \approx \frac{1}{2\kappa^2} \int d^4x \sqrt{-G}(R^G + 4D_\mu\Phi D^\mu\Phi)$$

Since $H_{\mu\nu\kappa}$ the torsion field must vanish at the space-times endowed with a Levi-Cevita connection, which is the classical limit, and this is related to Cartan first structure equation $H = \Theta = D\theta = DX_{1,\pm} = 0$, something that we inutioned as version of the classical equation of motion of a quantum system in part II (See the Hilbert Einstein appendix in Part II, where it was related to the Dirac equation) and we easily see that this intution was correct for vanishing B-fields. For the general case on the other hand we were wrong, what we should have done is to move over H field to the other side of the equality sign and simply written $\nabla_Y Z - \nabla_Z Y - [Y, Z] - H(Y, Z) = 0$, but since we guessed that vanishing torsion was probably an on-shell condition not totally wrong in the general case. So that is a case where someone elses idea is better, and the part II intution sufficed quite far, but not far enough. Luckily it is has no consequence whatsoever for vanishing B-fields for the above stringy formulas, which, because of the reasons listed above, still may be wrong.

We digress a little bit further on comparisions with Part II, and in this vein have that the above is exactly what one gets by setting $X_0^a \sim \Phi$, so that one only retains the scaling of a vector and discards the rest, that is the $SO(n, m)$ behaviour, in the formulas of Part II, and that this coincides with what we have previously heuristically derived in the Hawking radiation problem of Part II. The field Φ is the the field u in the spin cobordism section in the section "Spin Cobordisms , $\cdots$" of Part IV, this in general in the subsection entitled "D3-branes/One-particle Space-times in General", which is supposed to be related to time-like evolution of branes, NC geometry in Alain Connes sense etc., where we in particular discover (By leaning on Hitchin and a doing conjecture of our own to generalize) that it is related to an Obata connection on a Hyperkhler space-time X_- satisfying

$$G_- = e^{-2u}(ds^2 + \eta_1^2 + \eta_2^2 + \eta_3^3)$$
$$\eta_1 = Ids, \eta_2 = Jds, \eta_3 = Kds$$

We also note that the above relations are also precisely what one obatins by using the formulas of Part II summarised in the introductory section of this part.

15.1. Summary.

- The β-functions in bosonic string theory were cited. These induced effective actions by integrating the variation, which, after naive dimensional reduction for vanishing B-field, coincided with the action given, the explicit factor of 4 inclusive.
- We pointed out in passing the form that the metric must have on helicity preserving reasons, modulo Wick rotations to Euclidean space-time, as well as that there are no extra dimensions in string theory interpreted in that manner.

16. Mass spectrum and Functional Determinant Relations in String Theory and Hilbert-Einstein Gravity.

This section is a brief summary of the calculations made in the appendix "An Example of a Spin Boundary with Relations to Elementary Analytic Number Theory, Combinatorics, Theta Function Theory and D-branes.", in particular the subsections entitled "Strings, Fields, Mass and Degeneracy", "Superstring Partition Function and Mass Spectrum in $D = 10$" and "Mass Spectrum and Degeneracy" of $D3$-branes. This appendix can be found in part IV, and we assume it understood that $D3$-brane in our usage is equivalent to what we call one-particle spacetime. In that section we use the parallellizability of space compatified as S^3, owing to the interpretation of the quaternionic unit sphere, to be able to calculate the naive free partition function of the standard $D3$-brane in a state of oscillatory motion in the time like direction, and compare to string theory result, and find equivalence. In particular this implies vanishing homomorphism anomaly at tree level[24].

16.1. Summary.

- In the appropriate appendix, we calculate explictly the "quantum" Hilbert-Einstein gravity partition function, and find it to correspond exactly to what we expect, and also generating the additional degenracy factor associated with the string theory factor of 16. Hence vanishing "homomorphism anomaly" of the first kind to tree level falls.

---

[24]It strikes me, as I write this, that I may be able use that to prove that to all levels by sewing constructions. Furthermore, for discrete normally convergent determinants I may be able to prove vanishing homomorphism anomaly at finite orders for all theories, in arbitrary admissibility, which would then be a theorem restricted to the category of closed D-branes. But since the solution spectra will not be obstructed by compactification of "nice" ends' since they vanish at infinity, compactification of manifolds with boundaries and ends will be admissible, hence extending it throughout the smooth manifolds, this as the Novikov construction of the D-folds is exhaustive, with pertaining sewing along knots and links. Furthermore, for the case of non-trivial ends, forming closed doubles of manifolds will resolve the issue that case too. Hence since that will be true for each element on a generalized superdiagram that includes finitely many admissibility types, that will be true for the entire diagram, hence even non-diagonally over the dimension spectrum of the D-folds, as long as we restrict to integer dimension. I may or may not be able to extend this to non-integer dimension, however, and it may be that there is a non-trivial obstruction that I do not see directly. Furthermore, that will only be true at most on a dense subspace of the Hilbert spaces involved on non-compact manifolds, which may not even be countably separable, although the closed ones are certainely always separable.

- As a consequence of the above, we find equivalent mass spectrum in our quantization of gravity as compared to string theory, this with the understanding that additional degeneracy pertaing to the factor 16 in the appropriate partition function also comes out.
- In the appropriate appendix, we find mass spectrum at high masses given by

$$\rho(m) = m^{-\frac{3}{2}} exp(\frac{m}{m_0}), \ m_0 = const.$$

where we empasize that this spectrum is discrete, as $S^1 \times S^3$ is compact, and has been turned to contiuum description in the above.

17. Introduction to Quantization of The Three theories.

In this very large section we shall be concerned with the quantization of the three theories involved. We shall start by a quite heuristic procedure, just to get a feel for things first. Then we shall briefly summarize a quantization procedure and way of calculating Feynman rules which was made for the special purpose and experience of quantizing gravity, which the author calls "geometric" (or "noncommmutative") quantization. These names are admittedley ill chosen, as they coincide with the names of other ideas proposed by other authors which are only vaguely related, but we feel we have other things to put our mind to work on than quibbling with ourselves over names an leave to the readers to choose whatever name they whish for. It is in effect a natural extension of the Baitlin-Villikovsky formalism, BRST cohomology, etc, encoded in the tensor algebra of an infinite dimensional (pseudohermitean, pseudoriemannian, or special cases of Banach) manifold and a relative to the "geometric quantization" in th book of Woodhouse, as well as with noncommutative quantization, although we shall not go into the links in this thesis, but merely have provide enough information to be able to systematicall obtain Feynman rules of any theory of any admissibility type, in particular the Feynman rules of our three theories.

Finally we end this section with a number of examples of diagrams.

17.1. Heuristic Amplitudes and Crossections of The Three Theories.

In these subsections we undertake the task of comparing amplitudes, invariant matrix elements, crossections etc between the relevant theories featuring in our conjecture. The hope is to as far as possible confirm that they give the same results. After briefly sketching the interrelations to the reader, the empahsis is instead put on calculations, as we feel that this says more. We shall however, omit non-relevant parts well known to string theorists and quantum field theorists in general, and instead try to concentrate on the non-trivial parts. To train ourselves in writing down amplitdes in general we shall also write down high loop amplitudes—although we may not make the computation, this in order to make the reader have enough examples to look at to understand the Feynman rules of string field theoretic gravity.

17.1.1. *Theoretical Developments and Relations.*

We shall, firstly, only opt to calculate the crossections for graviton-graviton and supergraviton-supergraviton scattering, and only at a much later stage begin to worry about more general circumstances, this because it is best to understand the relevant Feynman rules and underpinnings first, and for now shall concentrate on doing so without dimensional overcounting in our test case. The main tool to pull this off will be the Chiral Splitting Theorem and then dimensional reduction of the last dimension via mass spectral density arguments, related to getting one space-time as a 4-dimensional space-time wave front in Kleins space-time. Of course the latter is naive, and would induce severe constraints on the homotopy type of a space-time if assumed to be embedded in Kleins original $\mathbb{R}^5$, but as we do not do so, rather merely assume that we can put it in an arbitrary five-dimensional space-time, e.g. in the form of the space-time factor factor in $X_+ \times S^1$, this is not a constraint. The diagrams we are looking at are of the type

on , say, X_- the on-shell leftmoving/holomorphic space-time in $\Sigma \times X_+ \times X_-$. Let us work with IIB on $X = \Sigma \times X_+ \times X_-$ as we want the SYM to pop out at the other end, and that theory is symmetrical in both chiralities. We remind the reader of our three theories, String $D = 10$, which is 2-admissible, Hilbert-Einstein, $D_\pm = 4$, which is also 2-admissible, and finally SYM, $D_\pm = 4$ which is 1-admissible. We want ot interrelate these. To this end we will need the

Theorem 17.1 (Chiral Splitting Theorem, As Stated By E. D'Hoker).
We begin by defining factorized vertex operators; A vertex operator V_i

$$ V_i = \int dz^2 d^2\theta W_i(z_i, \theta, \bar{z}_i, \bar{\theta}_i) $$

238

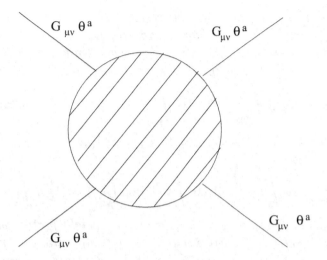

$\mathrm{G}_{\mu\nu}\,\theta^a$ $\mathrm{G}_{\mu\nu}\,\theta^a$

$\mathrm{G}_{\mu\nu}\,\theta^a$ $\mathrm{G}_{\mu\nu}\,\theta^a$

FIGURE 15. Two arbitrarily colliding gravitons.

is factorized provided V_i is the product of holomorphic/leftmoving/positive helicity and antiholomorphic/rightmoving/negative helicity factors,

$$W_i(z_i, \theta_i, \bar{z}_i, \bar{\theta}_i) = W_{i,L}(z_i, \theta_i) W_{i,R}(\bar{z}_i, \bar{\theta}_i)$$

e.g for $NS - NS$ massless states we have

$$W_i(z_i, \theta_i, \bar{z}_i, \bar{\theta}_i) = \zeta_{i,\,\mu} D_+ X^\mu e^{ikX_L} \bar{\zeta}_{i,\,\mu} D_- X_+^\mu e^{ikX_R}$$

the correlation function at fixed internal loop momenta p_I,

$$p_I = \oint_{A_i} dz d\theta D_+ X$$

A_I *a basis for $H_1(\Sigma, \mathbb{C})$ the first homology group, $dim(H_1(\sigma, \mathbb{C})) = 2g$, $g = h$ the genus or number of handles, is given by*

$$< W_1 \cdots W_N > (p_I) = \int DX \int DBDC W_1 \cdots W_N$$
$$= \Pi_{I=1}^{n=2g} \delta(p_I^\mu - \oint_{A_i} dz d\theta D_+ X^\mu) \Pi_{k=1}^{dim \mathcal{M}_s} | < \mu_k, B > |^2 e^{-(S_X + S_B C)}$$

Brackets denote insertion—for notational conventions see the references. The first part of the "Chiral Splitting Theorem" theorem states that the RNS amplitudes decompose as

$$< W_1 \cdots W_N > (p_I) = \delta(k) C_\nu^F \bar{C}_\nu^F$$

where $C_\nu^F = C_\nu^F(z_i, \theta_i, \zeta_k, m_k, p_I, k_I)$

is a complex analytic function of the supermoduli $m_k, k = 1, \cdots, dim \, s\mathcal{M}$, of the insertion points (z_i, θ_i), and of the leftmoving factors $\zeta_{i,\mu}$ of the polarization vectors.[Brackets denote note added by the author: i.e, the

239

parts only depend on respective space-times.] $\bar{C}_\nu^F$ *is the complex conjugate of* $C_{\nu'}^F$ *with the same spin structure, [So, it corresponds to the reflected path in the reflected copy of a space-time, i.e* X_+^5 *if was the original leftmoving* X_-^5 *copy.] The second part states for IIA/IIB*

$$A_k = \delta(k) \sum Q_{\nu\bar{\nu}} \int_{\mathbb{R}^{10h}} d^{10h} p_I \Pi_k \int_{s\mathcal{M}_k} dm_k d\bar{m}_k \Pi_{i=1}^N \int_\Sigma d^2 z d^2 \theta C_\nu^F \bar{C}_\nu^F$$

and for heterotic $E_8 \times E_8$, $spin(32)/\mathbb{Z}_2 = SO(32)$,

$$A_k = \delta(k) \sum Q_\nu \int_{\mathbb{R}^{10h}} d^{10h} p_I \Pi_k \int_{s\mathcal{M}_k} dm_k \int_{M_k} d\bar{m}_k \Pi_{i=1}^N \int dz d\theta \int d\bar{z} C_\nu^F \bar{C}^B$$

Here $Q_{\nu,\bar{\nu}}, Q_\nu$ *are spin structure dependent weight factors wich realize the GSO projection. If carried out in the supermoduli picture,* $\nu, \bar{\nu}$ *should run over only even or odd, but if the integrals are understood in components, then one should sum over all individual spin structures in each class as well. In general* $Q_{\nu\bar{\nu}} = \pm 1$, $Q_\nu = \pm 1$.

We need to adapt this theorem to the situation at hand, with two stacked space-times, so that we gan get accurate crossections in the physical dimension $D_\pm = 4$. This is done like this, notice that what is necessary to make this string amplitude become a probability is to factorize, just as in $dx^{10} = d^5 x_- d^5 x_+$

$$d^{10} p = d^5 p_+ d^5 p_-$$

on the two off-shell space-times, and do similarly with the worldsheet degrees of freedom z, $\bar{z}$(which are thus to be treated independently). Then we simply have to project out the states of the appropriate mass in usual Källen-Lehmann spectral density way for correlators.

We remind ourselves of, e.g. for a free scalar ϕ,

$$< \phi(x_1)\phi(x_2) >$$
$$= \int \frac{dm^2}{2\pi} \rho(m^2) < \phi_m(x_1)\phi_m(x_2')) >$$
$$= \int \frac{dm^2}{2\pi} \rho(m^2) D_F(x_1 - x_2, m^2)$$

D_F the usual Feynman propagator. Time ordering (or rather, conformal) ordering is always implicit when not otherwise stated, here and elsewhere in this thesis. One generalizes this easily to

Lemma 17.1.

$$< \phi(x_1) \cdots \phi(x_n) >$$
$$= \int (\frac{dm^2}{2\pi})^n \rho(m_1^2, \cdots, m_n^2) < \phi_{m_1}(x_1) \cdots \phi_{m_n}(x_2')) >$$

where we have $\rho(m_1^2, m_2^2) = \rho(m_1^2)(2\pi)\delta(m_1^2 - m_2^2)$.

240

Proof.

$$< \Pi_{i=1}^n \phi_i >$$
$$= \{\phi_i = \int dm_i^2 \phi_{m_i} \rho(m_2^2),$$
$$\phi_{m_i} = \int \frac{dm_i^2}{2\pi} \frac{dm_{i+1}^2}{2\pi} |m_i^2><m_i^2|\phi|m_{i+1}^2><m_{i+1}^2|$$
$$= \{<m_i^2|\phi|m_{i+1}^2> = (2\pi)\delta(m_i^2 - m_{i+1}^2)f(m_i^2)\}$$
$$= |m_i^2><m_i^2|\phi|m_i^2><m_i^2|\}$$
$$= <\Pi_{i=1}^n \int dm_i^2 \rho(m_i^2)\phi_{m_i} >$$

where we recover the familiar spectral density function by, e.g for complex scalars,

$$< \phi(x_1\bar{\phi}(x_2)) >$$
$$= \int \frac{dm_1^2}{2\pi} \frac{dm_2^2}{2\pi} \rho(m_1^2)\bar{\rho}(m_2^2) \underbrace{<\phi_{m_1}(x_1)\bar{\phi}_{m_2}(x_2)>}_{=\Delta(m_1^2, x_1, x_2)(2\pi)\delta(m_1^2 - m_2^2)}$$
$$= \int \frac{dm^2}{2\pi} |\rho_1(m^2)|^2 <\phi_m(x_1)\phi_m(x_2)) >$$

with $|\rho_1(m^2)|^2 = \rho_{<2>}(m^2) = \rho(m^2)$ the classical spectral density function.

$\square$

We can use the naïve expectation that all multiparticle spectral density functions for $1, 2, \cdots$ respectively have the looks

$$\rho_{<1>}(m^2) = (2\pi)\delta(m^2 - m_\infty^2) + \rho'_{<1>}(m^2),$$
$$\rho_{<2>}(m^2) = (2\pi)\delta(m^2 - m_\infty^2) + \rho'_{<2>}(m^2),$$
$$\cdots$$

with support $supp(\rho'_2(m^2)) \in [4m_{pair}^2, \infty)$, where m_{pair} is the lowest individual mass in the lightest pairs which can be created from the ϕ particle. m_∞ is the interacting mass at infinity of the S-matrix, i.e the physical mass. (This is, of course, also subject to following, e.g, the Feynman rules of the vertices of the lagrangean, which we go through later.). In our predicament we can simply neglect $\rho'(m^2)$, which enables to return to our previous amplitudes and know what to do with them when we have reduced the first 5 dimensions stemming from the space-time with opposite helicity. We shall use this to reduce the expressions to the correct masshell, thereby retaining the 4-dimensional wave-front in the range of our higher dimensional sigma model and so have reduced on the domain correctly as well by taking into account the various other mechanisms as well. It is an accurate observation on behalf of the reader at this point to claim that we are thus doing field theory on a wave front. And then we map this free (background) wave front to the interacting wave front corresponding to an interacting physical system, so that this is what the sigma model map corresponds

241

to classically. Due to the nature of wave fronts this means that our D-branes more look like objects from algebraic geometry than the theory of manifolds, think e.g of how Hamilton-Jacobi equations give wave fronts as smooth varieties. This wave front mechanism is actually responsible for most dimensional reductions I know of and is in a first naive approximation best modelled by taking a direct limit over sheaves of pertaining sections for trivial transversal monodromy to the appropriate D-brane and associated algebraic objects in various topological theories. In particular the mutiplicity of such varieties is coincident with the concept of N-admissibility as inspected by functional determinants in Part II and IV. One should point out that due to various singularity blow-up theorems of algebraic geometry etc we can always locally model our D-brane as a collection of covering spaces of manifolds with different dimension. Usually, we shall do as in this thesis and simply naively think of it as a manifold. The general situation is contained in the generalized super (D-brane) diagrams, and we will not touch this further. Let us go back to our pressing matters.

Hence by the above we obtain, for type II, which is what we concentrate on,

$$A_h = \delta(k) \sum_{\nu,\bar{\nu}} Q_{\nu\bar{\nu}} \int_{(\mathbb{R}^{5h})_+ \times (\mathbb{R})^{5h}_-} (d^5 p_I^+ d^5 p_I^-)^n \int_{s\mathcal{M}} dm_k d\bar{m}_k \Pi_{i=1}^N \int_\Sigma d^2 z d\theta^2 C_\nu^F \bar{C}_\nu^F$$
$$= \delta(k) \sum_{\nu,\bar{\nu}} Q_{\nu\bar{\nu}} \int_{(\mathbb{R}^{4h})_+ \times (\mathbb{R}^{4h})_-} \int_{(\mathbb{R})_- \times (\mathbb{R})_+} dp_I^{\Sigma-} dp_I^{\Sigma+} \int_{s\mathcal{M}} dm_k \Pi_{i=1}^N \int d^2 z_i d\theta_i^2 C_\nu^F \bar{C}_\nu^F$$
$$= \delta(k) \sum_{\nu,\bar{\nu}} Q_{\nu\bar{\nu}} | \int_{(\mathbb{R}^{4h})_+} \int_{s\mathcal{M}} dm_k dp_I^{\Sigma-} \Pi_{\nu=1}^N \int \Sigma_- dz_i d\theta_i C_\nu^F |^2$$
$$= \{Set\ P_I^{\Sigma-} = m\ \} = \delta(k) \sum_{\nu,\bar{\nu}} Q_{\nu\bar{\nu}} | \int_{(\mathbb{R}^{4h})_+} \frac{dm^2}{2m} \int_{s\mathcal{M}} dm_k \Pi_{\nu=1}^N \int \gamma_- dz_i d\theta_i C_\nu^F |^2$$

In the above the loop momentum is assumed on the fivedimensional mass-shell, this so being because the loop momentum is assumed transversal to the evolution direction in five dimensions, thus restrainted to live in four dimensions, as in usual $D = 4$ field theory. Assuming our delta like-spike in the spectral density and throwing away the off-shell part from interactions and other Hilbert spaces than the one corresponding to the relevant mass states at infinity, i.e discarding the ρ' part and possible other spikes except for the one at the relevant mass-shell, we get

$$\delta(k) \sum_{\nu,\bar{\nu}} Q_{\nu\bar{\nu}} | \underbrace{\int_{(\mathbb{R}^{4h})_+} \frac{1}{2m} \Pi_k \int_{s\mathcal{M}} dm_k \Pi_{\nu=1}^N \int \gamma_- dz_i d\theta_i C_\nu^F}_{:=\mathcal{M}} |^2$$

with γ_- the SYM path in the negative helicity space-time. But this $\mathcal{M}$ we also recognize as a very good candidate for the SYM invariant

242

matrix element in a sewing construction in the worldline formulation of field theory, if we are dearing enough to interpet the above as a crossection that missies some simple factors. But since

$$W_\pm = \zeta_\pm D_\mp X_\pm^\mu e^{ikX_\pm}$$

we have, if we include integration over the superpath $\gamma_\pm$ in either space-time

$$\int_{\gamma_\pm} dz_i^\pm d\theta_i^\pm D_\mp X_\pm^\mu e^{ikX_\pm}$$

and by $D_\pm X_\mp^\mu = \{X_-^\mu = X^{AA'}\} = \bar\Phi_\pm D_\pm \Phi_\mp$ or, more explictly,

$$D_\mp X_\pm^a = D_\mp \bar\Phi_\mp^A \sigma_{AA'}^{a,\pm} \Phi_\pm^{A'} = \bar\Phi_\mp^A D_\mp \sigma_{AA'}^{a,\pm} \Phi_\pm^{A'}$$

with $D_\pm = D_\pm + \omega_\pm + \mathcal{A}_\pm + \cdots$ the covariant space-time superderivative pulled back to the appropriate space-time path, and hence, letting $\bar\Phi D_{X_\pm} \Phi$ be belonging to the usual Dirac operator we get for the free part

$$\bar\Phi(\partial_{\gamma_+} \cdot \Gamma_+ + \partial_{\gamma_-} \cdot \Gamma_-)\Phi$$

with the two parts projected to the world sheet clearly visible in free form, each part giving the antiholomorphic/holomorphic sector, depending on each rightmoving/leftmoving space-time. Geometrically, we can thus think of the n-loop Riemann surface with $2g = g_+ + g_-$ cohomology classes in first cohomology group, thus by Hurewicz isomorphism being of homotopy type $\pi_1(S^1 \times S^1 \cdots S^1) \cong \bigoplus_{2g=g_++g_-} \pi_1(S^1)$ in the middle line of the Hodge decomposition(Σ is compact Kähler, hence Hodge's decompositon of $H^1(X, \mathbb{C}) \cong H_1(\Sigma, \mathbb{C})$ applies) in a Hodge diamond of the free paths (or rather, background) paths in each helicty space-time we expand about

$$(H_{\bar\partial}^{(p,q)}(\Sigma_{h=h_\pm}, \mathbb{C})) = \begin{pmatrix} & \mathbb{C} & \\ \oplus_{h_+}\mathbb{C} & & \oplus_{h_-}\mathbb{C} \\ & \mathbb{C} & \end{pmatrix}$$

which is perfect considering the fact that the $\partial_{\bar z}$-operator is the Dirac operator on a worldline, and is the "usual" part in $\bar\partial + i_{\bar\theta} = \bar\theta \partial_{\bar z} + \frac{\partial}{\partial\bar\theta} = D_+$, $D_+ := \slashed{D}_+ = \partial_{\bar z} + \bar\theta\frac{\partial}{\partial\bar\theta}$, $d\bar z = \bar\theta$, i.e the free antiholomorphic superderivative, where we used the identies that in Part II permitted us to recognize that D^2 was a Lie derivative, hence generating admissible lagrangeans. In the above the cohomology classes to the left/right are

243

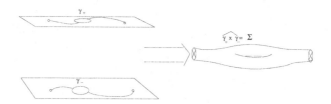

FIGURE 16. Paths γ_+, γ_- in the two space-times, and the string they produce. The above is an example for $h = 1$. Hat denotes compatification in the above.

thus part of the loop body of a SYM diagram in 5/4-dimensional space-time $X_\pm$ of the appropriate helicity whilst the other generator comes from the second D-brane of opposite helicty stacked onto the other. Of course we can always measure mass and helicity at infinity of the S-matrix, so we must stick to the $D = 4$ space-time at infinity.

But then we have established that, modulo summation over spin structures and moduli, since we have recoginzed the SYM worldline vertex operators in the above construction,

$$
\begin{aligned}
A_{h=L,Strings} \\
= \delta(k) \sum_{\nu,\bar{\nu}} Q_{\nu\bar{\nu}} \int_{(\mathbb{R}^{5h})_+ \times (\mathbb{R})^{5h}_-} (d^5 p_I^+ d^5 p_I^-)^n \int_{s\mathcal{M}} dm_k d\bar{m}_k \Pi_{i=1}^N \int_\Sigma d^2 z d\theta^2 C_\nu^F \bar{C}_\nu^F \\
= \delta(k) \sum_{\nu,\bar{\nu}} Q_{\nu\bar{\nu}} | \underbrace{\int_{(\mathbb{R}^{4h})_+} \frac{1}{2m} \int_{s\mathcal{M}} dm_k \Pi_{\nu=1}^N \int \gamma_- dz_i d\theta_i C_\nu^F}_{:=\mathcal{M}} |^2 \\
= \delta(k) \sum_{\nu,\bar{\nu}} Q_{\nu\bar{\nu}} |\mathcal{M}_{SYM,\,L}|^2
\end{aligned}
$$

where L stands for the loop order of the SYM diagrams in the two space-times, and $\gamma_\pm$ the approriate superpaths in the two space-times, with masses accoring to their curvature. $\mathcal{M}$ stands for invariant matrix element in the above. Notice how well this fits in with our claim that string theory deals with full densities while SYM with halfdensities. We have

$$
C_{\nu,m}^F = M_\nu \Pi_{i=1}^N W(z_i, \theta_i)
$$

M_ν from determinants. What, however, is certainly the case is that this is a little bit too sketchy, and, furthermore, we did not even touch the issue of summation over spin structures and moduli, at least not from the SYM perspective of paths in space-time and stackings. Having seen this, and admitted the shortcomings to ourselves, we have all the reasons and most components to state a theorem, and try to attempt it's proof. We include some things that will be included later in this

244

theorem, so that we obtain a fairly large reference theorem that we can refer back to. We sum up so far before going on to the theorem.

17.2. Summary.

- We displayed above, sketchedly but quite clearly via elementary methods, how to reduce the string pertubation (or , rather, when only retaing component fields of eack space-time, Hilbert-Einstein) theory series to SYM on the two space-times that make up, via stacking and two additional mass off-shell dimensions, the string space-time. We stated that our reasoning, however, was a little bit too sketchy, and that we wished to opt for a theorem with proof in the usual sense instead.
- We failed to provide a mechanism that yields summation over moduli and spin structures in the above line of reasoning, but claim that we shall remedy this later.

18. The Main Theorem

Theorem 18.1. *Let the IIA/B pertubation theory series be given in $D = 10$, and let SYM be given in $D_\pm = 4$ be given, as well as Hilbert-Einstein gravity in $D_\pm = 4$. Then*

(1) *Effective Dimension Coincidence: All of these theories are Weyl invariant in their respective dimension, thus they are in their effective dimensions, to which they tend to when being in other dimensions.*

(2) *Probability interpretation $\longrightarrow$ strings/Triality statement: There are, pointwise in the moduli, bijections*

$$IIA/B \iff (Q.)H.E \iff SYM$$

in terms of diagrams of the theories in the various admissibility types, then not incuding the overcounting of diagrams caused by different helicity and color quantum numbers, which causes matehmatically identical copies, which may or may not differ by values of physical constants etc. In field components it is given by

$$(\mathbb{X}^\mu) = (\mathbb{X}^{\mu+}_+) \oplus (\mathbb{X}^{\mu-}_-), \ (D = 10 \mapsto D =_\pm= 5/4)$$
$$\mathbb{X}^\mu_\pm = \bar{\Phi}_\mp \sigma^\mu_\pm \Phi_\pm, (H.E, \ D_\pm = 4 \mapsto SYM, \ D_\pm = 4)$$
$$\mathbb{X}^{\mu, \ \alpha\beta}_\pm = \bar{\Phi}_\mp \sigma^\mu_\pm T^{\alpha\beta} \Phi_\pm$$

and the space-time vectors thus obtained are $T_p X \otimes U(N_C)$ valued, i.e take values in the Lie algebra over a space-time point, this by tensor product formulae.[25]

(3) *How to Compute Ampiltudes and Crossections: A mass-fixed string diagram and amplitude in $D = 10$ Matrix space-time corresponds to the square of an invariant matrix element, i.e a crossection of SYM in $D_\pm = 4$.*

---

[25]The reader is implored to notice that the last formula is related to the color averaged formulas by traces in Lie algebras. This does not, however, necessarily imply that one generically has vanishing average fields since the generic gauge group is not $U(N)$, but rather $GL(N_C, \mathbb{C})$ for finite time S-matrices. When considering S-matrices at infnty we reduce to $U(N_C)$, which corresponds to the case above, but must be careful not take a trace when discussing physical currents at any time, this as color is observable. This is like usual probability currents, if we average over colors naively by a trace then we get zero probability current, but this does not imply inexistance of physically measurable quark currents. The reader who insists at taking averages over $U(N_C)$ in particular for S-matrices at infinity for some valid purpose may take averages in other ways at infnty, e.g the obvious, just sum them and divide by the rank of $U(N_C)$. E.g, the space-time we see is the color average in this second sense in $U(1) \times SU(2) \times SU(3) \times \cdots$.

246

(4) *Field Interpretation:* With fields as follows; $\mathbb{X}_0$ the embedding via exponentiation, $\mathbb{X}_1 = \theta$ the vielbein as a vector valued form, which corresonds to Dirac matrices, Φ_0 the squarks, corresponding to sigma model maps on and into the twistor/spinor brane, locally having tangent $\mathbb{C}^4 \cong \mathbb{T}$, and Φ_1 the vielbein in twistor/spinor space, also simply called Dirac field. A twistor brane is called a Y-brane in the following, as in $X, Y, \cdots$. Furthermore, this is compatible.

(5) *Maldacena Statement:* If regarding the off-shell string spacetime obtained by the usual SYM topology $S^5 \sim AdS^5$ on the factors, then we obtain the statement of the classical Maldacena conjecture on $AdS^5 \times S^5$.(This is a conjecture of Juan Maldacena, and only the proposed underlying mechanism is the authors.)

(6) *D-brane/C^*-algebra correspondence to NC Geometry:* On relevant twistor integrable space-time backgrounds (vacua) with reasonable physical properties the Euclidean space-times are Weyl selfdual and hyperkähler, with metric

$$G_{X_-} = e^{-2u}(ds^2 + \eta_1^2 + \eta_2^2 + \eta_3^2)$$

, u being a harmonic dilaton field, corresponding to scalings of the $\mathbb{X}$ field (this statement is due to Hitchin and not to the author). (The authors claim, which is a conjecture, now follows). This gives a NC geometry correspondence by the conjecture (See the appropriate appendix):

Conjecture 18.1 (NC-Geometry/ D-brane Correspondence). *Let det* : $\mathcal{A} \mapsto \mathbb{C}$ *be the determinant homomorphism from an appropriate C^* algebra to an appropriate field, here taken as $\mathbb{C} = \mathbb{R}^{\mathbb{C}}$. Then, after continuation to the Euclidean region so that the unitary action of the Hamiltonian is a scaling of the determinant instead of preserving it, we have our spatial sets induced by varieties of the form $det(O) = c$, $c \in \mathbb{C}$, where it is understood that we act by the determinant on appropriate initial value data on a spatial slice at some a priori given instant on a D-brane.*

This last conjecture gives a natural interconnection to Morse theory, thus in a special case (the above) foliating a space-time by the dilaton. Another example is in five dimensional spacetime where we have foliation by the Ricci scalar to obtain uusual space-times corresponding to various masses.

(7) *Physical Predictions:* and this suffices to make eight physical predictions.

Proof. The proof of this theorem is spread all over this thesis. We shall only be concerned with the parts that fit into this context presently and instead refer to the proper proofs in appendices and elsewhere when we can do so.

(1) Weyl Invariance: Is trivial/well known for the Lagrangean in $D = 10$ string theory at the classical and quantum level respectively, as manifisted in Weyl invariance of the world sheet lagrangean and vanishing Weyl anomaly in that order. For string theory we cannot regard the invariance dimension lagrangean as the space-time invariance dimension of the theory since the lagrangean lives on only the world seet and not the entire stringy space-time. Hilbert-Einstein gravity, on the other hand, is obiously dealing with self maps in terms of higher dimenisional sigma model maps in a space-time lagrangean, so it suffices to look at the classical invariance dimension of the lagrangean if it is consistent with an assumed invariance dimension at the quantum level, which is easily seen to be $D_{\pm} = 4$, where we again remind the reader that the Riemann curvature with co-efficents in the various gauge groups is functionally dependent only on the connection a priori. That the Yang-Mills part in SYM is well-known to be Weyl invariant, as well the interaction part of the SYM has Weyl invariance is also trivial by normalization conditions of spinors, which gives them their dimension. Now we must prove this to hold at the quantum level too. In the operator field formulation of quantum field theory the amplitudes will, however, be the result of the time-ordered exponential action of the interaction lagrangean, which is trivially seen to retain the Weyl invariance of it's classical counterpart. Hence, remembering that the vacuum normlization will scale in the same manner as the unnormalized ampliude, Weyl invariances of the ampliudes in Hilbert-Einstein gravity and SYM must fall. To ty it up we must show the existance of an operator formulation of Hilbert-Einstein gravity, which is mentioned in the subsequent section "Geometic Quantization of D-branes", which is there shown to exist under the axiom of existance of path integrals, an axiom which we, as previously stated, are subject to in our reasoning.

(2) By the Chiral Splitting Theorem, we have, indeed,
$\Longrightarrow$:

248

$$A_h = \delta(k)Q_{\nu\bar{\nu}}\Pi \int dm_k d\bar{m}_k C_\nu^F C_{\bar{\nu}}^{\tilde{F}} = \delta(k) \sum_{\substack{spin\ stuctures,\ moduli}} |C_\nu^F|^2$$

so with $C_\nu^F = M_\nu F$,

$$F = \Pi_{i=1}^N \int dz_i d\theta_i W_+(z_i, \theta_i) == \Pi_{i=1}^N \int d\bar{z}d\bar{\theta}\zeta_{\mu,\ +}D\mathbb{X}_+^\mu$$

The above corresponds to $\mathcal{L}_{M-theory/II} = |D\mathbb{X}|^2 = \mathbb{X}^*(D_+D_- \oplus D_+D_-)\mathbb{X}$. Looking at the various chiral parts, we recognize the SYM world line vertex by

$$\zeta_{\mu,\ +}D_+\mathbb{X}_-^\mu = \zeta_\mu\bar{\Phi}_+\sigma_+^\mu D_+\Phi_-$$

the left corresponds to $\mathcal{L}_{Q.H.E,\ D_\pm=4} = \mathbb{X}_+^* D_+D_-)\mathbb{X}_+$, while the right to $\mathcal{L}_{SYM} = \bar{\Phi}_- D - \Phi_+$. Reducing dimension in this way by fixing the mass to localize the front, which we can by using using the spectral density, e.g

$$\phi(x_i) = \int \frac{dm^2}{2\pi}\phi_m(x_i)\rho(m^2)$$

ϕ an arbitrary field, we have SYM on $D_\pm = 4$. Hence $\mathcal{L}_{SYM} = \bar{\Phi}_\mp D_\mp\Phi_\pm$ comes out.

$\Longleftarrow$: Conversely we have for SYM/Gauge theory, setting

$$\mathbb{X}_+^{\mu\alpha\beta} = \bar{\Phi}\sigma_+^\mu T^{\alpha\beta}\Phi$$

the components of a space-time supervector, e.g the usual embedding

$$\mathbb{X}_{0,\ +}^{\mu\alpha\beta} = \bar{\Phi}_0\sigma_+^\mu T^{\alpha\beta}\Phi_0$$

for squark fields Φ_0, we obtain in SYM treated in worldline formulation in a PI vertices

$$\zeta_\mu^a D_+\mathbb{X}_-^\mu e^{ik\cdot X}$$

by the related

$$\sigma^a\bar{\Phi}_-\sigma_a D_-\Phi_+ = \sigma^a D_-\bar{\Phi}_-\sigma_a\Phi_+$$

i.e in the terms of exterior forms, which gives us when projected onto the worldline

$$\zeta_\mu D_+\mathbb{X}_-^\mu$$

We must now show that $\mathbb{X}_\pm$ truly are $X_\pm \otimes U(N_C)$ valued (via exponentiation, that is), but this is easy since the defining formula

249

$$\mathbb{X}_\pm = T^{\alpha\beta}\sigma^a\bar{\Phi}_\mp\sigma_a T_{\alpha\beta}\Phi_\pm$$

gives so directly.

(3) This was proved already in the course of the proof of II.

(4) $\underline{\mathbb{X}_0, \mathbb{X}_1}$:

We expand the $\mathbb{X}$ field on, e.g, the negative helicity one-particle space-time. Then, retaining only the first two terms,

$$\mathbb{X} = \mathbb{X}_0 + \mathbb{X}_1 + \cdots = \mathbb{X}_0 + \mathbb{X}_{1,\mu}\theta^\mu + \cdots$$

we see that $\mathbb{X}_0$ is a vector. Hence in the integrated picure, i.e after applying the exponential map, this is a—possibly singular—"embedding". $\mathbb{X}_{1\mu}\theta^\mu$ on the other hand, by being a vector valued form of the appropriate type, is seen to be a vielbein. We prove that these $\mathbb{X}_{1\mu,\pm}\theta^\mu_\pm$ also define gamma matrices, by using $\mathbb{X}_{1\mu,\pm} = \delta^a_\mu$ in flat, or rather the background, space-time. In the operator picture this becomes—as usual we include the dual field in the operator picture as differing from the picture with classical fields in a lagrangean—

$$\Gamma^\mu_\pm = \theta^{\mu,\pm} + \theta^{\mu,*,\pm}$$

which gives

$$
\begin{aligned}
&[\Gamma^\mu_\pm, \Gamma^\nu_\mp] \\
&= (\theta^{\mu,\pm} + \theta^{*,\mu,\pm})(\theta^{\nu,\mp} + \theta^{*,\nu,\mp}) + (\theta^{\nu,\mp} + \theta^{*,\nu,\mp})(\theta^{\mu,\pm} + \theta^{*,\mu,\pm}) \\
&= 2\delta^{\mu\nu}
\end{aligned}
$$

$\underline{\Phi_0, \Phi_1}$

There is no doubt for the interpretation of $\Phi_{1, A}\underbrace{dz^A}_{=\theta^A}$ the classical spinor, and it is much studied in particle physics as the 4-component Dirac spinor, and in view of $\Phi_0 = [\Phi^A_0]$ taking the same form as $\mathbb{X}_0$ but in another space, considering consistency requirements among formulations, it has to have the same interpretation. Now it remains to show that this is compatible with the definition of $\mathbb{X}$. We have

$$
\begin{aligned}
\mathbb{X}^a_+ &= \bar{\Phi}_+\bar{\sigma}^a\Phi_- \\
&= (\bar{\Phi}_{0,\,+} + \bar{\Phi}_{1,\,+})\bar{\sigma}^a(\Phi_{0,\,-} + \Phi_{1,\,-}) \\
&= \bar{\Phi}_{0,+}\bar{\sigma}^a\Phi_{0,-} + \bar{\Phi}_{1,+}\bar{\sigma}^a\Phi_{0,-} \\
&\quad + \bar{\Phi}_{0,+}\bar{\sigma}^a\Phi_{1,-} + \bar{\Phi}_{1,+}\bar{\sigma}^a\Phi_{1,-}
\end{aligned}
$$

Now anyway we want to handle this, the trick lies in recoginzing that an X field is but a current(Currents are dual to positions,

250

so that is ok and not contradictory.) We could, e.g, directly notice that the appropriate superintegration, something that we must perform to obtain the usual current, would scrap the terms with mixed grading, hence giving us the consistency we opt for when comparig component formalism with non-component formalism. We must also show that the zero points of the different fields have been consistently chosen. Since we decided to have a field redefinition, see part II above from the definition of the operator fields,

$$e^a_\mu \mapsto \delta^a_\mu + e^a_\mu$$

this, e.g, so that that interacting vielbein components be vanishing at infinity and hence amenable to Fourier analysis. We see, directly, that since interacting spinors have to vanish at spatial infinity, that we at least have consistency in our choice of zero point at spatial infinity. Hence consistency of choice of field zero point also falls. We must also show that $\mathbb{X}_1$ indeed defines a vielbein when written in components, i.e a graded vector field of the appropriate kind. But using the map provided by the vector (*not* algebraic) space isomorphism $Cl \mapsto \bigwedge$ the Clifford algebra of some space and it's exterior cotangent algebra, we note by (before projection to the exterior algbra of the Y-brane)

$$\bar{\Phi}_-\sigma^a\Phi_+\sigma_{a,AA'}dz^A \otimes d\bar{z}^{A'}$$

that we get

$$\bar{\Phi}_-\sigma^a\Phi_+\theta^a$$

just as should be. Hence we are done proving consitency among field definitions.

(5) Now we wish to prove Maldacena correspondence in string space-time. Take off-shell space-times of various helicities to be simply $\mathbb{R}^5_\pm$ compactified, then we have $S^5_+ \times S^5_-$. Notice that the dimension of the spinors on the SYM's of the two helicities on the two space-times is appropriate. By continuation we get $AdS^5 \times S^5$ as claimed, and since we know that the string amplitudes are to coincide with the Yang-Mills amplitudes summed over moduli and spin structures we are done.

(6) Please see the appropraite appendix in Part IV. Interpreting a u field staisfying

$$G_{\mu\nu,\,-} = e^{-2u}(ds^2+\eta_2^2+\cdots\eta_3^2) = \theta^0\otimes\theta^0+\theta^1\otimes\theta^1+\theta^2\otimes\theta^2+\theta^3\otimes\theta^3 = \delta_{ab}\mathbb{X}_1^a\otimes\mathbb{X}_1^b$$

as a dilaton the result follows by theorems of Hitchin and our conjecture based on his results, elementary C*-algebaric theory, and field theory.

(7) See the pertainings solutions to the problems that give, e.g,

(a) Hawking Entropy.

(b) Hawking Radiation.

(c) Mass in terms of curvature.

(d) The entire classical limit, this as we are dealing directly with the Hilbert-Einstein lagrangean. Independently, we have also checked occurrence of post-newtonian limit.

(e) Gravitational crossections as given above and below, in particular for graviton-graviton scattering.

(f) The entire SYM sector, in particular QCD, Electroweak theory and QED, with crossections. In particular standard model physics must fall, and hence the experiments that have been confirmed in that sector for more than 50 years.

(g) Expansion rate of the universe.

(h) The physical dimension $D_\pm = 4$, consitent with Weyl invariance of string theory in $D = 10$ with two off-shell dimensions.

(i) Mass spectrum, in particular coinciding with string theory dimensionally reduced in the appropriate way.

(j) Coinciding effective actions and β-functions in the physical dimesion with vanishing torsion.

(k) Cosmic censorship of naked singularities, as generated by BPS bounds on black branes in string theory.

(l) Thermodynamic flows of equilibria of interstellar molecular clouds in weak inhomogenous background gravitational fields. In particular, detectable by spectral emissitivity checks in the appropriate domains.

(m) Standard mass additivity formulas of black holes, corrected to the dependency of particle physics mass additivity formulas by our quantum considerations.

(n) Gravitational quantum noise deviations from the classical path in weak inhomogenous gravitational background fields. In particular this can be used to calculate the gravitational noise in a laser signal in targeting a detector in space or in a inhomogenous graviational field affecting electrons in a metal.

□

We also have the lemma, that generalizes the concept of 2-admissibility and it's underlying mechanism here, and sums up problem 16 and 20 in a clean way.

Lemma 18.1. *We have an isomorphism*

$$\Pi_{i=0}^{N}\phi_i(x_i) \mapsto *\sigma_N$$

the Poincare duals of N-simplices of the appropriate kind.

Proof. By the Feynman diagram rules, we do not, apriori, treat the way we graphically represent the Feynman rules in a different way for boson fields as compared to fermion fields. We note that the above is an obvious isomorphism for classicall fermions , given by the fermion $\Phi_1(x) = \psi(x)$ from space-time to a graded algebra. Hence we conclude that this must be so by symmetry of interpretations under the assumption that the rules exist. But since the existance of Feynman rules and the worldline formulation is here treated as an axiom, this must fall. □

Lemma 18.2. *∃ unique lagrangean that has generalized supersymmetry and is diagonal over admissibilities $1/D$, and it is given by*

$$\mathcal{L} = \mathbb{X}_{G,\pm}^* D_{\mp} D_{\pm} \mathbb{X}_{G,\mp}$$

furthermore, when the fields contracted with the vertices $\mathcal{V}_u$ of this lagrangean are disidentified they give terms of admissibility type N, where $N + 1$ is the number of attached fields, as $*\mathcal{V}$ the Poincare dual is the graphic representation of the vertex—or simplex, we are, after all, dealing with Hilbert spaces—

$$\mathcal{V}_N =< \Pi_{i=1}^{N}\phi_i(x_i) >$$

Proof. By problem 20, part II, the first statement falls. Thus in view of the previous lemma the second statement falls. □

19. Geometric Quantization of D-branes and D-brane Feynman Rules

In this section we go through—in brief manner— a quantization procedure that will give us the Feynman rules in a particularly handy way.

19.1. Geometric Quantization.

Let $\mathcal{H}_1$ be a local coordinate chart (the direct sum of one-particle Hilbert spaces) on $\mathbb{X}$ an infinite dimensional manifold, of pseudhermitean, pseudoriemannian, Hilbert or other appropriate type such that the fiber of a tangent bundle is a identified with a quadratic vector space, with the understanding that

$$T^*\mathbb{X} \cong \overset{1}{\bigwedge} T^*\mathbb{X} \cong \mathcal{H}^* \cong Cl_1(\mathbb{X})$$

We have field operators, which one maps the classical fields to by the vector space isomorphism $\bigwedge \mapsto Cl$,

$$\mathbb{X} := \hat{\mathbb{X}}_i = \mathcal{O}_i \subset Cl(\mathbb{X}) \subset Hom_{\mathbb{C}}(\mathcal{H}, \mathcal{H})$$

and tensors

$\omega \in \mathcal{H}$

, $\mathcal{H} = \oplus_{\forall \text{ orders } p \in \mathbb{Z}, \forall \text{ possible orderings and factors at each order } p} \otimes_{\text{relevant factors}} H_i$, $\mathcal{H}_i^{-1} = \mathcal{H}_i^*$

We also have a symplectic form,

$$\omega^{AB} = \omega(\mathbb{X}_1^A, \mathbb{X}_2^B) = < \mathbb{X}_1^A \mathbb{X}_2^B - (1)^{pq} \mathbb{X}_2^B \mathbb{X}_1^A >$$

with the usual identites of symplectic geometry and field theory relating to this as usual. More generally,

$$\omega^{A_1 A_2 \cdots A_n}$$
$$= \omega(\mathbb{X}_1^{A_1}, \cdots, \mathbb{X}_n^{A_n})$$
$$= < \mathbb{X}^{A_1} \otimes \cdots \otimes \mathbb{X}_n^{A_n}, \omega >$$

from which we deduce that ω^{AB} must be antisymmetric. The above satisfy, depending on which day of the week it is and the mood of the reader what are called Dyson-Schwinger, Slavnov-Taylor, Baitlin-Villikovsky, BRST conditions or suchlike (They are all the same when generalized) by setting $D^2 = \mathcal{L} = Q$ a charge operator in a calculus of full densities (e.g. the $\mathbb{X}$-fields in string theory) which can be written as

$$D^2\omega = \mathcal{L}\omega = 0$$

or for halfdensites (e.g Φ-fields in SYM),

$$Dw = 0$$

. In the above we truly mean the covariant derivatives and nothing else, so this automatically includes the mass term that otherwise would have appeared in the charge operators. In a calculus of full densities this works out, e.g, as follows: Since $\mathcal{L}$ is a Lie derivative,(that is why it generates admissible transfrmations, i.e flows) it satisfies the Leibniz rules of Lie derivatives, so

$$
\begin{aligned}
&\mathcal{L}\omega_{Strings}(\mathbb{X}^1, \mathbb{X}^2, \cdots, \mathbb{X}^n) \\
&= \ (n := 1) \\
&= (\mathcal{L}\omega)(\mathbb{X}^1) + \omega([\mathbb{X}_2, \mathbb{X}_1]) \\
&= (\mathcal{L}\omega)(\mathbb{X}^1) + \omega_C \mathbb{X}^{2,A} \mathbb{X}^{1,B} \Delta_{AB}^C \\
&= (\mathcal{L}_A\omega)_B \mathbb{X}^{2,A} \mathbb{X}^{1,B} + \omega_C \mathbb{X}^{2,A} \mathbb{X}^{1,B} \Delta_{AB}^C \\
&= ((\mathcal{L}_A\omega)_B + \Delta_{AB}^C) \mathbb{X}_1^A \mathbb{X}_2^B
\end{aligned}
$$

which gives, since the linearity of the Lie derivative means that we can move out the index of the operating vector field,

$$-(\mathcal{L}\omega)_{AB} = \omega_C \Delta_{AB}^C$$

as an operator equation. In the same vein we have for halfdensities, where we for simplicity stick to the same notation as we have previously used for halfdensities,

$$
\begin{aligned}
D\omega_{SYM}(\Phi_2) &= (D\omega)(\Phi^2) + \omega([\Phi^1, \Phi^2]) \\
&= (D_A\omega)_B \Phi^{1,B} \Phi^{2,B} + \omega_C S_{AB}^C \Phi^A \Phi^B = <\Phi^1 | D\omega_{SYM} + \omega_A S^A | \Phi^2 >
\end{aligned}
$$

which gives the statement that

$$-D\omega_{SYM} = \omega_A S^A$$

which we recognize to be the Euclidean Dirac equation, the latter corresponding to the general lagrangean for half densities

$$\mathcal{L}_{SYM, G} = \bar{\Phi}_G D \Phi_G$$

while the former to

$$\mathcal{L}_{M, G} = \mathbb{X}_G^* D^2 \mathbb{X}_G$$

for full densities.

19.2. Relation to Feynman Rules. One can use the above to calculate Feynman Rules in a more systematic way. We have

$$S = \int d^D x \mathcal{L} = \sum_{A_1 > \cdots > A_n} \omega_{A_1 \cdots A_N} \mathbb{X}_1^{A_1} \cdots \mathbb{X}_N^{A_N} = \sum \frac{1}{|A|!} \omega_{A_1 \cdots A_N} \mathbb{X}_1^{A_1} \cdots \mathbb{X}_N^{A_1}$$

directly the vertices and propagators in a lagrangean as the tensor component entires $\omega_{A_1 \cdots A_N}$. [26]

i.e from amplitudes and expressions of the form

$$\omega(\mathbb{X}^1, \cdots, \mathbb{X}^N) = < [\mathbb{X}^1, \cdots, \mathbb{X}^N] >$$

$[\cdots]$ denotes the appropriate symmetrizer of fields. E.g in a multiparticle function related to the Generalized Clebch-Gordan coefficients of the reduction of parallell transport of the tensor product of isometry subalgbras of the automorphism algebra of the relevant Hilbert spaces in the fiber over space-time, so that the paralell transport is diagonal in multiparticle Hilbert space classified by the total charges $Q = \sum Q_i$, e.g the degree charge $Q_i = \frac{deg(\Phi)}{2}$, $Q_i = deg(\mathbb{X})$, something that we really in field theory never have to worry about, since we only look at propagation on the vacuum of lowest total polarization Q of total charge, where the functions, heuristically, are antisymmetric for fermion entries and symmetric for boson entries. For a simple example of a non-trivial vacuum, take the total spin state $l = 1, m = 0$ of two fermions coupling, it figures a multiparticle amplitude symmetric in fermion factors. Trivial vacuum means lowest vacuum here. Our procedure is general enough to include all vacua.

We must now include the basic elements of computations of Feynman rules in one way or another. Usually, one isolates the terms in the lagrangean of each definite type, and then one (anti)symmetrizes carefully over identical particle entries, thereby remembering to appropraitely move around the various indices. However, we shall do the same thing in a slightly different way. We use the entries of a lagrangean directly as rules by putting $< \cdots >$ around the terms as above, while keeping outgoing particles in a definite Fourier mode. By noticing that a trace tr has both the property that it appears in or Lagrangeans and is cyclic, i.e has a certain symmetry property under certain permutations, we generalize this to str, which has the propery that it is of the appropriate symmetry between identical fields, so that it becomes a symmetrizer of fields of the appropriate kind. Finally, we take the amplitudes that we then have as vertices and contract the fields within the traces to make up a diagram. That was a little bit brief, so let's take it again: We have correlators and we contract the field in a correlator with fields—which are of the same kind usually,

---

[26]Remark: We see here that one has identification of space-times to only retain one integral over one-space-time at the end. Concretely, to give an example of a similar situation, one sometimes uses this the other way, by introducing delta functions so that one has to integrate over additional space-times, e.g when calculating position space propagators.

this since the symplectic structure usually is diagonal over different field species—outside of that correlator to make up the final diagram. So one can read each contraction as 'Goes to', then meaning that the particle went to that second place in space-time from the original place, where the original field was located. Since the space-times are anyway copies of each other when identified that still makes sense. So that is the way that many $D3$-branes are a multiparticle space-time, which can then be measured as perturbations on a standard copy, namely usual background space-time, and that is also the way that we are going to calculate our rules. Finally, since each contraction etc can be expressed in usual canonical sewing of D-brane partition functions(with appropriate ghost factors included in them, e.t.c.) that means that we get the D-brane Feynman rules directly. A picture in section 10, strings and $D_\pm = 4$ CFT, Part I, illustrates this sewing.

19.3. Informal Questions and Answers.

(1) How many D-brane diagrams/manifolds are generated in this way?
Answer: By the Novikov construction of the D-folds, this procedure is exhaustive. However, overcounting of homeomorphism type is a recurrent phenomemon, so one cannot use this to enumerate the D-folds by generation.

(2) In higher dimension there are exotic structures-
should one sum over them?
Answer: Naively, one would guess so. But it's hard to say, really. The best I can give in that direction is to calculate specific examples and see how they come out. So , really, there is not much I can say about that. Please alo see the below on spain structure and moduli.

(3) I have noticed that one and the same collections of spaces in D-brane diagrams, i.e leaves in a foliation, can be cobordant to a lot of different diagrams of different topology. Are those diagrams the generalization of "multiloop"?
Answer: Yes.

(4) I do not understand what "cobordism" means—is it equivalent to bulk?

Answer: In the present usage, which is admittedley a little bit sloppy, yes. Usually one differs between different cobordism types, such as KO, K or spin cobordisms.

(5) In the section on generalized supersymmetry (e.g, Problem 20) it said that we are supposed to look at all theories of all admissibility type, so are we supposed to look at all the propagators of various admissibility types?

Answer: Precisely!

(6) Ok. So I get it. But I still think it's sloppy. Why don't you improve the rigour?

Answer:

1) Because if I did, you would loose focus. You'd think we are her to discuss Feynman rules. We are not. We are here to quantize gravity a couple of times, and we need to concentrate on that task.

2) Because it is much better to calculate an entire bunch of examples to check the various hypothesis to see if we are right, at least in terms of giving us confidence in our lofty ideas. At least I think so. And when we do the calculations we may discover things which are important to a systematic theoretical development. We shouldn't try a systematic development unless we are absoloutly sure that we know what we are doing. Are you absolutley sure of what we are doing, in the sense of each technical germ?

19.4. Some Examples of Feynman Rules.

Above all, we are interested in graviton-graviton scattering in our three theories. So we focus on the rules relevant to those interactions, and furthermore do not mention the rules for SYM. Then is then used to generate a diagram buy contracting fields, e.g. in the diagrams (CONTRACT)

$$A_{h=0} = str(\mathbb{X}_\pm^* \not\Omega \ \mathbb{X}_\mp) str(\mathbb{X}_\pm^* \not\Omega \ \mathbb{X}_\mp)$$

usually we substiute out the smooth coefficient of the outgoing particles, so that $\mathbb{X}^a \Gamma_a$ is used when $\mathbb{X}^a$ is dropped. Because of the identity

$$\int [d\mathbb{X}]_{\mathcal{H}_1 \oplus \mathcal{H}_2} = \int [d\mathbb{X}]_{\mathcal{H}_1} [d\mathbb{X}]_{\mathcal{H}_2}$$

which in particular implies that integrals factorize in independent degrees of freedom for independent multiplying factors in an integrand, the above can be used to calculate arbitrary diagrams of arbitary admissibility type (even mixed) from the components of these diagrams. E.g, sticking to string diagrams we would expect (of course, naively)

$$\mathbb{X}^{1,A} \ CONTRACT \ \mathbb{X}^{2,B} =< [\mathbb{X}^{1,A}, \mathbb{X}^{2,B}] >$$
$$= \int_{cylinder, \ or \ carpet} [d\mathbb{X}][dC][dB] e^{-S[\mathbb{X}] - S_{BC}[B,C]}$$
$$(\mathbb{X}^a(z, \bar{z}, \theta, \bar{\theta}) \mathbb{X}^b(z', \bar{z}', \theta', \bar{\theta}') - \mathbb{X}^a(z', \bar{z}', \theta', \bar{\theta}') \mathbb{X}^b(z, \bar{z}, \theta, \bar{\theta}))$$

the above then yield the string field theory Feynman rules for the three theories, and in particular the matrix notation goes well hand in hand with matrix valued space-time. I must, however, be staated that rigorously the entire procedure of calculating the PI to ghost form on the relevant D-brane world volume must be repeated for each case, but which hopefully goes without saying.

20. Explicit Checks of Ampliudes.

20.1. Tree Level Amplitudes. We start off with the string amplitudes, then compare them to the Q.H.E amplitudes and SYM amplitudes, $D = 10$, $D_\pm = 4$ respectively. We work excusively with NS-NS massless states in IIA/B theories—up to some tachyon amplitudes used to exemplify the calculus in the beginning. Again, we stress that $\omega_{\mathcal{A}} = \omega + \mathcal{A}$ and $\Omega = \mathcal{R}_{\mathcal{A}} = D\omega_{\mathcal{A}} = d\omega_{\mathcal{A}} + \omega_{\mathcal{A}}^2$ are functionally independent of the metric G and/or vielbein $\theta = \mathbb{X}_1$. This is important, as this provides vertices—to both the Q.H.E, the SYM and something similar in II A/B string theory.

20.1.1. N Tachyons. We start off by the simplest, but totally artificial case, of N tachyons on loop order $h = g$. Then we have

$$< \Pi_{i=1}^N : e^{ik\cdot X(z_i, \theta_i)} :>$$
$$= e^{-\Delta^{\mu\nu, AB}\partial_{\mu,A}\partial_{\nu,B}} <: \Pi_{i=0}^N e^{ik\cdot X(z_i,\theta_i)} :>$$
$$= e^{-\Delta^{\mu\nu} ik_{i,\mu} ik_{j,\nu}} <: \cdots :>$$
$$= \{\Delta^{\mu\nu} = \eta^{\mu\nu}(ln|E(z_i, z_j)| + S_\nu(z_i, z_j)\theta_i\theta_j)\}$$
$$= \Pi_{1 \leq i < j \leq N}(E(z_i, z_j)exp(S_\nu(z_i, z_j)\theta_i\theta_j))^{k_i \cdot k_j}$$

Where E is the prime form (Lecture 5, E. D'Hoker) and S_ν is the usual Szegö kernal on the spin structure indicated by ν on the actual Riemann surface of genus $g = h$. Hence for the sphere, by $ln\Delta(z, z')_{S^2 \cong \mathbb{C}P^1} = z - z' - \theta\theta'$, we get

$$(2\pi)^D \delta^D(\sum k_i)\Pi_{i<j}|z_{12}|^{k_i \cdot k_j}$$

$z_i - z_j - \theta_i\theta_j$. This is not to be squared further, since string theory deals with full densities, according to our "strings in quantum theory $\leftrightarrow$ full densites in a formalism for half densities"- philosophy, as exemplified by the first theorem of this part.

20.1.2. N Gravitons. We set vertex operators

$$\epsilon_{\bar\mu\nu}D_+\mathbb{X}_-^\mu D_-\mathbb{X}_+^{\bar\nu} e^{ik\cdot X}$$

Where we use bars to denote quantities belonging to the positive helicity space-time, this by analogy to complex notation, although dealing with real (although perhaps complexified) geometry.

By covariance

$$\eta_{\bar a b}D_+\mathbb{X}_-^a D_-\mathbb{X}_+^{\bar b} = \eta_{\bar a b}e_{\bar\mu}^a e_\nu^{\bar b} D_+\mathbb{X}_-^\mu D_-\mathbb{X}_+^{\bar\nu} = e_{\bar\mu}e_\nu D_+\mathbb{X}_-^\mu D_-\mathbb{X}_+^{\bar\nu}$$

where we skipped the degrees of freedom transversal, or, equivalently, codimensional, to the world sheet $\Sigma = \gamma_+ \times \gamma_-$ in the last line. Hence, expanding by

260

$$G_{\mu\nu} = \eta_{\mu\nu} + \epsilon_{\mu\nu}e^{ik\cdot X}$$

We recognize, since only helicity preserving second variations can remain for a vielbein expansion from an onshell background of our particular lagrangean, that

$$\epsilon_\mu \epsilon_{\bar\nu} = \epsilon_{\mu\bar\nu}$$

We can use, e.g, $\zeta := e_\mu^+ e_-$ to disitinguish better in the following, where e_- is the ON holomorphic vector paralell to the world sheet, so that $\zeta, \bar\zeta$ are odd grassmannians. Then we obtain, by the Chiral Splitting Theorem,

$$
\begin{aligned}
&< \Pi_{i=1} : \zeta_{i,\mu} D_+ \mathbb{X}_+^\mu e^{ik_i\cdot X_{i,+}} :> \\
&=< \Pi_{i=1}^N : \zeta_{i,\mu} D_+ \tfrac{1}{i}\tfrac{\partial}{i}\tfrac{\partial}{\partial k_i} e^{ik_i\cdot \mathbb{X}_{i,+}} :> \\
&= \Pi_{i=1}^N \zeta_{i,\mu} D_+ \tfrac{1}{i}\tfrac{\partial}{\partial k_{i,\mu}} <: e^{ik_i\cdot X_{i,+}} :> = \Pi_{i=1}^N \zeta_{i,\mu} D_+ \tfrac{1}{i}\tfrac{\partial}{\partial k_{i,\mu}} \Pi_{i=1}^N <: e^{ik_i\cdot X_{i,+}} :> \\
&= \Pi_{i=1}^N \zeta_{i,\mu} D_+ \tfrac{1}{i}\tfrac{\partial}{\partial k_{i,\mu}} \Pi_{i<j} e^{ln|z_{12}|k_i\cdot k_j} \\
&= \Pi_{i=1}^N \sum_{i<j} \zeta_{i,\mu} S_{ij,+}^{\mu\nu} k_{j,\nu} \Pi_{i<j} |z_{12}|^{k_i\cdot k_j}
\end{aligned}
$$

There remains determinants etc. In the above $S_+^{\mu\nu} = D_+\Delta^{\mu\nu}$.

But, let us— before proceeding—consider if this is physically realistic, and if so what this corresponds to. Now, in the ON frame, our favourite frame, we have

$$\mathcal{L} = \eta_{ab} D_+\mathbb{X}_-^\mu D_-\mathbb{X}_+^\mu$$

in order to obtain our chirally split amplitde, we used $\mathbb{X}^a = e_\mu^a \mathbb{X}_+^\mu$ a change of frame from the coordinate frame, and this gave us us our appropriate term. For this to hold, however, e_μ^a need be superholomorphic— and this is may (or may not) be reasonable. What more, the above amplitude implies that several fermions—without intermediate bosons— are interacting with each other. To understand the puzzle, and how our Feynman rules resolve the issue, we shall halt a little bit and see that we have intermediate particles as usual—it just a little bit hard to see them right now. To be precise, the above corresponds to a diagram with a $1P1$-blob in the middle and two external legs. This certainly not explicitely the kind of diagram we want to look at, for we need one where what is going on in the centre is visible. We need vertices, in brief, and where to put them. So let us get them. We have

$$\mathcal{L}_{D=10} = \mathcal{L}_{Q.H.E//\Sigma} = \eta_{cd}\mathbb{X}_-^c (D_+D_- + R_{ab}\frac{[\Gamma^a, \Gamma^b]}{2}P_-)\mathbb{X}_+^d$$

where we have identified $\gamma_+ \times \gamma_- = \sigma \subset X_+^{(5)} \times X_-^{(5)}$ and put not bars on the indices, since we now assume it understood that the various

261

gravitons can have different polarizations. $R_{ab} = D_{[a}\omega_{b]}$ is functionally independent of the metric. We can expand around $\omega = \omega_{\mathcal{A}} = 0$, i.e the pure gauges. Then, by

$$R_{ab} = \partial_{[a}\omega_{b]} + \omega_{[a}\omega_{b]}$$

we get, including an i from the coefficent in front of the action,

$$\eta_{ab}\mathbb{X}_+^a i R_{ab}\Gamma_-^{ab}\mathbb{X}_-^b$$
$$= \eta_{ab}\mathbb{X}_+^a i(-ik_{[a}\omega_{b]}\Gamma_-^{ab} + \omega_{[a}\omega_{b]}\Gamma_-^{ab})\mathbb{X}_-^b$$
$$= \eta_{ab}\mathbb{X}_+^a i(-ik_{[a}\omega_{b]}^\alpha T_\alpha\Gamma_-^{ab} + \delta_{[a}^c\delta_{b]}^d\omega_c^\alpha\omega_d^\beta[T_\alpha, T_\beta]\Gamma_-^{ab})\mathbb{X}_-^b$$

We need to calculate the result of the *str*'s. We do it on the lowest vacuum.

For the first term this gives

$$\eta_{ab}\delta_b^d\delta_{[a}^c\mathbb{X}^a k_{c]}\omega_d^\alpha T_{alpha}\mathbb{X}_-^b$$
$$= \eta_{ab}\mathbb{X}_+^{a,\alpha} k_c\omega^\alpha C_{\alpha\beta\gamma}\mathbb{X}^{b,\gamma}$$

Hence

$$\eta_{ab}C_{\alpha\beta\gamma}k_c\delta_{[a}^c\delta_{b]}^d$$
$$= \eta_{ab}C_{\alpha\beta\gamma}k_c(\delta_a^c\delta_b^d - \delta_b^c\delta_a^d)$$

$$= 2\,worldonlegs + 1\,connectionleg$$

In the same manner, for the second term we have

$$i\eta_{ab}\mathbb{X}_+^a\omega_a^\alpha\omega_b^\beta[[T_\alpha, T_\beta], \cdot]\mathbb{X}_-^b$$
$$= \mathbb{X}_+^{a,\kappa} i\eta_{ab}C_{\alpha\beta\gamma}C_{\gamma\delta\kappa}\mathbb{X}_-^{b\delta}\omega_c^\alpha\omega_d^\beta\Gamma^{cd}$$
$$= [i\Gamma^{cd}\eta_{ab}C_{\alpha\beta\gamma}C_{\alpha\beta\gamma}]\mathbb{X}_+^{a,\kappa}\mathbb{X}_-^{b\delta}\omega_c^\alpha\omega_d^\beta$$

Hence

$$i\Gamma^{cd}\eta_{ab}C_{\alpha\beta\gamma}C_{\gamma\delta\kappa} = 2\,worldonlegs + 2\,connectionlegs$$

That actually completes the non-YM part of the quantization, with Feynman rule

$$\Delta^{ab} = \frac{-i\eta^{ab}}{\Box + m^2}$$

inserted, which is obtained by inverting the quadratic part, and $S_\pm = D_\pm\Delta^{ab}$.

We now know the SYM part. Then, with $\mathbb{X}_-^{\mu\alpha} = \bar{\Phi}_-\sigma^\mu T^\alpha\Phi_+$ as usual relating the string fields to the SYM fields via Hilbert-Einstein Fields, we have the "usual" Feynman rules—which can be looked up in any textbook—for the component fields. But then we have to remeber to square the amplitudes at the end since we are deling with halfdensities in SYM. Thus we seem to have the necessary components for our purposes, nmaely graviton-graviton scattering at low loop. The process with a 4-graviton diagram at tree level, resembeling the one we decided to doubt slightly since we could not see any intermediate interaction bosons. We shall see that the ampliudes are equivalent and they actually, just as one would suspect, are the same diagram. The right diagram above is in the SYM picture while the above left diagram is in the string picture.

The ω-field can be set to be only $SL(2, \mathbb{C})$-valued, so that we skip the Gauge space-time degrees of freedom, and we can retain only the $\Phi_{1,\pm}$ component of $\Phi_\pm$, i.e the usual Dirac fields. Then-please note that the Diagrams needed correspond to each other directly, so we can compute, for electrons in one-particle space, and consequently make calculations on the Vielbeins, the invariant matrix element as, in SYM,

$$
\begin{aligned}
\mathcal{M} &= \bar{u} - g_G(M) T_\alpha \Gamma_a u \frac{-i}{k^2 - m_\omega^2} \bar{u} - i g_G(M) T_\beta u \\
&= (-i)^3 g_G(M)^2 \bar{u}(k_1) T_\alpha \Gamma_a u(k_2) \frac{1}{\underbrace{k^2 - m_\omega^2}_{:=0}} \bar{u}(k_3) T^\beta \Gamma^a u(k_4)
\end{aligned}
$$

$$
, k = k_1 + k_2 = k_3 + k_4
$$

which implies, if we let $\overline{\sum}$ mean spin average and supress both the coupling factor and the spin normalization factor,

$$
\begin{aligned}
\overline{\sum}|\mathcal{M}|^2 &= \bar{u}_s(k_1) T_\alpha \Gamma_a u_{s'}(k_2) \frac{1}{k^2} u_{s''}(k_3) T^\alpha \Gamma^a u_{s'''}(k_4) \bar{u}_{s'''}(k_4) T^\beta \\
&\Gamma^c u(k_3) \frac{1}{k^2} \bar{u}(k_2) T_\alpha \Gamma^c u(k_1) \\
&= \cdots \bar{u}(k_3) \Gamma^a T^\alpha (k\!\!\!/_4 - m) T^\beta \Gamma^a u(k_3) \cdots
\end{aligned}
$$

and thus should yield, as the calculations here after coincide, the result listed in previous calculations.

We notice that the above is quadric in the boson propagator as we have to look at both string end points propagating in string space-time. The SYM cannot give any other crossections of the type we are looking for at order $\alpha^2 = (\frac{g_G(M)^2}{4\pi})^2$. We can now check the stringy crossection. We should, if we are right, be able to recognize the chiral stringy parts in the SYM crossection as the matrix elements $\mathcal{M}$,

$$
\mathcal{M} = \bar{u}_s(k_1) T_\alpha \Gamma_a u_{s'}(k_2) \frac{1}{k^2} \bar{u}_s(k_3) T^\alpha \Gamma^a u(k_4)
$$

263

a clearly restrainted condition. Setting

$$D_\pm = \theta^\pm \partial \cdot \Gamma_\pm + \frac{\partial}{\partial\theta^\pm}$$
$$\Delta = \frac{i}{\Box+m^2}\delta^4_\theta(\theta - \theta')$$
$$S_\pm = D_\pm\Delta$$

we can check the SUSY answer too, and we see that it would not change. We can go back to the string version then— with the understanding that we will have to include vertices in the series, which are to be doing what the ChanPaton rules usually do. Hence we obtain

$$= \underbrace{\frac{g^4}{4}}_{includes\ spin\ factor} \frac{1}{(k_1+k_2)^4}tr[D_+\mathbb{X}_+D_-\mathbb{X}_1]tr[D_+\mathbb{X}_1D_-\mathbb{X}_1]$$

$$= \frac{g^4}{4}\frac{1}{(k_1+k_2)^4}tr[(\slashed{k}_1 + m_1)\Gamma^a(\slashed{k}_2 - m)\Gamma_a]tr[(\slashed{k}_3 + m)\Gamma^a(\slashed{k}_4 - m)\Gamma_a]$$
$$= \frac{g^4}{4}\frac{1}{(k_1+k_2)^4}tr[\slashed{k}_1\slashed{k}_2 - m] \cdot (perm.k_1 \mapsto k_3\ and\ k_3 \mapsto k_4)$$
$$= \frac{g^4}{4}(4k_1 \cdot k_2 - 4m) \times \cdots$$
$$= \frac{4g^4}{(k_1+k_2)^4}(k_1 \cdot k_2 - m)(k_3 \cdot k_4 - m)$$

We have thus that the Matrix formulation is identical to the Feynman trace trick. Let us proceed; Let us try $\mathbb{X}^{\mu,\alpha}$-field formulation in a spacetime with Gauge degrees of freedom-we can try a loop correction with fixed masses.

20.1.3. *Summary and Important Points.*

- In the above, we had relations

$$\eta_{ab}\mathbb{X}^a_- D_+D_-\mathbb{X}^b_+ = \eta_{ab}e^a_\mu e^b_\nu \mathbb{X}^\mu_{B,+} D_+D_-\mathbb{X}^\nu_{B,-} = (G_{\mu\nu,B}+\bar\zeta_\mu\zeta_\nu e^{ik\cdot X})\mathbb{X}^\mu_{B,-}D_+D_-\mathbb{X}_{B,+}$$

  B indicating background field, that made it possible for us to switch freely between the metric graviton and vielbein graviton picture.
- We formulated the various rules in terms of traces that naturally arise in the various pictures, e.g by matrix space-time in string theory and Feynman technology in SYM, and coincide.
- In order to write down an amplitude correctly in H.E gravity one has to remember the chirally split nature of amplitudes. In particular that means that one has to follow both ends of intermediate particles when associating propagators to them, so that the final crossection becomes quadratic in that propagator, with one contribution from each chirality, which are supposed to be mutually independent.

264

20.2. **1-Loop Amplitudes.** We try 1-loop calculations. We have learned our lesson, the stringy diagram is the square of the inavariant matrix element, again we use our subdivision space-times of various helicities. Thus we know what to we expect. For the above, we have, if we only use SYM theory and the usual Feynman trace trick, that

$$\overline{\sum}|\mathcal{M}|^2$$
$$= tr[(\slashed{k}_1 - m_1)\Gamma^\mu T^\alpha (\slashed{k}_2 + m_2)\Gamma^\nu$$
$$T^\beta]\tfrac{1}{k^4}\Pi_{1-LOOP}(k)tr[(\slashed{k}_3 - m_3)\Gamma^\mu T^\alpha (\slashed{k}_4 + m_4)\Gamma^\nu T^\beta]$$
$$, \Pi_{1-LOOP}(k)$$
$$= \int d^4p_+ d^4p_- tr(T^\alpha \tfrac{i}{\slashed{p}_+ - m} T^\beta \tfrac{i}{\slashed{k}_+ - \slashed{p}_+ - m}) tr(T^\alpha \tfrac{i}{\slashed{p}_- - m} T^\beta \tfrac{i}{\slashed{k}_- - \slashed{p}_- - m})$$
$$= \int d^4p_+ d^4p_- |tr(T^\alpha \tfrac{i}{\slashed{p}_+ - m} T^\beta \tfrac{i}{\slashed{k}_+ - \slashed{p}_+ - m})|^2$$

and so on, just as in the chiral splitting theorem, hence we recognize that we are truly dealing with the same thing—up to the issue of spin structures and moduli. So a string ampliude is just a clever way never to have take the square explictely. We can write down the rules that we have collected so far:

21. Spin Structures and Moduli of Complex Strucures on $\Sigma_{h=L}$

We have long postponed the discussion of necessary summation over moduli of spin strucutres and complex structures, which is necssary to ty together our previous reasoning.

Insofar as spin strucures are concerned, we notice these to be included in the Feynman rules for intermediate loops, with usual fermionic (-1) -factors arising from the rules, hence there is really not much we can do there but realizing this.

The matter of complex structures, however, is more interesting, and is thus what we shall concentrate on. Our objecive is to show, in whatever sense it may be true, that the SYM diagrams and the string diagrams must be equivalent(or rather, as it turns out, at least a singular limit, but we shall go beyond this).

By theorem 1.1 we know that SYM and usual field theoretic interpretations lead to two-dimensional conformal field theories. And, as pointed out in that same theorem, this does not necessarily imply summation over moduli. One can see that this can be deduced in some circumstances. Assume for example that we started in a real formulation of superstrings, that is what woul have resulted by the patching construction of the two field theories on the two intervals in theorem 1.1, then we know by assumed consistency that we must obtain the same result as in th complex formalism. Hence summation over moduli must fall. Naively, such a reasoning would (?) perhaps suffice, but there are truly no guarantees w.r.t to consistency, hence we do feel dubious about this line of reasoning, although it certainly at least contains some grain of truth. Purely field theoretically we are obviously looking at inequivalent Dirac vacua of solutions in the kernel of the Dirac operator, so one might then reason that one is to sum over these degenerate vacua when obtaining the solutions involved. However, this again fails, since we want to precisely furthen ourselves from such heuristic lines of reasoning, which are presicely what we try—in whatever sense we can—to avoid in this part. Thus remains to either to deduce a theorem our to define ourselves around it.

We try a theorem, where we warn the reader not to believe that the assumptions are realistic. We go into remedying this after the theorem.

Theorem 21.1 (Equivalence of Amplitudes Under Some Circumstances).
Assume that the string amplitudes, which are assumed to be modular invariant, are either holomorphic in their dependence of the moduli or e.g in the kernel of the laplacian on the moduli. Assume furthermore that the moduli can be obtained as a quotient of an open subset of $\mathbb{C}$

w.r.t to the action of some discret group, e.g a Fuchsian group, and that this typical fundamental region is of compact closure, for which it suffices that it is bounded. Then, under the assumption that the amplitudes are holmorphic up to and including these edges of the moduli, the SYM and string ampliudes must coincide.

Proof. We can display the line of reasonig for a torus first. We have, by modular invariance,

$$A_h(\tau + 1) = A_h(\tau),$$
$$A_h(\tfrac{\tau}{\tau+1})$$

But since

$$A_{h=L,SYM}(\tau_0) = A_{h=L,Strings}(\tau_0)$$

at the relvant corner τ_0 of the moduli, we have, remembering that a constant function is also a periodic function, that

$$A_{h,SYM}(\tau) - A_{h,Strings}(\tau)$$

is periodic with the above modular periodicities. A holomorphism on a compact complex manifold or a harmonic function on a compact manifold is a constant, which we use, since the periodicity implies that we have a compact scenario. Since we know that this diffirence, which now know to be a constant, attains the value 0 at τ_0, which we could extend without troubles to the edges according to hypothesis in the theorem, it falls that this difference is identically zero throughout the moduli. But then then the amplitudes are trivially equivalent throughout the moduli. Thus, in view of this, normalizing the modular measure with the volume of moduli space we have

$$A_{h,SYM} = \int \frac{d^{dims\mathcal{M}}}{Vol(s\mathcal{M})} A_{h,Strings}(\tau) = A_{h,Strings}$$

since the string ampliude would have to be a constant of the moduli under such circumstances. $\qquad\square$

The main trouble, now is that although this is certainly in one regard a beautiful scenario because it would have given us exactly what we would have opted for, and thus in effect finishing the proof of the main theorem quoted before, we cannot realistically expect that our amplitudes are non-singular at the boundary of the moduli. Indeed, given either the infinities in field theory or the explicit representations of modular forms in terms of Eisenstein series etc it would quite remarkable if they were holomorphic/harmonic at the edges of the moduli. That is where the trouble lies. Furthermore, to make the Lebsgue integral over the moduli defined in the existance of singular behaviour

267

in the moduli, we would have been forced to take a principal value over exhaustions from the enterior of the moduli. One can however, use the latter, to combine the both to retain another beauty—the nonsingularity of string theory with world sheets as a smooth cutoff. Thus, instead of quibbling over weather we should sum over the moduli or not, we define the SYM answer to be the moduli averaged for comparisions to string theory, this with the understanding that we regard structures as any other observable that we have taken an average over, and which may or may not be measurable. Reasoning as to the effect of measurability need not concern us at all, it suffices to know how to compute both answeres and then state them, and then let the notion of averaging over moduli be something settled in the laboratory, this as it seems not deducible, and we are not, logically, in posession of a train of though that would permit us to choose. Why choose when we do not need to choose? That would also be a satisfactory point of view when considering previous reasoning, that told us that we are anyway dealing with a two-dimensional CFT as generated by the SYM's.

Finally, we must ask ourselves, if in the predicament that the structure is observable, where would most of the excitations be? The answer to this comes from a generalization of the same mechanism that makes long distances on a world sheet become long distances in space-time; Bloch bounds from below and above on magnitudes of holomorphisms on a Riemann surface. Since we know thaat a reasonable string would have to be of proportions, say vaguely,

$$\frac{l_{Transversal\ to\ worldline}}{l_{Parallell\ to\ worldline}} = \frac{l_{Planck}}{l_{Worldline}} \sim 10^{-33} \sim 0$$

we also deduce that the string is very near the assumed SYM excitation of the amplitude. So taking the singular SYM limit in string theory would not be a totally unmotivated affair.

21.1. Summary.

- We concluded that we should, in view of the above, compare moduli averaged ampliudes when comparing with standard stringy formulas. Furthermore the amplitudes in A_{SYM} and $A_{strings}$ are at least to coincide at an edge of the moduli.
- In the predicament that only a small part of the moduli is excited, the part of the moduli that will be the relevant is the one very near the SYM singular worldline like point(s) at the edge of the moduli, this by growth estimates on holomorphisms.

22. Strings and $D_{\pm} = 4$ CFT, Part I: Links to 4-Dimensional Topology and Knot Theory

This is one of two sections on strings and 4-dimensional space-time conformal field theory. By consistency, in view of how we interpret the S-matrix operations on string worldsheets as gluing and sewing constructions and our brief section on quantization and sewing of D-branes we must interpret the to chirality parts of the string S-matrix as doing topological operations on usual $D_{\pm} = 4, 5$ space-times. This is also in good agreement with the homotopy type, before compactification, of the space-times we are considering, which is half the real dimension of string space-time. Thus, e.g in the 4-dimensional case, we regard the propagators and vertices dually as topological operations in space-time. That is, mathematically stated, we are sewing 4-folds along knots and links in worldlines with three-dimensional space codimensional to them. This is in good parallell with the Morse theory construction of a foliation of space-time in instances of equal time by an appropriate homomorphism from a C^*-algebra (indeed, as some reader might remark, this would lead in the proper context to the appeareance of the Alexander polynomial, and actually by the same determinant mechanism as we foliate our space-time, albeit used in different context). This behaviour of 4-dimensional topology has been known at least since the mid 80's to the mid 90's, this due to developments in topological field theory in interconnection with knot theory in 4-dimensional setting. In that context, and perhaps in similar vein, some focusing on something called the fundamental group of the complement (of the knot) gives information on the knot itself, i.e information about time gives information about space and, in some sense, conversely. There is a wonderful J.W.Gibbs lecture by M.F. Atiyah from 1990 on video at some libraries in this subject, which we use as a more competent yet not too tecnichal reference for the on wishing with a first encounter. Actually, the grounds of hypermathematics(of course not "done" or "good" or "perfect"in any sense) were developed among other things with the hope that it may reach some aspects 3/4 dimensional topology from the back door, namely using this typical foliated function theoretic behavior in $D = 4$. In particular the S^3 Poincare conjecture felt interesting as a problem, as it seems to be at the very centre of happenings. No signs yet of any sucess whatsoever in that direction.

22.1. Summary.

FIGURE 17. Sewing along knots of D-branes to implement contractions. The above is just about the most trivial situation one can have.

- We pointed out that, by consistency, the S-matrix must be generating topological operations on space-time, mathematically interpretable as sewing along knots and links.

23. THE SYMMETRIES OF THE GRAVITATIONAL FIELD

As already pointed out in Part II, we observe macroscopically certain symmetries in gravity in nature. Indeed we required as a *criterion* to even discuss a quantum theory of gravity that it at least macroscopically and classically reproduce these symmetries, and we tried, with the small resources that were to our disposition during our quantization in part II, to touch and understand these imperative aspects of gravity, this with varying failure an success. Our favourite scenario, with a spin 2 field being responsible for gravity, wich would have engulfed the entire dicussion and made it unnecssary, was obstructed by the fact that we chose the vielbein and connection, functionally unrealted, as our fundamental fields. Let us thus reassume this discussion now that we know (slightly) more.

As previously pointed out, the determining object for how the action is to affect the free fields is it's interaction term, then including the sign related to the free term. In our scenario, we have for the Hilbert Einstein term in a calculus of full densities, with pertaining strings,

$$\mathcal{L}_{X_\mp} = \mathbb{X}_\mp^* \Omega_{\ \mp} \mathbb{X}_\pm$$

Ω the Riemann tensor of space-time with coeffcients in various gauge groups. We have not called the Ω-field anything in particular, although it is it is one of the two fundamental fields of gravity. We notice, however, that it is of conformal weight 2, hence, remebering, following the heuristics that a rotation is a rescaling with antiselfadjoint exponent, we know that we have our candidate for such a spin 2 field. Rewriting the relevant part in the above to obtain a symmetric tensor, which is the usual case when considering such spins, we have

$$\mathcal{L}_{X_\pm} = Ric_{ab}G^{ab} = Ric_{ab}(\mathbb{X}_{1,\mu}^a \mathbb{X}_{1,\nu}^b \eta^{\mu\nu})$$

Hence we see that this lagrangean can really be rewitten as two components, which although can both be set to be functionally independent by a redefnition of the fundamental fields, still are both symmetric spin two fields—just as should be. Thus our symmetries are saved.

Furthermore, in the on-shell limit at the Levi-Cevita connection our lagrangean can be written as a function of a single symmetric field, namely the metric,

$$\mathcal{L}_{X_\pm}(G^{ab}) = Ric_{ab}G^{ab}$$

hence the symmetries at the classical macroscopical limit fall in twofold a manner, and are retained in the quantum theory by the first argument.

23.1. Summary.

- It was shown, on the basis of elementary arguments, that the symmetries at the classical level of our gravities are those that should be, and are retained in the quantum theory.

24. Formulae For The Effective Riemann and Stress-Energy

By the formulae of Part II, repeated in the introductury section of Part III, we realize directly, e.g for the positive helicity Riemann,

$$\not{R}_+ = \frac{\partial^2}{\partial \mathbb{X}_+^* \partial \mathbb{X}_-} \ln Z_{int}[\mathbb{X}_+^*, \mathbb{X}_-]$$

which becomes "quantum corrected" when other terms than only the classical part of the interaction term are included. This should be compared to the classical formula of Hermitean complex geometry

$$\mathfrak{Ric} = \bar{\partial}\partial \ln |G|$$

for the Ricci form on Hermitean complex manifolds. This can, actually, be obtained from the former by taking a trace tr over space-time indices,

$$
\begin{aligned}
&tr(\not{R}_+) \\
&= \partial_{\bar{\mu}}\partial_\nu tr \ln Z_{int}[\mathbb{X}_+^*, \mathbb{X}_-]^{\frac{[\Gamma^\mu, \Gamma^\nu]}{2}} \\
&= \{tr \ln = \ln \det\} \\
&= \partial_{\bar{\mu}}\partial_\nu \ln \det Z_{int}[\mathbb{X}_+^*, \mathbb{X}_-]^{\frac{[\Gamma^\mu, \Gamma^\nu]}{2}} = \{G = Z_{int}\} \\
&= \mathfrak{Ric}_{\bar{\mu}\nu}\frac{[\Gamma^\mu, \Gamma^\nu]}{2}
\end{aligned}
$$

i.e using $Cl \mapsto \bigwedge$ our usual vector space isomorphism,

$$tr R_{+, \bar{\mu}\nu}\theta^\mu \wedge \theta^{\bar{\nu}} = \frac{\mathfrak{Ric}_{\bar{\mu}\nu}}{2}\theta^\mu \wedge \theta^{\bar{\nu}}$$

just as in the defining relations of the complex Ricci form. We note in the above that

$$< \mathbb{X}_+^*(x', t')|\mathbb{X}_-(x, t) >$$

is a metric on a Hilbert space. Although this is implied by the above this does not, of course, neccarily imply that we have complex manifolds, (albeit this is certainly so locally, see the Stein manifold discussion in part II.). Rather we are dealing with a scenario of Riemannian geometry, as we even before noted falles by the Oka-Grauert principle, see Part I and II. Compactification of our scenarios, which is allowed by the restrainted cohomologies we like to regard, then yields the compact models we physicists like to consider—admittedley with some imprudence at times. The above formula for the Riemann also gives the stress-energy and usual real Ricci tensor by Einstein onshell-conditions and usual bastard trace respectively. It should, however, be pointed out that it is better conceptually to envisage the Riemann itself as a kind of stress-energy, since the Einstein field equation is much less an equation than an identity in our Einstein-condition is no-condition quantization of gravity, and all mass and kinematics is but geometrical in nature in it. One also notices the above to imply a generalized Schroedinger/Klein-Gordon equation $(-\partial_+\partial_- + \not{R})Z = 0$, from which, e.g., the usual Klein-Gordon equation can be drawn as consequence by taking traces, although we shall not be going further into that particular matter in this thesis.

24.1. Summary.

- We gave a simple formula for the Riemann, from which remaining standard entities of gravity can be deduced.

25. Noncommutative Moduli of ASD YM Vacua and β-functions

We may be interested in the moduli of vacua of the YM theory(we can drop the S for now) on $X_\pm$ and compare to it's stringy counterpart. Again our strive to quantization that is a unification of string theory, twistor/spinor geometry and NC geometry pays off, namely by the anticipated mechanism—someone else has made the work for us. According to Seiberg and Witten, String Theory and Noncommutative Geometry, IASSNS-HEP-99/74, hep-th/9908142, the classical moduli of instantons, i.e ASD $U(N_C)$ gauge connections, on $X_\pm$ the one-particle space-time, i.e. equivalently in their language $D3$-branes, coincides for zero B-field $B_{\pm,\mu\nu} = 0$ [27]. For nonzero B-field, this is supposed to be equivalent to NC geometry, which can according to the same paper also be realized by usual YM with higher dimensional correction terms, with an explicit transformation from the two theories NC YM and YM to each other. Please see section 3.1 in the above mentioned article. Under the hypothesis that their results are accurate, it would fall that also β-functions are to coincide. We apologize again for our briefness, and that we have to refer to authors better than ourselves, but we simply cannot expand all details, as we are delaing with all of physics in our thesis, and thus must(and are truly glad when we can) trust others with details we cannot see to but which we know anyway are *more* than relevant.

25.1. **Summary.**

- Instantons on $X_\pm$ with a B-field are described by NC YM. Usual YM with higher dimensional corrections terms is equivalent to NC YM, with an explicit change of variables.

---

[27]Note: Of course, S. & W. do not specifically look on their D3-branes as space-times of various helicities, thus this is under the assumption that no non-trivial obstruction arises from our identification of $D3$-branes as such in their analysis.

26. STANDARD STRINGY DEVELOPMENTS

26.1. **Strong/Weak Coupling Limits and Exact Results.** Again, we largely rely on authors more competent than ourselves. The weak coupling limit is what we usually consider, when we have a meaningful pertubation theory series, thus the references D'Hoker and Polchinski below should do. For strong coupling limits we use the reference being chapter 14 in Polchinski II, in particular section 14.1 on type IIB. References on exact results are also found in Polchinski.

26.2. **T,S,U Duals.** Again we rely and refer to Polchinski, chapter 14.

26.3. **Strings and $D_{\pm} = 4$ CFT, Part II: Stringy Developments.** Again we rely and refer to Polchinski, this time to chapter 18, "Physics in four dimensions", where ideas on $D_{\pm} = 4$ of more standard stringy type, which are never the less relevant to this thesis, are expounded.

27. BPS Bounds and Naked Singularities on Black D-branes

A generalization of the usual Birkhoff theorem from elementary general relativity states that the unique axisymmetric solution of the vacuum Einstein field equation with charge Q, angular momentum $L = J = L_z$ and mass M in Boyer-Lindquist coordinates is given by

$$G_- = ds^2 = -(1 - \tfrac{2Mr}{\Sigma})dt^2 - 4Mra\tfrac{sin^2(\theta)}{\Sigma}dtd\phi$$
$$+(r^2 + a^2 + \tfrac{2M^2ra^2sin(\theta)^2}{\Sigma})sin(\theta)^2d\phi^2 + \tfrac{\Sigma}{\Delta}dr^2 + \Sigma d\theta^2$$
$$\Delta = r^2 - 2Mr + a^2 + Q^2,$$
$$\Sigma = r^2 + a^2cos(\theta)^2,$$
$$a = \tfrac{L_z}{M}$$

where the last is the angular momentum per unit mass. The event horizon—which is a spherical surface of revolution— is located at

$$R_- = M + \sqrt{M^2 - Q^2 - a^2}$$

while the static limit— another surface of revolution of the hole— is at

$$R_+ = M + \sqrt{M^2 - a^2cos(\theta)^2}$$

The region between the static limit and the horizon is called ergosphere. This above is illustrated in the picture below:

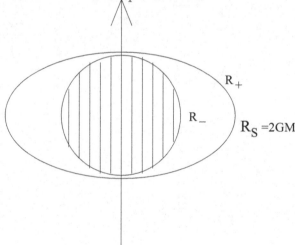

FIGURE 18. A charged rotating black hole, with it's event horizon at R_- and static limit at R_+. The ergosphere is the region between the horizon and static limit.

27.1. **Naked Singularities.** As is apparent from the above formula for the event horizon that $a^2 + Q^2 \leq M^2$ must be satisified in order for it to make sense. If, on the other hand the latter is not true, then this is interpreted as that the event horizon will disappear, leaving the singularity in the middle of the hole visible to external observers. This is, quite pictorially of course, called a naked singularity. Some people feel for some metaphysical reason or the other that a naked singularity is an impossibility, although this has actually not been proved rigorously with in classical general relativity. (This last statement is based on the book "Blackholes: Theory and References , Lecture Notes in Physics, Springer, 1998", and may or may not have become inaccurate since then.)

In string theory, a brane endowed with a vacuum Einstein metric, in particular one belonging to a black hole, in some of it's dimensions, is called a black brane. Consequently, one calls a $D3$-brane with such a metric a black $D3$-brane. Because of BPS bounds on the mass of such branes, precisely the condition of prohibition of vanishing of the horizon falls in several places in string theory. Please see Polchinski II, chapter 14, for further information.

28. Hyperkähler Geometry and Cosmic Inflation

We can use our observation of the form that the metric must have at the hyperkähler limit, please see the relevant appendix in Part IV for notation and suchlike,

$$G = e^{-2u}(ds^2 + \eta_1^2 + \eta_2^2 + \eta_3^2)$$

the volume form om the hypercomplex manifold is

$$\theta^0 \wedge \theta^1 \wedge \theta^2 \wedge \theta^3 = e^{-4u}(ds \wedge \eta_1 \wedge \eta_2 \wedge \eta_3)$$

and the natural—and preserved—volume form on $M_s = \Sigma$ the spatial slices is

$$e^{-2u}\eta_1 \wedge \eta_2 \wedge \eta_3$$

Hence from the latter we deduce that no expansion of a volume element can be predicted, classically, for times $s \in U, U \times M \cong \mathbb{X}_-$. And thus vanshing expansion rate of the universe as a first approximation must fall.

29. Particle Physics and The Three Theories Above

As we have previously emphasized at some point or another, our three theories are but aspects of each other. In particular, if we look only at interactions mediated by the the $U(N_C)$ degrees of freedom of space-time we retain the interactions of the usual standard model $U(1) \times SU(2) \times SU(3) \cdots$. Thus all interactions of particle physics itself are but a mere consequence of our gravity, wether we we wish to prefer the SYM, String, or H.E. formulation does not matter, and indeed it would be hard to seperate them from one another, as one has to reduce the string answer to H.E. form to make it useful in the natural setting of the physical dimension. This, that it predicts the right kind of physics, and in the right dimension $D_\pm = 4$, must be envisaged as a strong support for our string field theory—or M-theory if one so whishes. Perhaps, all the better, simply quantum theory.

30. The Systematic Classical Limit

To take the classical limit in the above model is particulary easy, this because of the sigma model structure, which takes background paths to interacting paths. Thus in any Fourier expansion of the fields

$$\mathbb{X} = \underbrace{\mathbb{X}_B}_{=constant} + \sum {}' X^A$$

one only has to retain the constant terms, which are the terms that correspond to the background congruences $\mathbb{X}_B$, hence turning the usual functional integral over a background subspace of the original infinite dimensional Hilbert space into an integral over the possible values of these constants. For example, if we are dealing with a black hole this would delimit the background position spectrum to be within the Schwarzild radius $R_S = 2GM$, while we could either take discrete or infinite momentum space, which gives us the usual statistical sum

$$Z[V] = (\int d^D p + \sum)(\int d^D x + \sum)e^{-S[x,p]}$$

If, on the other hand, we are dealing with a gas of volume V, we may simply use the same formula for that gas. This is, in fact, usual classical statistical mechanics, but it is remarkable that it should follow so directly as a generality when we have the right picture. Stating it differently; Hence classical thermodynamics also follows.

References

[1] Black Holes: Theory and Observation, Lecture Notes in Physics; 1999.

[2] Quantum Fields and Strings; A Course for Mathematicians, volumes I and II, Editors Pierre Deligne, Pavel Etingof, Lisa C. Jeffery, Daniel S. Freed, David Khazdan, David K. Morrison, John W. Morgan, Edward Witten.

In particular

[3] Eric D'Hoker, String theory lecture notes from the 1996-1997 year; Institute for Advanced Study, e.g. in vol. II, Quantum Fields and Strings; A Course for Mathematicians, Editors Pierre Deligne, Pavel Etingof, Lisa C. Jeffery, Daniel S. Freed, David Khazdan, David K. Morrison, John W. morgan, Edward Witten.

[4] Green, Schwarz, Witten, Superstring Theory, vol. I and II, Cambridge; 1987.
and we relied heavily on

[5] J.Polchinski, String Theory, vol. II, Cambridge, 1998.
as a reference better than ourselves for sections $15 - 17$.
The article referred to regarding β-functions and Y.M. moduli is

[6] String Theory and Noncommutative Geometry, N. Seiberg and E. Witten, IASSNS-HEP-99/74, hep-th/9908142.

Also

[7] Notes on D-branes, J. Polchinski, S. Chaudhuri, C. V. Johnson

[8] Dirichlet-Branes and Ramond-Ramond Charges, J. Polchinski
make interesting and available reading on D-branes,
and

[9] Black Hole Entropy in M-Theory, J. Maldacena , A. Strominger and E. Witten, hep-th/ 9711053
it too accessible by and large.

Part IV: Quantum Gravity, String Field Theory and Physics:

Appendices

31. Introduction

The appendices in this section are of varying linkage to the topics discussed in this thesis. In particular they contain a fairly blunt, but non-iterative, evaluation of the general field theory diagram for a restricted class of field theories where the diagram can be expressed as a sum of higher dimensional Euler integrals in complex D. Subsequent to this follows a formula for connections on bundles of good use in a wide variety of cases(as it is simple and general, this becaue we in this thesis express a metric by expressing it's ON-frame, i.e it's "vielbein".) and a discussion related to D-branes and strings, namely a brief and not very detailed discussion of the singular Cauchy problem and spin cobordisms. We also compute the mass spectrum of $D3$-branes for what we assume to be a standard case. Finally maps, pertaining to part II and Part III, illustrating the relation of the different parts of physics to each other appear.

31.1. Appendix; Evaluating the General Quantum Field Theory Diagram in Terms of a Closed Formula for a Restricted Class of Theories.

In this appendix we shall evaluate the general Q.F.T. diagram for a large class of theories in a non-iterative way invented by the author during the winter 1998. The main tool for this will be elementary pluricomplex/complex analysis, used to evaluate Feynman diagrams non-iteratively for special cases when the diagram can be expressed a sum of higher dimensional Euler integrals. Although we do not enter into this here, this can also be used to provide different and more consistent renormalization schemes at high loop order. The expression of the Feynman integrals directly as a function of complex dimension allows us to directly implement the economical method of t'Hooft-Veltman dimensional regularization. We begin with a definition

Definition 31.1. *A theory is called Mickelsson if*

(1) *It has propagators of type $\frac{p(k_1)}{k_1^2+m^2}$, $p(k_1)$ a polynomial in $k_1 \in \mathbb{R}^n$ and vertices of type $g(k_1, \cdots, k_d)$ a matrix-valued polynomial in attached momenta $k_\sigma \in \mathbb{R}^n$.*

(2) *All expressions of the theory can be continued to complex space-time dimension.*

Example 31.1. *QED, tensor boson gravity, and phi-fourth are Mickelsson. Any non handed part of the standard model is also Mickelsson. GWS is however not Mickelsson since it involves expressions with Γ^5, something that cannot be extended in a straightforward way to complex space-time dimension.*

We shall now use this definition to prove that in Mickelsson theories there is a renormalized formula for the arbitrary diagram. We first do a little detour to provide the necessary tool;

31.1.1. *A Generalized Euler Integral of Higher Complex Dimensions.* We will use the observation that the integrals of quantum field theory tend to be higher dimensional analogues of Euler integrals. In particular that means that some families of (complex) deformations of these definite integrals are expressible in terms of gamma functions. We define the function [28] $\mathcal{B}_2$: $\mathbb{C}^d \to \mathbb{C}$

$$\mathcal{B}_2(\zeta, d, z, \Delta) \equiv \int_{\mathbb{E}^d} \frac{r^\zeta}{(r^2+\Delta^2)^{\frac{z}{2}}} d^d x$$

where $r = ||(x_1, x_2, \cdots, x_d)||$ is the euclidean norm function in $\mathbb{E}^d, d \in \mathbb{C}$, analytically continued to $\mathbb{C}^d$. In the region $Re[\zeta + d] > 0, Re[z - \zeta - d] > 0, Re[d] > 0$ this integral is a holomorphism of several complex variables and can be expressed in terms of gamma functions, and in $\mathbb{C}^d$ it is a meromorphism expressible in ditto functions. If we let $\mu(S^{d-1}) = \frac{2\pi^{\frac{d}{2}}}{\Gamma(d/2)}$ denote the formal (Lebesgue)mass of the d-dimensional sphere we have

---

[28]Called beta-two.

$$\mathcal{B}_2(\zeta, d, z, \Delta)$$
$$= \mu(S^{d-1}) \frac{B(\frac{\zeta+d}{2}, \frac{z-\zeta-d}{2})}{2\Delta^{z-\zeta-d}} = \frac{2\pi^{\frac{d}{2}}}{\Gamma(d/2)} \frac{B(\frac{\zeta+d}{2}, \frac{z-\zeta-d}{2})}{2\Delta^{z-\zeta-d}}$$

with B the ordinary beta function. In particular this implies by symmetry on the regions we are interested in for

$$x = (-ix_0, x_1, x_2, \cdots, x_d)$$

$$\int_{\mathbb{E}^d} \frac{x_\mu x_\nu}{(r^2+\Delta^2)^{\frac{z}{2}}} = -\frac{g_{\mu\nu}}{d} \mathcal{B}_2(2, d, z, \Delta)$$
$$\int_{\mathbb{E}^d} \frac{x_\mu x_\nu x_\sigma x_\rho}{(r^2+\Delta^2)^{\frac{z}{2}}} = +\frac{[g_{\mu\nu}g_{\sigma\rho}+g_{\mu\sigma}g_{\nu\rho}+g_{\mu\rho}g_{\nu\sigma}]}{d(d+2)} \mathcal{B}_2(4, d, z, \Delta)$$
$$\cdots$$

or in full generality for non-vanishing cases of the last type with α a multiindex, $\chi_{0 \ mod \ 2}$ a characteristic function of the even integers and $\{\}$ the unnormalized symmetrizer of $|\alpha|$ symbols,

$$\int_{\mathbb{E}^d} \frac{x_\alpha}{(r^2+\Delta^2)^{\frac{z}{2}}} = \chi_{0 \ mod \ 2}(|\alpha|) \frac{(-1)^{\frac{|\alpha|}{2}}}{2^{\frac{|\alpha|}{2}}} \frac{g_{\{\alpha_1}\cdots g_{\alpha_n\}}(d-2)!!}{(d+|\alpha|)!!} \mathcal{B}_2(|\alpha|, d, z, \Delta)$$
$$\equiv \mathcal{B}_{2, \alpha}(|\alpha|, d, z, \Delta), \quad \alpha \in \mathbb{N}^d.$$

making integrals at arbitrary order of perturbation theory easy to evaluate by t'Hooft-Veltman dimensional regularization.

31.1.2. *The Main Theorem.* We now recollect the pieces of the theorem and give a proof.

Theorem 31.1. *Assume that a theory is Mickelsson, then the general diagram is expressible through one and the same closed formula.*

We postpone the proof in order to give an of example that will illustrate the vital parts of the proof and formula.

We shall evaluate the above diagram using a slightly unusual technique, instead of using iterated integration we shall integrate directly in $2d = 8 + 2\epsilon$ dimensional momentum space. We have vertices $-ie\gamma^\mu$, and propagators $\frac{i}{\not{k} - m}, \frac{-i}{k^2}$. Let the integration over internal momenta be implicit and Tr be the usual trace over the Dirac algebra, then with m electron mass and $e < 0$

285

Example 31.2.

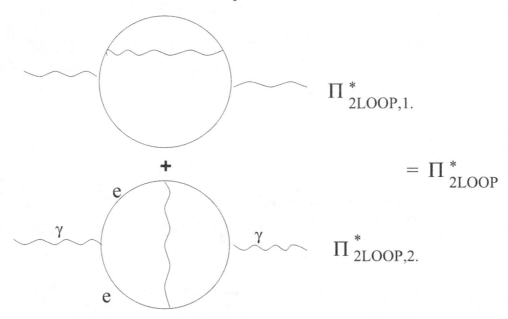

$$\Pi^*_{2\text{LOOP},1}.$$

$$= \Pi^*_{2\text{LOOP}}$$

$$\Pi^*_{2\text{LOOP},2}.$$

FIGURE 19. A Feynman diagram of simple 2-loop radiative corrections in QED. We evaluate the latter one, $\Pi^*_{2LOOP,2}$, as our first illustrative example.

the fundamental charge we have have from physics kindergarten

$$\Pi^*_{2LOOP,2} = Tr[-ie\gamma^\mu \frac{i}{\not{p}-\not{k}-m} - ie\gamma^\nu \frac{i}{\not{p}-\not{k}-\not{q}-m} + ie\gamma_\mu \frac{i}{\not{k}+\not{q}-m} + ie\gamma_\nu \frac{i}{\not{k}-m}\frac{-i}{q^2}]$$

$$= \underbrace{-ie^4}_{supress} Tr[\gamma^\mu \frac{\not{p}-\not{k}+m}{(p-k)^2-m^2}\gamma^\nu \frac{\not{p}-\not{k}-\not{q}+m}{(p-k-q)^2-m^2}\gamma_\mu \frac{\not{k}+\not{q}+m}{(k+q)^2-m^2}\gamma_\nu \frac{\not{k}+m}{k^2-m^2}\frac{1}{q^2}]$$

$$= Feynman \ parameters \ \{x,y,z,w\}$$

$$= \int \frac{dk^d}{(2\pi)^d}\frac{dq^d}{(2\pi)^d} \underbrace{\int_{[0,1]^4=I^n} \underbrace{dxdydzdw}_{=d\varsigma^n}}$$

$$\underbrace{}_{supress}$$

$$Tr[\frac{\gamma^\mu(\not{p}-\not{k}+m)\gamma^\nu(\not{p}-\not{k}-\not{q}+m)\gamma_\mu(\not{k}+\not{q}+m)\gamma_\nu(\not{k}+m)}{(x((p-k)^2-m^2)+(y-x)((p-k-q)^2-m^2)+(z-y)((k+q)^2-m^2)+(w-z)(k^2-m^2)+(1-w)q^2)^5}]$$

We now combine the sum of denominators-it is seen to be a quadric of momenta, thus elementary algebra gives that this quadric can be written as [29] $||\Lambda^{-1}(k \oplus q)+T||^2 - \Delta^2$ *with Λ, T a Feynman parameter dependent matrix and $|| \cdot ||$ the norm induced by the metric $h = g_1 \oplus g_2$, g_i usual Minkowski space metrics. Hence a simple change of variables gives, setting $t = \Lambda^{-1}(k \oplus$*

---

[29]Orthogonal diagonalization of the quadric and a little algebra suffices to show this for example.

$q)|_{k_0 \mapsto -ik_4, q_0 \mapsto -iq_4}$ *together with setting the numerator to* $\sum A^\alpha t_\alpha$ [30] *that we can finally see the use of the previous section;*

$$\Pi^*_{2LOOP,2} = (-i)^2 det(\Lambda) \frac{\sum A^\alpha t_\alpha}{(t^2 + \Delta^2)}$$
$$= (-i)^2 det(\Lambda) \sum A^\alpha \mathcal{B}_{2,\,\alpha}(\zeta, d, z, \Lambda)|_{\zeta := |\alpha|,\, d := 2d = 8 + 2\epsilon,\, z = 10}$$
$$= (-i)^2 \frac{-ie^4}{(2\pi)^d} \int_{I^n} det(\Lambda) \sum_{0 \le |\alpha| \le 4}$$
$$[A^\alpha \chi_{0 \bmod 2}(|\alpha|) \frac{(-1)^{\frac{|\alpha|}{2}}}{2^{\frac{|\alpha|}{2}}} \frac{g_{\{\alpha_1} \cdots g_{\alpha_n\}}(d-2)!!}{(d+|\alpha|)!!} \mathcal{B}_2(\zeta, d, z, \Lambda)]|_{\zeta := |\alpha|,\, d := 2d = 8 + 2\epsilon,\, z = 10}$$

In the above A must be transformed covariantly with Λ, this is something that we assume from now on. The calculation of the A^α in the example above can be made and consits of some tedious algebra [31]. The main point is that we now have an answer- that was reached without intermediate use of counterterms. Also, the procedure outlined has great generality, for we can evaluate the Laurent series expansion of $\mathcal{B}_2$ once and for all for all cases. Thus t'Hooft-Veltman regularization, the above tricks and the formula for the above Euler integrals will give us our answers.

Let us give the general formula for the Laurent expansion. It might not be understandable why we choose to give such a formula, but we will need the exhaustiveness to achieve generality. In even background dimension $d_0 \in \mathbb{N}$ the following case covers the non-trivial cases I have seen so far [32]

$$\frac{1}{(2\pi)^d} \mathcal{B}_2(\zeta, d, z, \Delta)$$
$$= \frac{1}{4\pi \Delta^{z-\zeta-d_0}} \frac{\Gamma(\frac{d_0+\zeta}{2})}{\Gamma(\frac{d_0}{2})} \frac{(-1)^{\frac{|z-d_0-\zeta|}{2}}}{\frac{|z-d_0-\zeta|}{2}} [\frac{-2}{\epsilon} - [ln(\frac{\Delta^2}{4\pi}) - \frac{\Gamma'(\frac{d_0}{2})}{\Gamma(\frac{d_0}{2})} + \frac{\Gamma'(\frac{\zeta+d_0}{2})}{\Gamma(\frac{\zeta+d_0}{2})}$$
$$+ \frac{(-1)^{\frac{|z-d_0-\zeta|}{2}}}{\frac{|z-d_0-\zeta|}{2}} (\gamma - \sum_{1 \le n \le \frac{|z-d_0-\zeta|}{2},\, n \in \mathbb{Z}} \frac{1}{n})] + O(\epsilon)] \,,$$
$$\frac{z-d_0-\zeta}{2} \in \mathbb{Z}_- \cup \{0\}, \frac{\zeta+d_0}{2} \in \mathbb{Z}_+$$

or in still greater generality

---

[30] The matrices A^α depend on dimension in the general case, although they do not do so in this specific case. Expressions in momenta before transformation must be expressed in the new variables t, so theis induces a covariant transformation of the matrices A^α by Λ, something that is supressed later.

[31] The main impetus for inventing my function was to have some device for writing down the regularized result for any Feynman integral at any order of perturbation theory. Of course there still remains the evaluation of the integrals over Feynman parameters, which is best made numerically. The above procedure still holds at high loop, and there it can be advisable to also handle the calculation of Λ and the Dirac algebra on computer too.

[32] However there are cases $z - d - \zeta \le 0$ where regularization is not needed. Direct evaluation in terms of gamma functions is then possible according to the previous section, so those cases are trivial in some sense.

$$\frac{1}{(2\pi)^d}\mathcal{B}_{2,\,\alpha}(\zeta,d,z,\Delta)$$

$$=\frac{1}{(4\pi)^{d_0/2}\Delta^{z-\zeta-d_0}}\frac{\Gamma(\frac{d_0+\zeta}{2})}{\Gamma(\frac{d_0}{2})\Gamma(\frac{z}{2})}\frac{(-1)^{\frac{|z-d_0-\zeta|}{2}}}{\frac{|z-d_0-\zeta|}{2}!}\chi_{0\ mod\ 2}(|\alpha|)\frac{(-1)^{\frac{|\alpha|}{2}}}{2^{\frac{|\alpha|}{2}}}\frac{g_{\{\alpha_1}\cdots g_{\alpha_n\}}(d_0-2)!!}{(d_0+|\alpha|)!!}$$

$$[\frac{-2}{\epsilon}-[ln(\frac{\Delta^2}{4\pi})-\frac{\Gamma'(\frac{d_0}{2})}{\Gamma(\frac{d_0}{2})}+\frac{\Gamma'(\frac{\zeta+d_0}{2})}{\Gamma(\frac{\zeta+d_0}{2})}$$

$$+\frac{\frac{|z-d_0-\zeta|}{2}!}{(-1)^{\frac{|z-d_0-\zeta|}{2}}}(\gamma-\sum_{1\leq n\leq\frac{|z-d_0-\zeta|}{2},\,n\in\mathbb{Z}}\frac{1}{n})-2\frac{\partial_d(d-2)!!}{(d-2)!!}|_{d=d_0}+2\frac{\partial_d(d+|\alpha|)!!}{(d+|\alpha|)!!}|_{d=d_0}]$$

$$+O(\epsilon)]\,,$$

$$\frac{z-d_0-\zeta}{2}\in\mathbb{Z}_-\cup\{0\},\frac{\zeta+d_0}{2}\in\mathbb{Z}_+\,.$$

We are finally ready to make the proof. Bearing in mind that the formulas of the Euler integral section apply to a wide variety of cases we shall anyway restrict ourselves to the physical case and only consider even $d_0\in\mathbb{N}$ for simplicity.

Proof. The proof is simple and uses canonical components. First we use Feynman parameters then we make a change of variables in the dl-dimensional integration space, $d\in\mathbb{C}$ being the space-time dimension and l the loop order.

Say we are given an arbitrary Mickelsson diagram(The name for a Feynman diagram in an arbitrary Mickelsson theory). We recall the Feynman parameter formula

$$\frac{1}{D^\beta}=\frac{1}{B(\beta)}\int_{I^n=I^{|\beta|}}\frac{\delta(\sum\zeta_i-1)\zeta^{\beta-n}}{(\sum\zeta_iD^i)^\beta}$$

β a multiindex [33], D the denominators. The measure used is $d\zeta^n$. It all now all falls down to using the above formula, Wick rotate, then make a change of variables and express the end result as a superposition of relevant Laurent series. Let us call this general diagram $\mathcal{M}$, p the degree of the polynomial in the numerator and set $dl=d_0l+\epsilon$, $\epsilon\in\mathbb{C}$, $|\epsilon|>0$ small enough.

Example 31.3. *In QED dl is the dimension times loop order, in dimension 6 at loop order 4 we have $dl=6\cdot4=24$.*

Then we have

---

[33]The generalization of the beta B function used above is defined by $B(\beta)=\frac{\Gamma(\beta_1)\Gamma(\beta_2)\cdots\Gamma(\beta_{|n|})}{\Gamma(|\beta|)}=\frac{\Gamma(\beta_1)\Gamma(\beta_2)\cdots\Gamma(\beta_{|n|})}{\Gamma(\sum\beta_i)}.$

$$\mathcal{M}$$

$$= (-i)^l det(\Lambda) \sum_{0 \le |\alpha| \le p} [\; [A^\alpha|_{d=d_0} + \partial_d A^\alpha|_{d=d_0 l}\epsilon + O(\epsilon^2)]$$

$$\frac{1}{(2\pi)^{dl}} \mathcal{B}_{2,\,\alpha}(\alpha, dl, z, \Delta)]$$

$$= (-i)^l det(\Lambda) \sum_{0 \le |\alpha| \le p} [\; [A^\alpha|_{d=d_0 l} + \partial_d A^\alpha|_{d=d_0 l}\epsilon + O(\epsilon^2)]$$

$$[\chi_{\mathbb{C}_{-0}}(z - d_0 l - \zeta)$$

$$\frac{1}{(4\pi)^{\frac{d_0}{2}} \Gamma(\frac{z}{2}) \Delta^{z-\zeta-d_0 l}} \frac{\Gamma(\frac{d_0 l + \zeta}{2})}{\Gamma(\frac{d_0 l}{2})}$$

$$\frac{(-1)^{\frac{|z-d_0 l-\zeta|}{2}}}{\frac{|z-d_0 l-\zeta|}{2}!} \chi_{0\ mod\ 2}(|\alpha|) \frac{(-1)^{\frac{|\alpha|}{2}}}{2^{\frac{|\alpha|}{2}}}$$

$$\frac{g_{\{\alpha_1}\cdots g_{\alpha_n\}}(d_0 l - 2)!!}{(d_0 l + |\alpha| - 2)!!}$$

$$[\frac{-2}{\epsilon} - [ln(\frac{\Delta^2}{4\pi}) - \frac{\Gamma'(\frac{d_0 l}{2})}{\Gamma(\frac{d_0 l}{2})} + \frac{\Gamma'(\frac{\zeta + d_0 l}{2})}{\Gamma(\frac{\zeta + d_0 l}{2})}$$

$$+ \frac{\frac{|z-d_0 l-\zeta|}{2}!}{(-1)^{\frac{|z-d_0 l-\zeta|}{2}}}$$

$$(\gamma - \sum_{1 \le n \le \frac{|z-d_0 l-\zeta|}{2},\ n \in \mathbb{Z}} \frac{1}{n}) - 2\frac{\partial_d (d-2)!!}{(d-2)!!}|_{d=d_0 l} + 2\frac{\partial_d (d+|\alpha|-2)!!}{(d+|\alpha|-2)!!}|_{d=d_0 l}]\;]$$

$$+ \chi_{\mathbb{C}_+}(z - d_0 l - \zeta)$$

$$\frac{1}{(2\pi)^{d_0 ll}} \frac{2\pi^{d_0 l}}{\Gamma(\frac{d_0 l}{2})}$$

$$\frac{\Gamma(\frac{|\alpha|+d_0 l}{2})\Gamma(\frac{z-\zeta-d_0 l}{2})}{2\Delta^{z-d_0 ll-\zeta}\Gamma(\frac{z}{2})} + O(\epsilon)\;],$$

$$= (-i)^l det(\Lambda) \sum_{0 \le |\alpha| \le p} [$$

$$[\chi_{\mathbb{C}_{-0}}(z - d_0 l - \zeta)\frac{1}{(4\pi)^{\frac{d_0}{2}} \Gamma(\frac{z}{2}) \Delta^{z-\zeta-d_0 l}}$$

$$\frac{\Gamma(\frac{d_0 l + \zeta}{2})}{\Gamma(\frac{d_0 l}{2})}$$

$$\frac{(-1)^{\frac{|z-d_0 l-\zeta|}{2}}}{\frac{|z-d_0 l-\zeta|}{2}!} \chi_{0\ mod\ 2}(|\alpha|) \frac{(-1)^{\frac{|\alpha|}{2}}}{2^{\frac{|\alpha|}{2}}} \frac{g_{\{\alpha_1}\cdots g_{\alpha_n\}}(d_0 l - 2)!!}{(d_0 l + |\alpha| - 2)!!}$$

$$[$$

$$\frac{-2A^\alpha|_{d=d_0 l}}{\epsilon} - A^\alpha|_{d=d_0 l}[ln(\frac{\Delta^2}{4\pi}) - \frac{\Gamma'(\frac{d_0 l}{2})}{\Gamma(\frac{d_0 l}{2})}$$

$$+$$

$$\frac{\Gamma'(\frac{\zeta + d_0 l}{2})}{\Gamma(\frac{\zeta + d_0 l}{2})}$$

$$+ \frac{\frac{|z-d_0 l-\zeta|}{2}!}{(-1)^{\frac{|z-d_0 l-\zeta|}{2}}}(\gamma - \sum_{1 \le n \le |z-d_0 l-\zeta|,\ n \in \mathbb{Z}} \frac{1}{n})$$

$$-2\frac{\partial_d (d-2)!!}{(d-2)!!}|_{d=d_0 l} + 2\frac{\partial_d (d+|\alpha|-2)!!}{(d+|\alpha|-2)!!}|_{d=d_0 l}]$$

$$-2\partial_d A^\alpha|_{d=d_0 l}\;]$$

$$+ \chi_{\mathbb{C}_+}(z - d_0 l - \zeta)\frac{1}{(2\pi)^{d_0 ll}} \frac{2\pi^{d_0 l}}{\Gamma(\frac{d_0 l}{2})} \frac{\Gamma(\frac{|\alpha|+d_0 l}{2})\Gamma(\frac{z-\zeta-d_0 l}{2})}{2\Delta^{z-d_0 ll-\zeta}\Gamma(\frac{z}{2})} \chi_{0\ mod\ 2}(|\alpha|) \frac{(-1)^{\frac{|\alpha|}{2}}}{2^{\frac{|\alpha|}{2}}}$$

$$\frac{g_{\{\alpha_1}\cdots g_{\alpha_n\}}(d_0 l - 2)!!}{(d_0 l + |\alpha| - 2)!!} + O(\epsilon)\;],$$

289

Where $\mathbb{C}_+$ are the complex numbers of positive real part and $\mathbb{C}_{-0}$ the complement in $\mathbb{C}$. Thus desupressing the integral over Feynman parameters we get for $d_0 l \geq 2$

$$\mathcal{M} = \frac{1}{B(\beta)} \int_{I^n = I^{|\beta|}} \delta(\sum \zeta_i - 1) \zeta^{\beta - n} (-i)^l \det(\Lambda) \sum_{0 \leq |\alpha| \leq p} \Big[$$

$$\Big[\chi_{\mathbb{C}_{-0}}(z - d_0 l - \zeta) \frac{1}{(4\pi)^{\frac{d_0}{2}} \Gamma(\frac{z}{2}) \Delta^{z - \zeta - d_0 l}} \frac{\Gamma(\frac{d_0 l + \zeta}{2})}{\Gamma(\frac{d_0 l}{2})} \frac{\frac{|z - d_0 l - \zeta|}{2}!}{(-1)^{\frac{|z - d_0 l - \zeta|}{2}}}$$

$$\chi_{0 \bmod 2}(|\alpha|) \frac{(-1)^{\frac{|\alpha|}{2}}}{2^{\frac{|\alpha|}{2}}} \frac{g_{\{\alpha_1} \cdots g_{\alpha_n\}}(d_0 l - 2)!!}{(d_0 l + |\alpha|)!!}$$

$$\Big[\frac{-2 A^\alpha|_{d = d_0 l}}{\epsilon} - A^\alpha|_{d = d_0 l} [ln(\frac{\Delta^2}{4\pi}) - \frac{\Gamma'(\frac{d_0 l}{2})}{\Gamma(\frac{d_0 l}{2})} + \frac{\Gamma'(\frac{\zeta + d_0 l}{2})}{\Gamma(\frac{\zeta + d_0 l}{2})}$$

$$+ \frac{(-1)^{\frac{|z - d_0 l - \zeta|}{2}}}{\frac{|z - d_0 l - \zeta|}{2}} (\gamma - \sum_{1 \leq n \leq \frac{|z - d_0 l - \zeta|}{2}}, \, n \in \mathbb{Z}$$

$$\frac{1}{n}) - 2 \frac{\partial_d (d - 2)!!}{(d - 2)!!}|_{d = d_0 l} + 2 \frac{\partial_d (d + |\alpha|)!!}{(d + |\alpha|)!!}|_{d = d_0 l}] - 2 \partial_d A^\alpha|_{d = d_0 l} \Big]$$

$$+ \chi_{\mathbb{C}_+}(z - d_0 l - \zeta) \frac{1}{(2\pi)^{d_0 l}} \frac{2\pi^{d_0 l}}{\Gamma(\frac{d_0 l}{2})} \frac{\Gamma(\frac{|\alpha| + d_0 l}{2}) \Gamma(\frac{z - \zeta - d_0 l}{2})}{2\Delta^{z - d_0 l - \zeta} \Gamma(\frac{z}{2})} \chi_{0 \bmod 2}(|\alpha|) \times$$

$$\frac{(-1)^{\frac{|\alpha|}{2}}}{2^{\frac{|\alpha|}{2}}} \frac{g_{\{\alpha_1} \cdots g_{\alpha_n\}}(d_0 l - 2)!!}{(d_0 l + |\alpha|)!!} + O(\epsilon) \Big],$$

the regularized formula for the general field theory diagram in the Mickelsson class of theories. Subtracting at a suitable subtraction point, or employing minimal subtraction, i.e projecting away the purely meromorphic germ at the origin in complex ϵ-space, we have our renormalized diagram. Some remarks are to be done: This evaluation of the general diagram for this class of theories may at first hand seem sensitive to the order of integration over loop dimensions, however this problem is taken care of by regarding the whole integral as a meromorphism over several variables. Slicing up the integration region as a hypercube, letting the side of the cube go to infinity and integrating iteratively over cubes pertaining to the various loop momenta we note that in euclidean time we have a holomorphism infitisimally transversally by Paley-Wieners theorem from several complex variables. Furthermore the integral is invariant under order of integration by Funbini-Tonellis theorem from integration theory for each succesive cube in the limit taken.

$\square$

32. Appendix; A formula for Connections on Some Bundles

32.1. Introduction. We will begin with the mathematics and then proceed to the physics. As for the physics this formula is of practical consequences, among other things it makes life more pleasing when calculating connections in some instances of general relativity, in particular for the quite common special case when the metric is diagonal.

32.2. Results and Proofs. Let the tangent bundle $(E, \pi, M, F, G, \{U_i, \phi_i\})$ be given, E total space, π the projection, M the base space, $F \cong K^n$ the fibre for $K = \mathbb{R}$ or $K = \mathbb{C}$, $G = O(n, m, K)$ in pseudo-orthogonal case or $U(n, m)$ in pseudo-unitary case, $\{U_i, \phi\}$ a open cover of M with the belonging local trivializations ϕ_i on E. Let $*$ be the dual map over G (i.e transpose for orthogonal case and hermitian adjoint for unitary case) and it's associated left module F. Let $S_\mathfrak{g}$ be the antisymmetrizing projection from $F \otimes F^*$ to $\mathfrak{g}$ with explicit realization over the 'defining' representation of G by $S_\mathfrak{g}(X^{ij} e_{ij}) = X^{ij} e_{ij} \pm X^{ij} e_{ji}$ in pseudo-orthogonal case and $S_\mathfrak{g}(X^{ij} e_{ij}) = X^{ij} e_{ij} \pm \bar{X}^{ij} e_{ji}$ where $e_{ij} \in M(n + m, \mathbb{C})$, $M(n + m, \mathbb{C})$ the vector space of all $(n + m) \times (n + m)$ matrices with it's associated multiplication, which is naturally associated to the defining representation of these classical groups. [34]. Then I claim, firstly for these tangent bundles with vector space fiber equipped with an appropriate quadratic form, that the torsion-less Cartan stucture equation for the connection one-form on E, $d\theta = -\omega \wedge \theta$, $\theta \in \Lambda(E, K) \otimes K^{(n+m)}$ being the local ON-vielbein corresponding to the fibre metric $h = h_{\mu\nu} dx_\mu \otimes dx_\nu$, $\{x_\sigma\}$ local coordinates on the fibre with explicit coordinate representaion

$$
\theta = \begin{pmatrix} (h_{00})^{\frac{1}{2}} dx_0 \\ (h_{11})^{\frac{1}{2}} dx_1 \\ ... \\ (h_{n+m})^{\frac{1}{2}} dx_{n+m} \end{pmatrix}
$$

in the diagonal case, which we can take to simplify computations in the following without enforcing any relevant constraint as we prove coordinate freedom later on, [35] has solution

$$
\omega|_{TM} = S_\mathfrak{g}(e\theta^*)
$$

---

[34]Einstein sums apply only form now on

[35]The formula given is also true in the non-diagonal metric case, however we will consider diagonal metrics first for simplicty. Two proofs for the general case are below.

where $e \in der(\mathcal{T} \otimes K^{(n+m)})$, $\mathcal{T}$ denoting the tensor algebra on F, is a derivation associated to the repére vector

$$e = \begin{pmatrix} (h_{00})^{-\frac{1}{2}}\partial_0 \\ (h_{11})^{-\frac{1}{2}}\partial_1 \\ ... \\ (h_{n+m})^{-\frac{1}{2}}\partial_{n+m} \end{pmatrix}$$

[36]whose action is locally defined on individual Fibre tensors $t \in \mathcal{T}$ by

$$et$$
$$= e(t^{ijk..}_{\sigma,\rho...}P(dx_\rho \otimes .....\partial\sigma..))$$
$$= e(t^{ijk..}_{\sigma,\rho...}t^{ijk..}_{\sigma,\rho...})P(dx_\rho \otimes .....\partial\sigma..)$$
$$= \mathcal{L}_e(t^{ijk..}_{\sigma,\rho...})P(dx_\rho \otimes .....\partial\sigma..)$$

P some permutation of tensor bases, $\mathcal{L}$ denoting partial diffrentiation on smooth coefficients $t^{ijk..}_{\sigma,\rho...} \in C^\infty(F,K)$. Let us define an 'abstract' connection $\omega \in \mathfrak{g} \otimes T^*P$ on P the frame bundle corresponding to E to be

- 1) A unique smooth separation $T_qP = H_qP \oplus V_qP$ satisfying
- 2) $H_{ug}P = R *_g HuP$, $u \in P$, R_g^* denoting the right translation on TP induced by $g \in G$.

where one defines $V_uP = Ran(\#)$, $\# : \mathfrak{g} \to V_uP \subset T_uP$ as the isomorphism of vector spaces pointwisley defined by $A^\# f(u) = \frac{d}{dt}f(ue^{tA})|_{t=0}$ for any $f : P \to \mathbb{R}$ nondegenerate at u and the horizontal subspace H_uP as the complement in T_uP of V_uP. My mathematical claims are then

- That the formula given is the solution to Cartan's torsion-less stucture equation for the structure groups mentioned.
- That this quantity transforms as a connection with respect to admissible transition functions $t_{ij} \in G$ where t_{ij} is such that $\phi_j(p,f) = \phi_i(p,t_{ij}f)$.
- That we are actually dealing with an 'abstract' connection in the sense above.

Before proceeding we warn the reader of not forgetting that to get $\omega|_{T^*E}$ * must operate on $\omega|_{TE}$ in explicit computations. We begin with a lemma

Lemma 32.1. *Let* $A \in G$ *or* $G = O(n,m,K)$ *or* $G = U(n,m,\mathbb{C})$, *then for* $B \in M(n,\mathbb{C})$. *Then*

$$S_\mathfrak{g}(Ad_A B) = Ad_A S_\mathfrak{g}(B)$$

i.e the symmetrizing and adjoint action commute, $[S_\mathfrak{g}, Ad] = 0$, *where* $Ad_A B = A^{-1}BA$.

---

[36]In (pseudo)unitary cases there should be a complex conjugation here because of the hermitian metric.

Proof. Set A to be e_{ij} times some (complex) constant. Then by $A^{-1} = A^*$ we get

$$L.S = S_{\mathfrak{g}}(A^{-1}BA) = A^{-1}BA \pm (A^{-1}BA)^*$$
$$= A^{-1}BA \pm A^*B^*A^{-1*} = A^{-1}(B \pm B^*)A = A^{-1}S_{\mathfrak{g}}(B)A = R.S$$

and by linearity of the adjoint action the asserion now follows. $\square$

Let us define $S_{\mathfrak{g}}(d) = d$, d denoting exterior diffrentiation on the fibre. The wisdom of this choice can be confirmed with an elementary computaion left to the reader. Then we get the theorem

Theorem 32.1. *Let $\omega = S_{\mathfrak{g}}(e\theta^*)$. Then ω transforms as a connection under admissible transformations $t \in C^\infty(E, G)$.*

Proof. Since $t^{-1} = t^*$ we get

$$L.S = S_{\mathfrak{g}}(e'\theta'^*)$$
$$= S_{\mathfrak{g}}((t^{-1}e)(t^{-1}\theta)^*) = S_{\mathfrak{g}}(e'\theta'^*)$$
$$= S_{\mathfrak{g}}(t^{-1}e(\theta^*t)) = S_{\mathfrak{g}}(t^{-1}(e\theta^*)t) + t^{-1}(\theta^*e)t))$$
$$= S_{\mathfrak{g}}(t^{-1}(e\theta^*)t + t^{-1}dt)$$
$$= t^{-1}(S_{\mathfrak{g}}(e\theta^*) + d)t = t^{-1}(d + \omega)t = Ad_t(d + \omega) = R.S$$

where we used the previous lemma and the fact that e was a derivation, i.e linear with a Leibniz rule. $\square$

Corollary 32.1. *ω is a connection.*

Proof. To satisfy criterion 1) define $H_uP = Ker(\omega)$, since ω is smooth that takes care of that criterion by letting V_uP be the complement. As for criterion 2), letting $X \in H_{ug}P$ we get $\omega(R_{*g}X) = R_g^*\omega(X) = Ad_g(\omega(X)) = 0$ where we noted theorem 1.1 and the concequence

$$\omega' = g^{-1}\omega g + \underbrace{g^{-1}dg}_{=0} = Ad_g(\omega).$$

$\square$

Theorem 32.2. *ω satisfies the structure equation $-d\theta = \omega \wedge \theta$ on T^*M.*

Proof. For illustrative purposes we shall first work with a diagonal metric and O(m) case. Then (No Einstein over i)

$$d\theta^i = dh_{ii}^{\frac{1}{2}}dx_i = \partial_\sigma h_{ii}^{\frac{1}{2}}dx_\sigma \wedge dx_i$$
$$= -h_{\sigma\sigma}^{-\frac{1}{2}}\partial_\sigma h_{ii}^{\frac{1}{2}}dx_i \wedge \theta^\sigma = -(e\theta^*)_i^\sigma dx_i \wedge \theta^\sigma = ((e\theta^*)^*)_j^i dx_i \wedge \theta^\sigma = -\omega_\sigma^i \wedge \theta^\sigma$$

gives when taking the symmetrization conditions into consideration $\omega|_F = S_{\mathfrak{g}}((e\theta^*)^*) = (S_{\mathfrak{g}}(e\theta^*))^*$ just as it should be. We can now take the general case

$$\omega|_F \wedge \theta = (S_{\mathfrak{g}}(e\theta^*))^* \wedge \theta = e(\theta) \wedge \theta + (e\theta)^* \wedge \theta = [e^i\theta^i \wedge \theta^i] + d\theta = d\theta$$

in the above brackets denote a vector indexed by i. The elements in that vector vanish since squares of one-forms are null. $\square$

Example 32.1. *On S^2 with polar coordinates and the usual metric*

$$g = d\theta \otimes d\theta + sin(\theta)^2 d\phi \otimes d\phi$$

we have

$$\omega = S_{\mathfrak{g}}(e\theta^*) = -S_{\mathfrak{g}}\begin{pmatrix} e_{\hat{\theta}} \\ e_{\hat{\phi}} \end{pmatrix}\begin{pmatrix} \theta^{\hat{\theta}} & \theta^{\hat{\phi}} \end{pmatrix} = -S_{\mathfrak{g}}\begin{pmatrix} \partial_\theta \\ \frac{1}{sin(\theta)}\partial_\phi \end{pmatrix}\begin{pmatrix} d\theta & sin(\theta)d\phi \end{pmatrix}$$

$$= -S_{\mathfrak{so}(2)}\begin{pmatrix} 0 & -cos(\theta)d\phi \\ 0 & 0 \end{pmatrix} = \begin{pmatrix} 0 & -cos(\theta)d\phi \\ cos(\theta)d\phi & 0 \end{pmatrix}$$

Calculations in higher dimension proceed in exactly equivalent manner, and are drastically simplified when the metric is already given in diagonal form by expressing it's ON basis—this is assumed in the projection onto the appopriate isometry subalgebra of the automorphism algebra of the relevant vector bundle fiber, and is a common enough case in physics.v

With good approximation at relevant backgrounds, $X_\pm$ are hyperkähler, that is, they are hypercomplex space-times with their Obata connection[37] preserving a metric in the conformal class. If one lets the measured helicity space-time be embedded in the stringy space-time, something that can always be done by a special case of Whitney's theorem, we can represent the remaining by a contractible set, this to have homotopy equivalent topology inducing an isomorphism of the sheaf cohomological theories of the corresponding background partial differential operators (i.e BRST operators), e.g free operators in the simplest case on $\mathbb{R}^4$, which we often choose to compactify. One can also work with two copies of the same space-time, as we chose to do in part III as opposed to part II, as long as one remembers to take this into account the physical results should not differ—or at least have not done so so far in this thesis. Several particles, say N, are then supposed to correspond to a stack of N standard copies of the appropriate space-time background, which works a little bit as a common zero level or reference point for the various space-times. Here is a simple explanation for the reader who wishes to understand this; Think about usual electrons in space-time, they are but fields; they can have different field values, for they correspond to different particles with roughly independent physics, this by cluster decomposition of the S-matrix. One models this by having N different position configuration spaces, or space-times in more common language. This does not necessarily mean that one truly does have several space-times but that is anyway the way we model it. So we have several space-times because we want to allow for electrons with different field values. Now, we all agree that vielbeins and connections are but usual fields, and we can, for the line of reasoning, say we agree that we can interpret these fields as particles. The values of connections and vielbeins on a space-time can be—should be— *different*, because after all, they are but particle way functions. Since a space-time is minimally and uniquely determined by it's smooth topology, it's connection and it's ON frame(or metric, if one insists on being impractical), and both the vielbein and connection are but usual particles, we deduce that we must be having N different copies of the smooth topology but with different connections and vielbeins in order to model the physics. So we have a heap of space-times, because of independent geometry,— as many as the number of particles—but with the same position spectrum. String theory, i.e the addition of dimensions is to drop the last restraint. The stack of branes, then, is what is called multiparticle space-time. Let us see how a one-particle space-time looks. According to us—and as has been checked indirectly in the mathematical literature for one very simple but relevant case— we conjecture that the spatial sets in a brane can, after continuation

---

[37]This is the name of a torsion-less connection that preserves the hypercomplex structures on a hypercomplex space-time.

to Eucidean metric, be identified with the level sets of the determinant homomorphism $det(1 + \mathcal{O})$, $\mathcal{O} \in \mathcal{A}$ a C^* algebra. One can easily intutively see how that comes about, for if we make the analytic continuation of the usual expressions to imaginary time β it is easy to see that the the Hamiltonian H does not generate unitary isometries of the relevant Hilbert spaces but, e.g instead generates a scaling of determinants by the possibly ill-defined determinant

$$det(e^{-2H\beta})$$

In the standard case, $\mathcal{O} = Q_{int}/Q$, Q the BRST charge operator, e.g the Klein-Gordon operator $Q = -k^2 + m^2$. In the mathematical literaure, see Hitchin[1] in the references, this figured and was proved two years ago in his paper "Hypercomplex Manifolds and the Space of Framings", who was simply considering the special case of harmonic functions as generating spatial sets and time in a hypercomplex space, something that corresponds to Weyl(massless) fermions after reduction to halfdensities in the SYM, and the case for general m seems thus to be suggested, by "translation invariance" of the physical laws, as it were, along the mass scale, if we only remember that we can write solutions of partial differential equations by using partition functions and pertaining determinants.

It would be interesting to know the behaviour of such possibly non-hyperkahler space-times, might they be the non-vacuum states one percieves/believes them to be, and if so, how many of them are tachyons? Such answers could give further checks that truly everything fits in the grand hypothesis of this thesis—something that we can of course not exhaustively check ourselves as it involves all of physics.

Traditionally the "spectral varieties" $det(1 + O) = c$ are called isospectral sets for finite dimensional operators in the context of integrable systems (See Hitichin[2]). In the infinite dimensional case, it can be hard to establish any meaning to such an equation, but since we are any way wanting information on a particular space-time this is remidied by observing the level sets of the determinant homomorphism acting as background partition functions.

Let us return and sum up. We draw a picture in string space-time to make it clear;

We sum up our thoughts on the determinant homomorphism by

Conjecture 33.1 (NC-Geometry/ D-brane correspondence, "Determinant Homomorphism Conjecture"). *Let $det : \mathcal{A} \mapsto \mathbb{C}$ be the determinant homomorphism from an appropriate C^* algebra to an appropriate field, here taken as $\mathbb{C} = \mathbb{R}^{\mathbb{C}}$. Then, after continuation to the Euclidean region so that the unitary action of the Hamiltonian is a scaling of the determinant instead of preserving it, we have our spatial sets induced by varieties of the form $det(O) = c$, $c \in \mathbb{C}$, where it is understood that we act by the determinant on appropriate initial value data on a spatial slice at some a priori given instant on a D-brane.*

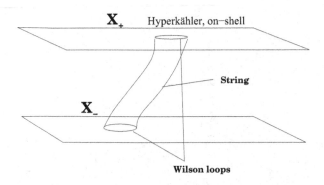

X₊ Hyperkähler, on–shell

String

X₋

Wilson loops

FIGURE 20. $X_\pm$ is a level surface of mass(i.e a foliation by the Ricci instead of the dilaton) in Kaluza-Klein space-time, $X_\pm = U_\pm \times \Sigma_\pm$, $\Sigma_\pm$ a level surface of time, and $det(1+\mathcal{O})$ gives the branes at various instances. The above corresponds to how a Wilson loop of open strings would look like. A usual free propagating string in gravity would like the opposite, with the end points in the various space-times joining, being able to propagate arbitrarily in either space-time. Hence the closed dimension and open dimension would reverse in the above picture.

and this gives a natural interconnection to Morse theory that will, up to the following mindmaps and flowcharts, end our thesis.

Remark: For the massless *classical* case of a scalar with only a Laplacian as charge operator— which is actually precisely what our gravitons and dilatons correspond to in this thesis—this has already been proven (by Hitchin in the references). Otherwise the question is open. Again we emphasize that the above includes illdefined objects (the determinants) whose cure may well prove to have some interesting effects on the evolutions of branes. As an example of the above scaling behaviour, theta functions are quasi-periodic, and they are partition functions on a cobordism in the Euclidean plane belonging to a parabolic differential equation.

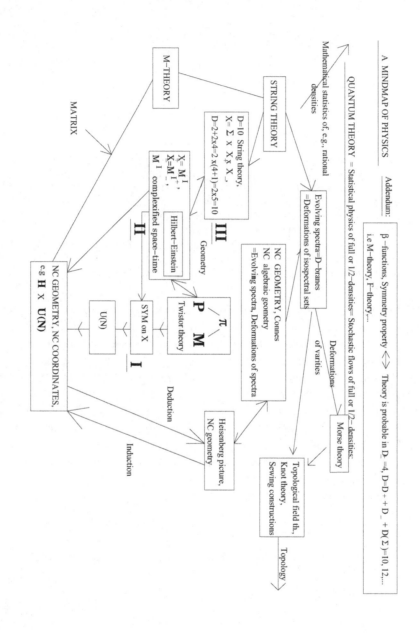

Remark on previous page; Our hypothesis consists of the interrelations illustrated in the mindmap above, and the three theories in our hypothesis have been denoted by I, II and III.

With ϵ the modulus of the imaginary part, $limdir_{\epsilon \to 0} X_{\pm} = Y = Re(X), D(X_{\pm}) = 4$. $X_{\pm}$ the holomorphic and antiholomophic part of a complex Stein manifold of complex dimension 4.

Fundamental Fields: Φ super Dirac field, ω a spin connection with gauge degrees of freedom, so $\omega := \omega + \mathcal{A}_0 + \mathcal{A}_1 + \cdots$, $\mathcal{A}_i \in \{U(N_C), SU(N_C), \cdots\}$-connections. A 1-admissible theory.

$$\mathcal{L}_{SYM,Dirac} = \bar{\Phi}_{\mp} D_{\mp} \Phi_{\pm},$$

Form $\mathbb{X}^a$ from Φ^A

II: H.E. on $X_{\pm}, D(X_{\pm}) = 4$ $\quad\Updownarrow$

Fundamental Fields: $\mathbb{X}_{\pm}^a$ a supervector in the ON-frame, $\mathbb{X}_{\pm,1}^a = e_{\pm,\mu}^a dx_{\pm}^{\mu} = e_{\pm,\mu}^a \theta_{\pm}^{\mu}$ a vielbein, $\mathbb{X}_{\pm,0}^a = X_{\pm}^a$ a vector which gives an embedding via exponentiation. $\omega_{\pm}$ a connection with extra gauge degrees of freedom.

$$\mathcal{L}_{H.E.} = *(D_{\mp}\mathbb{X}_{\pm}^* D_{\pm}\mathbb{X}_{\mp}), \quad Hilbert - Einstein$$

Stack X_+ onto X_-, include a worldsheet to be able to deform mass states. $X = X_+^{(5)} \times X_-^{(5)} = \Sigma \times X_+ \times X_-$, $X_{\pm}^{(5)}$ now each include one degree of freedom more than usual, as well as gauge degrees of freedom. The extra degree of freedom is by effectively killed by world sheet on-shell(BRST) conditions.

III: Strings on $X, D(X) = 10$ $\quad\Updownarrow$

Fundamental Fields: $\mathbb{X}^a$ a NC supervector with gauge degrees of freedom. This is also a 2-admissible theory.

$$\mathcal{L} = *(\mathbb{X}^* D^2 \mathbb{X}), \quad STRING/MATRIX - THEORY$$

This is a theory with total space a Stein infitisimal transversal complexification of $X^{4/5}$, $D(X) = 5/4$ complex dimnsion, Mathematically a cohomological problem(such as finding solutions to field equations with some symmetry) is then by the Stein property[38] a problem on a real analytic 5 or 4-manifold, so ten dimensional Matrix-theory is equivalent to 5-dimensional Gauge Theory with no mass fixed or alternatively fixed mass, which is the same as prescribing the curvature Ricci scalar , and then do 4-dimensional gauge-theory. In $D = 10$ String Theory we fix the dimension and kill the

---

[38]See the Oka-Grauer Principle in the mathematical part. Incidentally this property also proves crossing symmetry, as well as well defined Wick rotations since a Stein manifold is necessaraly pseudoconvex.

off-shell dimensions so that this becomes the same as effective dimenion $D_{effective} = 8$ and this is then equivalent to the physics of Gauge Theory in $X_{\pm}$ with $D(X_{\pm}) = 4$.

A Picture Series Of The Above;

<u>I: SYM on $Y_\pm(X_\pm)$</u>

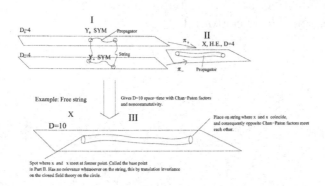

I

$D_i=4$ Y, SYM   Propagator

$D=4$ Y, SYM   String

π_+ **II**   X, H.E., D=4

π_- Propagator

Example: Free string   Gives D=10 space–time with Chan–Paton factors and noncommutativity.

X   **III**

D=10

Place on string where x and x coincide, and consequently opposite Chan–Paton factors meet each other.

Spot where x and x meet at former point. Called the base point in Part II. Has no relevance whatsoever on the string, this by translation invariance on the closed field theory on the circle.

301

Regarding the previous page; The points I,II and III above are the vartious steps and pertaining theories on the previous page. Both the 4 dimensional space-time and the 10 dimensional space-times above are noncommutative, and are in particular hyperkähler for a variety of more or less sensible backgrounds.

Our Hypothesis Features;

(1) Branes, strings, SYM, H.E. gravity, in particular string theory in $D = 10$ as the inevitable consequence of field theory in $D_\pm = 4$.

(2) Spinors(Twistors)

(3) NC geometry, both in Connes sense and in NC space-time meaning. For the former case, the determinant homomorphism conjecture provided the link to NC algebraic geometry.

(4) Toplogical fluctuations of arbitrary dimension, i.e. N-admissibility and generalized SUSY.

(5) The various dimensions as results of Weyl invariance, which makes theories in such dimensions more probable (they would, e.g., tend to flow to such dimensions when being in other dimensions). We have our stringy results directly in the physical dimension $D_\pm = 4$ as a consequence of this, and we called the correctly reduced string theory in $D_\pm = 4$ H.E. gravity.

It is hopefully visible that these interlock most naturally and are indispensible and common to each other.

References

[1] Hugett, Mason,Tod, Tsou and Woodhouse, The Geometric Universe, Oxford.

In particular

[2] "Hypercomplex Manifolds and the Space of Framings" by Nigel Hitchin in the above book.

[3] "Noncommutative Differential Geometry and the Structure of Space-Time" by Alain Connes in the above book.

[4] Daniel S. Freed, Five Lectures on Supersymmetry, American Mathematical Society, 1999.

[5] Quantum Fields and Strings; A Course for Mathematicians, volumes I and II, Editors Pierre Deligne, Pavel Etingof, Lisa C. Jeffery, Daniel S. Freed, David Khazdan, David K. Morrison, John W. morgan, Edward Witten.

In particular

[6] Eric D'Hoker, String theory lecture notes from the 1996-1997 year; Institute for Advanced Study, e.g. in vol. II, Quantum Fields and Strings; A Course for Mathematicians, Editors Pierre Deligne, Pavel Etingof, Lisa C. Jeffery, Daniel S. Freed, David Khazdan, David K. Morrison, John W. morgan, Edward Witten.

[7] Green, Schwarz, Witten, Superstring Theory, vol. I and II, Cambridge; 1987.

In particular volume I, where the material which we refer and lean on in the mass spectrum calculation we have done is located. Comparative material in string theory which we used for proving vanishing homomorphism anomaly of the first kind to tree level is also there.
Finally

[8] Hitchin, Segal, Ward, Integrable Systems, Oxford Science Pubications; 1999. which we was our inspiration to regard our infinite dimensional determinants from a partly integrable systems perspective (beside the C^*-algebra perspective), in particular using the same nomenclature to describe some features of infinite dimensional systems such as our D-branes.

Epilogue and Retrospect

35. A Retrospect to the History of Dual Models

In our thesis, we derived string theory as the inevitable consequence of our wish to deal with space-time entities in quantum gravity, which are full densities rather than half. Thus, in effect, it was the probability interpretation of quantum mechanics that implied string theory. We did this by using the path formulation of quantum field theory, applied in a novel way, but still within the usual rules or quantum mechanics, thus yielding a derivation of string theory out of axiomatized principles. This gives space-time a noncommutative structure, with e.g. $U(N)$ degrees of freedom on space-time vectors, and a natural interpretation of Chan-Paton factors in terms of quarks on the ends of strings. But it also gave the underlying mechanism, namely the probability interpretation, to the existence of string theory, as well as why it takes the form it does. As examples of how close this touches and founds the early ideas that founded string theory, that were based on heuristics and not direct and inevitable deduction as in this thesis, we quote directly GSW, Part I, on two places with figures.

Thus it is hopefully seen that we truly have come up with the *raison d'être* of string theory, which is as fundamental as quantum mechanics itself.

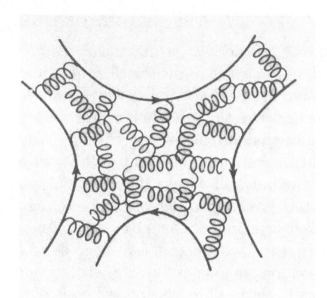

FIGURE 21. This is a figure on page 52, Part I of GSW. It has text below it as follows; "Meson scattering in the large-n limit of QCD is described by 'planar' Feynman diagrams with quarks on the boundary. Describing the flavor state of a meson by the flavor matrix λ_i, the planar amplitude of M mesons in the cyclic order 123...M involves a contraction of each quark with the antiquark of the adjoining meson, and so involves the Chan-Paton factor $tr(\lambda_1\lambda_2\cdots\lambda_M)$."

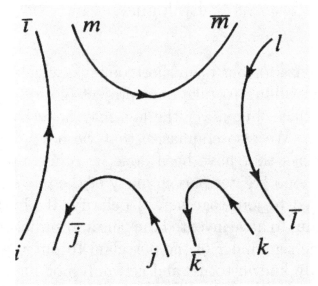

FIGURE 22. The above is the element (c) of a picture series (a)-(d) in GSW Part I, page 48, with a figure text as follows; "We can suppose that an oriented open string has a 'quark' at one end and an 'antiquark' at the other end, as sketched in (a). They can be assumed to transform, respectively, in the n and $\bar{n}$ representation of a $U(n)$ symmetry group. When open strings join, the quark and antiquark charges are required to match as in (b). The group-theory factor associated with general planar open-string amplitude, sketched in (c), is then $tr(\lambda_1 \lambda_2 \cdots \lambda_M)$. In a more general string diagram, as in (d), a similar group theory factor is assigned to each boundary component."

Epilogue

We apologize for our deep shortcomings, which we believe that anyone willing to take on a project of this kind must, in the end, face, but never the less may have handled better than us. We also emphasize that we have many times touched things that have been done by other authors, and that we have no greater part in many of the parts of physics we have tried to join together. To clarify; If physics was a vase, we claim to have invented the glue and sufficiently many pieces of the remainder of the porcelain to put together the parts already known to us, and certainly not have invented the greater part of the large pieces of our precious shattered porcelain, thus feeling very obliged, since the main part of the porcelain is not our creation. In this last vein we feel sincere gratitude and admiration in view of the contributions of J.Polchinski(D-branes), J. Maldacena(AdS-CFT correspondence), Hitchin(The article on hypercomplex framings and harmonic level sets, which drew our attention to a reformulation in the sense of a determinant homomorphism conjecture) and finally A. Connes (Noncommutative Algebraic Geometry). We also acknowledge the strong incentive that the dual work of A.Polyakov (Stochastic flows via CFT) and R. Jackiw (Stochastic flows via Y.M.) as a motivation for believing in early stages in the hypothesis presented. None of the hypothesis set forth could have been done, in particular neither the two extra quantizations of gravity presented nor the unification of physics, were it not for their brave contributions to physics and mathematics.

36. ERRATA

We set

$$D_+ = \not{D}\,^{\mathbb{C}}_+,$$

so the Dirac operator we use is the holomorphic complexified version in complex dimension 4 or 5. This is to among other things make the reasoning about Stein manifolds and QFT in half the dimension trivial via the Oka-Grauert Principle.

37. REMARK

The equivalence

$$\left(\int \rho(m_1^2, \cdots, m_n^2) + \sum_m\right)|Dirac - Yang - Mills, D = 4|^2$$

$$= \left(\int \rho(m_1^2, \cdots, m_n^2) + \sum\right)Gravity, (D = 4)$$

$$= Re(SteinGravity, (D = 8))$$

$$= \left(\int d^n m_+ dm_-^n \rho(m_1^2, \cdots, m_n^2) + \sum_m\right)M_{strings, D_{eff}=8 \equiv D=10}$$

, can be seen in the Feynman rules of, e.g, QED. Indeed we have $u(k) \otimes \bar{u}(k) = \not{k} - m$, and $e_\mu e_\nu^* = g_{\mu\nu}$, so the square of the field/variable pieces give the correct new dynamical variables/fields in the tensored theory which is gravity. Obviously we see that we have two fields or variables, $[X_\pm = \not{k}_\pm \pm m = \frac{1}{2\pi i}\int_\gamma \partial_\pm E^\mu_\pm \Gamma_{\mu,\pm} \pm m = \frac{1}{2\pi i}\int_\gamma \partial_\pm \bar{\psi}\Gamma^\mu_\pm \psi \Gamma_{\mu,\pm} \pm m]$ the asymptotic worldon at infinity and $[g = (g_{\mu\nu})]$ the metric in our gravity. Please note that the e.g the worldon has gauge theory algebra degrees of freedom as well, which are exactly the same as including Chan-Paton rules or stacking D-branes.